BASIC PROBABILITY THEORY

Robert B. Ash

Department of Mathematics
University of Illinois

DOVER PUBLICATIONS, INC.
Mineola, New York

Bibliographical Note

This Dover edition, first published in 2008, is an unabridged republication of the work originally published in 1970 by John Wiley & Sons, Inc., New York. Readers of this book who would like to receive the solutions to the exercises not published in the book may request them from the publisher at the following e-mail address: editors@doverpublications.com

Library of Congress Cataloging-in-Publication Data

Ash, Robert B.
    Basic propability theory / Robert B. Ash. — Dover ed.
        p. cm.
    Includes index.
    Originally published: New York : Wiley, [1970]
    ISBN-13: 978-0-486-46628-6
    ISBN-10: 0-486-46628-0
        1. Probabilities. I. Title.

QA273.A77 2008
519.2—dc22

200800473

Manufactured in the United States by Courier Corporation
46628005   2013
www.doverpublications.com

Preface

This book has been written for a first course in probability and was developed from lectures given at the University of Illinois during the last five years. Most of the students have been juniors, seniors, and beginning graduates, from the fields of mathematics, engineering and physics. The only formal prerequisite is calculus, but an additional degree of mathematical maturity may be helpful.

In talking about nondiscrete probability spaces, it is difficult to avoid measure-theoretic concepts. However, to develop extensive formal machinery from measure theory before going into probability (as is done in most graduate programs in mathematics) would be inappropriate for the particular audience to whom the book is addressed. Thus I have tried to suggest, when possible, the underlying measure-theoretic ideas, while emphasizing the probabilistic way of thinking, which is likely to be quite novel to anyone studying this subject for the first time.

The major field of application considered in the book is statistics (Chapter 8). In addition, some of the problems suggest connections with the physical sciences. Chapters 1 to 5, and Chapter 8 will serve as the basis for a one-semester or a two-quarter course covering both probability and statistics. If probability alone is to be considered, Chapter 8 may be replaced by Chapter 6 and Chapter 7, as time permits. An asterisk before a section or a problem indicates material that I have normally omitted (without loss of continuity), either because it involves subject matter that many of the students have not been exposed to (for example, complex variables) or because it represents too concentrated a dosage of abstraction.

A word to the instructor about notation. In the most popular terminology, $P\{X \leq x\}$ is written for the probability that the random variable X assumes a value less than or equal to the number x. I tried this once in my class, and I found that as the semester progressed, the capital X tended to become smaller in the students' written work, and the small x larger. The following semester, I switched to the letter R for random variable, and this notation is used throughout the book.

Fairly detailed solutions to some of the problems (and numerical answers to others) are given at the end of the book.

I hope that the book will provide an introduction to more advanced courses in probability and real analysis and that it makes the abstract ideas to be encountered later more meaningful. I also hope that nonmathematics majors who come in contact with probability theory in their own areas find the book useful. A brief list of references, suitable for future study, is given at the end of the book.

I am grateful to the many students and colleagues who have influenced my own understanding of probability theory and thus contributed to this book.

I also thank Mrs. Dee Keel for her superb typing, and the staff of Wiley for its continuing interest and assistance.

Urbana, Illinois, 1969   *Robert B. Ash*

Contents

BASIC
PROBABILITY
THEORY

1

Basic Concepts

1.1 INTRODUCTION

The origin of probability theory lies in physical observations associated with games of chance. It was found that if an "unbiased" coin is tossed independently n times, where n is very large, the relative frequency of heads, that is, the ratio of the number of heads to the total number of tosses, is very likely to be very close to 1/2. Similarly, if a card is drawn from a perfectly shuffled deck and then is replaced, the deck is reshuffled, and the process is repeated over and over again, there is (in some sense) convergence of the relative frequency of spades to 1/4.

In the card experiment there are 52 possible outcomes when a single card is drawn. There is no reason to favor one outcome over another (the principle of "insufficient reason" or of "least astonishment"), and so the early workers in probability took as the probability of obtaining a spade the number of favorable outcomes divided by the total number of outcomes, that is, 13/52 or 1/4.

This so-called "classical definition" of probability (the probability of an event is the number of outcomes favorable to the event, divided by the total number of outcomes, where all outcomes are equally likely) is first of all restrictive (it considers only experiments with a finite number of outcomes) and, more seriously, circular (no matter how you look at it, "equally likely" essentially means "equally probable," and thus we are using the concept of probability to define probability itself). Thus we cannot use this idea as the basis of a mathematical theory of probability; however, the early probabilists were not prevented from deriving many valid and useful results.

Similarly, an attempt at a frequency definition of probability will cause trouble. If S_n is the number of occurrences of an event in n independent performances of an experiment, we expect physically that the relative frequency S_n/n should coverge to a limit; however, we cannot assert that the limit exists in a mathematical sense. In the case of the tossing of an unbiased coin, we expect that $S_n/n \to 1/2$, but a conceivable outcome of the process is that the coin will keep coming up heads forever. In other words it is possible that $S_n/n \to 1$, or that $S_n/n \to$ any number between 0 and 1, or that S_n/n has no limit at all.

In this chapter we introduce the concepts that are to be used in the construction of a mathematical theory of probability. The first ingredient we need is a set Ω, called the *sample space*, representing the collection of possible outcomes of a random experiment. For example, if a coin is tossed once we may take $\Omega = \{H, T\}$, where H corresponds to a head and T to a tail. If the coin is tossed twice, this is a different experiment and we need a different Ω, say $\{HH, HT, TH, TT\}$; in this case one performance of the experiment corresponds to two tosses of the coin.

If a single die is tossed, we may take Ω to consist of six points, say $\Omega = \{1, 2, \ldots, 6\}$. However, another possible sample space consists of two points, corresponding to the outcomes "N is even" and "N is odd," where N is the result of the toss. Thus different sample spaces can be associated with the same experiment. The nature of the particular problem under consideration will dictate which sample space is to be used. If we are interested, for example, in whether or not $N \geq 3$ in a given performance of the experiment, the second sample space, corresponding to "N even" and "N odd," will not be useful to us.

In general, the only physical requirement on Ω is that *a given performance of the experiment must produce a result corresponding to exactly one of the points of* Ω. We have as yet no mathematical requirements on Ω; it is simply a set of points.

Next we come to the notion of *event*. An "event" associated with a random experiment corresponds to a question about the experiment that has a yes or no answer, and this in turn is associated with a subset of the sample space. For example, if a coin is tossed twice and $\Omega = \{HH, HT, TH, TT\}$, "the number of heads is ≤ 1" will be a condition that either occurs or does not occur in a given performance of the experiment. That is, after the experiment is performed, the question "Is the number of heads ≤ 1?" can be answered yes or no. The subset of Ω corresponding to a "yes" answer is $A = \{HT, TH, TT\}$; that is, if the outcome of the experiment is HT, TH, or TT, the answer

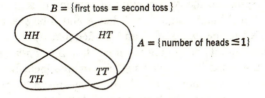

FIGURE 1.1.1   Coin-Tossing Experiment.

to the question "Is the number of heads ≤ 1?" will be "yes," and if the outcome is *HH*, the answer will be "no." Similarly, the subset of Ω associated with the "event" that the result of the first toss is the same as the result of the second toss is $B = \{HH, TT\}$.

Thus an *event* is defined as a subset of the sample space, that is, a collection of points of the sample space. (We shall qualify this in the next section.)

Events will be denoted by capital letters at the beginning of the English alphabet, such as A, B, C, and so on. An event may be characterized by listing all of its points, or equivalently by describing the conditions under which the event will occur. For example, in the coin-tossing experiment just considered, we write

$$A = \{\text{the number of heads is less than or equal to 1}\}$$

This expression is to be read as "A is the set consisting of those outcomes which satisfy the condition that the number of heads is less than or equal to 1," or, more simply, "A is the event that the number of heads is less than or equal to 1." The event A consists of the points *HT, TH*, and *TT*; therefore we write $A = \{HT, TH, TT\}$, which is to be read "A is the event consisting of the points *HT, TH*, and *TT*." As another example, if B is the event that the result of the first toss is the same as the result of the second toss, we may describe B by writing $B = \{\text{first toss} = \text{second toss}\}$ or, equivalently, $B = \{HH, TT\}$ (see Figure 1.1.1).

Each point belonging to an event A is said to be *favorable* to A. The event A will occur in a given performance of the experiment if and only if the outcome of the experiment corresponds to one of the points of A. The entire sample space Ω is said to be the *sure* (or *certain*) event; if *must* occur on any given performance of the experiment. On the other hand, the event consisting of none of the points of the sample space, that is, the empty set $\varnothing$, is called the *impossible event*; it can *never* occur in a given performance of the experiment.

1.2   ALGEBRA OF EVENTS (BOOLEAN ALGEBRA)

Before talking about the assignment of probabilities to events, we introduce some operations by which new events are formed from old ones. These operations correspond to the construction of compound sentences by use of the connectives "or," "and," and "not." Let A and B be events in the same sample space. Define the *union* of A and B (denoted by $A \cup B$) as the set consisting of those points belonging to *either A or B or both.* (Unless otherwise specified, the word "or" will have, for us, the inclusive connotation. In other words, the statement "p or q" will always mean "p or q or both.") Define the *intersection* of A and B, written $A \cap B$, as the set of points that belong to *both A and B.* Define the *complement* of A, written A^c, as the set of points which do *not* belong to A.

▶ **Example 1.**   Consider the experiment involving the toss of a single die, with $N =$ the result; take a sample space with six points corresponding to $N = 1, 2, 3, 4, 5, 6$. For convenience, label the points of the sample space by the integers 1 through 6.

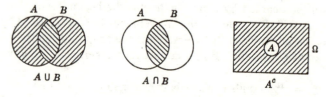

FIGURE 1.2.1   Venn Diagrams.

Then
$$\text{Let } A = \{N \text{ is even}\} \quad \text{and} \quad B = \{N \geq 3\}$$

$$A \cup B = \{N \text{ is even or } N \geq 3\} = \{2, 3, 4, 5, 6\}$$
$$A \cap B = \{N \text{ is even and } N \geq 3\} = \{4, 6\}$$
$$A^c = \{N \text{ is not even}\} = \{1, 3, 5\}$$
$$B^c = \{N \text{ is not} \geq 3\} = \{N < 3\} = \{1, 2\} \blacktriangleleft$$

Schematic representations (called *Venn diagrams*) of unions, intersections, and complements are shown in Figure 1.2.1.

Define the *union of n events* $A_1, A_2, \ldots, A_n$ (notation: $A_1 \cup \cdots \cup A_n$, or $\bigcup_{i=1}^{n} A_i$) as the set consisting of those points which belong to *at least one* of the events $A_1, A_2, \ldots, A_n$. Similarly define the union of an infinite sequence of events $A_1, A_2, \ldots$ as the set of points belonging to at least one of the events $A_1, A_2, \ldots$ (notation: $A_1 \cup A_2 \cup \cdots$, or $\bigcup_{i=1}^{\infty} A_i$).

Define the *intersection of n events* $A_1, \ldots, A_n$ as the set of points belonging to *all* of the events $A_1, \ldots, A_n$ (notation: $A_1 \cap A_2 \cap \cdots \cap A_n$, or $\bigcap_{i=1}^{n} A_i$). Similarly define the intersection of an infinite sequence of events as the set of points belonging to all the events in the sequence (notation: $A_1 \cap A_2 \cap \cdots$, or $\bigcap_{i=1}^{\infty} A_i$). In the above example, with $A = \{N \text{ is even}\} = \{2, 4, 6\}$, $B = \{N \geq 3\} = \{3, 4, 5, 6\}$, $C = \{N = 1 \text{ or } N = 5\} = \{1, 5\}$, we have

$$A \cup B \cup C = \Omega, \qquad A \cap B \cap C = \varnothing$$
$$A \cup B^c \cup C = \{2, 4, 6\} \cup \{1, 2\} \cup \{1, 5\} = \{1, 2, 4, 5, 6\}$$
$$(A \cup C) \cap [(A \cap B)^c] = \{1, 2, 4, 5, 6\} \cap \{4, 6\}^c = \{1, 2, 5\}$$

Two events in a sample space are said to be *mutually exclusive* or *disjoint* if A and B have no points in common, that is, if it is impossible that both A and B occur during the *same* performance of the experiment. In symbols, A and B are mutually exclusive if $A \cap B = \varnothing$. In general the events A_1, $A_2, \ldots, A_n$ are said to be mutually exclusive if no two of the events have a point in common; that is, no more than one of the events can occur during

FIGURE 1.2.2 $A \cap (B \cup C) = (A \cap B) \cup (A \cap C)$.

the same performance of the experiment. Symbolically, this condition may be written

$$A_i \cap A_j = \varnothing \qquad \text{for } i \neq j$$

Similarly, infinitely many events $A_1, A_2, \ldots$ are said to be mutually exclusive if $A_i \cap A_j = \varnothing$ for $i \neq j$.

In some ways the algebra of events is similar to the algebra of real numbers, with union corresponding to addition and intersection to multiplication. For example, the commutative and associative properties hold.

$$A \cup B = B \cup A, \qquad A \cup (B \cup C) = (A \cup B) \cup C$$
$$A \cap B = B \cap A, \qquad A \cap (B \cap C) = (A \cap B) \cap C \qquad (1.2.1)$$

Furthermore, we can prove that for events A, B, and C in the same sample space we have

$$A \cap (B \cup C) = (A \cap B) \cup (A \cap C) \qquad (1.2.2)$$

There are several ways to establish this; for example, we may verify that the sets of both the left and right sides of the equality above are represented by the area in the Venn diagram of Figure 1.2.2.

Another approach is to use the definitions of union and intersection to show that the sets in question have precisely the same members; that is, we show that any point which belongs to the set on the left necessarily belongs to the set on the right, and conversely. To do this, we proceed as follows.

$$x \in A \cap (B \cup C) \Rightarrow x \in A \quad \text{and} \quad x \in B \cup C$$
$$\Rightarrow x \in A \quad \text{and} \quad (x \in B \text{ or } x \in C)$$

(The symbol $\Rightarrow$ means "implies," and $\Leftrightarrow$ means "implies and is implied by.")

CASE 1. $x \in B$. Then $x \in A$ and $x \in B$, so $x \in A \cap B$, so $x \in (A \cap B) \cup (A \cap C)$.

CASE 2. $x \in C$. Then $x \in A$ and $x \in C$, so $x \in A \cap C$, so $x \in (A \cap B) \cup (A \cap C)$.

Thus $x \in A \cap (B \cup C) \Rightarrow x \in (A \cap B) \cup (A \cap C)$; that is, $A \cap (B \cup C) \subset (A \cap B) \cup (A \cap C)$. (The symbol $\subset$ is read "is a subset of"; we say that $A_1 \subset A_2$ provided that $x \in A_1 \Rightarrow x \in A_2$; see Figure 1.2.3. Notice that, according to this definition, a set A is a subset of itself: $A \subset A$.)

Conversely: Let $x \in (A \cap B) \cup (A \cap C)$. Then $x \in A \cap B$ or $x \in A \cap C$.

CASE 1. $x \in A \cap B$. Then $x \in B$, so $x \in B \cup C$, so $x \in A \cap (B \cup C)$.

CASE 2. $x \in A \cap C$. Then $x \in C$, so $x \in B \cup C$, so $x \in A \cap (B \cup C)$.

Thus $(A \cap B) \cup (A \cap C) \subset A \cap (B \cup C)$; hence

$$A \cap (B \cup C) = (A \cap B) \cup (A \cap C)$$

As another example we show that

$$(A_1 \cup A_2 \cup \cdots \cup A_n)^c = A_1^c \cap A_2^c \cap \cdots \cap A_n^c \qquad (1.2.3)$$

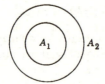

FIGURE 1.2.3 $A_1 \subset A_2$.

The steps are as follows.

$$x \in (A_1 \cup \cdots \cup A_n)^c \Leftrightarrow x \notin A_1 \cup \cdots \cup A_n$$

$\Leftrightarrow$ it is not the case that x belongs to at least one of the A_i

$\Leftrightarrow x \in$ none of the A_i

$\Leftrightarrow x \in A_i^c \quad$ for all i

$\Leftrightarrow x \in A_1^c \cap \cdots \cap A_n^c$

An identical argument shows that

$$\left(\bigcup_{i=1}^{\infty} A_i \right)^c = \bigcap_{i=1}^{\infty} A_i^c \tag{1.2.4}$$

and similarly

$$\left(\bigcap_{i=1}^{n} A_i \right)^c = \bigcup_{i=1}^{n} A_i^c \quad \text{i.e. } (A_1 \cap \cdots \cap A_n)^c = A_1^c \cup \cdots \cup A_n^c \tag{1.2.5}$$

Also

$$\left(\bigcap_{i=1}^{\infty} A_i \right)^c = \bigcup_{i=1}^{\infty} A_i^c \tag{1.2.6}$$

The identities (1.2.3)–(1.2.6) are called the *DeMorgan laws*.

In many ways the algebra of events differs from the algebra of real numbers, as some of the identities below indicate.

$$
\begin{array}{ll}
A \cup A = A & A \cup A^c = \Omega \\
A \cap A = A & A \cap A^c = \varnothing \\
A \cap \Omega = A & A \cup \varnothing = A \\
A \cup \Omega = \Omega & A \cap \varnothing = \varnothing
\end{array} \tag{1.2.7}
$$

Another method of verifying relations among events involves algebraic manipulation, using the identities already derived. Four examples are given below; in working out the identities, it may be helpful to write $A \cup B$ as $A + B$ and $A \cap B$ as AB.

$$1. \quad A \cup (A \cap B) = A \tag{1.2.8}$$

PROOF.

$$A + AB = A\Omega + AB = A(\Omega + B) = A\Omega = A$$

$$2. \quad (A \cup B) \cap (A \cup C) = A \cup (B \cap C) \tag{1.2.9}$$

PROOF.

$$(A + B)(A + C) = (A + B)A + (A + B)C$$
$$= AA + AB + AC + BC \quad \text{(note } AB = BA\text{)}$$
$$= A(\Omega + B + C) + BC$$
$$= A\Omega + BC$$
$$= A + BC$$

3. $A \cup [(A \cap B)^c] = \Omega$ $\hspace{2cm}$ (1.2.10)

PROOF.

$$A + (AB)^c = A + A^c + B^c = \Omega + B^c = \Omega$$

4. $(A \cap B^c) \cup (A \cap B) \cup (A^c \cap B) = A \cup B$ $\hspace{1cm}$ (1.2.11)

PROOF.

$$AB^c + AB + A^cB = AB^c + AB + AB + A^cB \quad \text{[see (1.2.7)]}$$
$$= A(B^c + B) + (A + A^c)B$$
$$= A\Omega + \Omega B$$
$$= A + B$$

(see Figure 1.2.4).

As another example, let Ω be the set of nonnegative real numbers. Let

$$A_n = \left[0, 1 - \frac{1}{n}\right] = \left\{x \in \Omega : 0 \le x \le 1 - \frac{1}{n}\right\} \qquad n = 1, 2, \ldots$$

(This will be another common way of describing an event. It is to be read: "A_n is the set consisting of those points x in Ω such that $0 \le x \le 1 - 1/n$." If there is no confusion about what space Ω we are considering, we shall simply write $A_n = \{x : 0 \le x \le 1 - 1/n\}$.) Then

$$\bigcup_{n=1}^{\infty} A_n = [0, 1) = \{x : 0 \le x < 1\}$$

$$\bigcap_{n=1}^{\infty} A_n = \{0\}$$

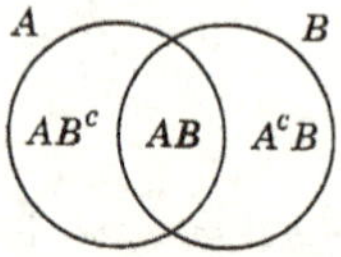

FIGURE 1.2.4   Venn Diagram Illustrating
$(A \cap B^c) \cup (A \cap B) \cup (A^c \cap B) = A \cup B$.

As an illustration of the DeMorgan laws,

$$\left(\bigcup_{n=1}^{\infty} A_n \right)^c = [0, 1)^c = [1, \infty) = \{x : x \geq 1\}$$

$$\bigcap_{n=1}^{\infty} A_n{}^c = \bigcap_{n=1}^{\infty} \left(1 - \frac{1}{n}, \infty\right) = [1, \infty)$$

(Notice that $x > 1 - 1/n$ for all $n = 1, 2, \ldots \Leftrightarrow x \geq 1$.) Also

$$\left(\bigcap_{n=1}^{\infty} A_n \right)^c = \{0\}^c = (0, \infty) = \{x : x > 0\}$$

$$\bigcup_{n=1}^{\infty} A_n{}^c = \bigcup_{n=1}^{\infty} \left(1 - \frac{1}{n}, \infty\right) = (0, \infty)$$

PROBLEMS

1. An experiment involves choosing an integer N between 0 and 9 (the sample space consists of the integers from 0 to 9, inclusive). Let $A = \{N \leq 5\}$, $B = \{3 \leq N \leq 7\}$, $C = \{N$ is even and $N > 0\}$. List the points that belong to the following events.

$$A \cap B \cap C, \quad A \cup (B \cap C^c), \quad (A \cup B) \cap C^c, \quad (A \cap B) \cap [(A \cup C)^c]$$

2. Let A, B, and C be *arbitrary* events in the same sample space. Let D_1 be the event that at least two of the events A, B, C occur; that is, D_1 is the set of points common to at least two of the sets A, B, C.
Let $D_2 = \{$exactly two of the events A, B, C occur$\}$
$D_3 = \{$at least one of the events A, B, C occur$\}$
$D_4 = \{$exactly one of the events A, B, C occur$\}$
$D_5 = \{$not more than two of the events A, B, C occur$\}$
Each of the events D_1 through D_5 can be expressed in terms of A, B, and C by using unions, intersections, and complements. For example, $D_3 = A \cup B \cup C$. Find suitable expressions for D_1, D_2, D_4, and D_5.

3. A public opinion poll (circa 1850) consisted of the following three questions:
(a) Are you a registered Whig?
(b) Do you approve of President Fillmore's performance in office?
(c) Do you favor the Electoral College system?
A group of 1000 people is polled. Assume that the answer to each question must be either "yes" or "no." It is found that:
550 people answer "yes" to the third question and 450 answer "no."
325 people answer "yes" exactly twice; that is, their responses contain two "yeses" and one "no."

100 people answer "yes" to all three questions.

125 registered Whigs approve of Fillmore's performance.

How many of those who favor the Electoral College system do not approve of Fillmore's performance, and in addition are not registered Whigs? HINT: Draw a Venn diagram.

4. If A and B are events in a sample space, define $A - B$ as the set of points which belong to A but not to B; that is, $A - B = A \cap B^c$. Establish the following.
   (a) $A \cap (B - C) = (A \cap B) - (A \cap C)$
   (b) $A - (B \cup C) = (A - B) - C$
   Is is true that $(A - B) \cup C = (A \cup C) - B$?

5. Let Ω be the reals. Establish the following.

$$(a, b) = \bigcup_{n=1}^{\infty} \left(a, b - \frac{1}{n}\right] = \bigcup_{n=1}^{\infty} \left[a + \frac{1}{n}, b\right)$$

$$[a, b] = \bigcap_{n=1}^{\infty} \left[a, b + \frac{1}{n}\right) = \bigcap_{n=1}^{\infty} \left(a - \frac{1}{n}, b\right]$$

6. If A and B are disjoint events, are A^c and B^c disjoint? Are $A \cap C$ and $B \cap C$ disjoint? What about $A \cup C$ and $B \cup C$?

7. If $A_n \subset A_{n-1} \subset \cdots \subset A_1$, show that $\bigcap_{i=1}^{n} A_i = A_n$, $\bigcup_{i=1}^{n} A_i = A_1$.

8. Suppose that $A_1, A_2, \ldots$ is a sequence of subsets of Ω, and we know that for each n, $\bigcap_{i=1}^{n} A_i$ is not empty. Is it true that $\bigcap_{i=1}^{\infty} A_i$ is not empty? (A related question about real numbers: if, for each n, we have $\sum_{i=1}^{n} a_i < b$, is it true that $\sum_{i=1}^{\infty} a_i < b$?)

9. If $A, B_1, B_2, \ldots$ are arbitrary events, show that

$$A \cap \left(\bigcup_i B_i\right) = \bigcup_i (A \cap B_i)$$

This is the distributive law with infinitely many factors.

1.3 PROBABILITY

We now consider the assignment of probabilities to events. A technical complication arises here. It may not always be possible to regard all subsets of Ω as events. We may discard or fail to measure some of the information in the outcome corresponding to the point $\omega \in \Omega$, so that for a given subset A of Ω, it may not be possible to give a yes or no answer to the question "Is $\omega \in A$?" For example, if the experiment involves tossing a coin five times, we may record the results of only the first three tosses, so that $A = \{$at least four heads$\}$ will not be "measurable"; that is, membership of $\omega \in A$ cannot be determined from the given information about ω.

In a given problem there will be a particular class of subsets of Ω called the "class of events." For reasons of mathematical consistency, we require that the event class $\mathscr{F}$ form a *sigma field*, which is a collection of subsets of Ω satisfying the following three requirements.

$$\Omega \in \mathscr{F} \tag{1.3.1}$$

$$A_1, A_2, \ldots \in \mathscr{F} \quad \text{implies} \quad \bigcup_{n=1}^{\infty} A_n \in \mathscr{F} \tag{1.3.2}$$

That is, $\mathscr{F}$ is closed under finite or countable union.

$$A \in \mathscr{F} \quad \text{implies} \quad A^c \in \mathscr{F} \tag{1.3.3}$$

That is, $\mathscr{F}$ is closed under complementation.

Notice that if $A_1, A_2, \ldots \in \mathscr{F}$, then $A_1^c, A_2^c, \ldots \in \mathscr{F}$ by (1.3.3); hence $\bigcup_{n=1}^{\infty} A_n^c \in \mathscr{F}$ by (1.3.2). By the DeMorgan laws, $\bigcap_{n=1}^{\infty} A_n = (\bigcup_{n=1}^{\infty} A_n^c)^c$; hence, by (1.3.3), $\bigcap_{n=1}^{\infty} A_n \in \mathscr{F}$. Thus $\mathscr{F}$ is closed under finite or countable intersection. Also, by (1.3.1) and (1.3.3), the empty set $\varnothing$ belongs to $\mathscr{F}$.

Thus, for example, if the question "Did A_n occur?" has a definite answer for $n = 1, 2, \ldots$, so do the questions "Did at least one of the A_n occur?" and "Did all the A_n occur?"

Note also that if we apply the algebraic operations of Section 1.2 to sets in $\mathscr{F}$, the new sets we obtain still belong to $\mathscr{F}$.

In many cases we shall be able to take $\mathscr{F} =$ the collection of *all* subsets of Ω, so that every subset of Ω is an event. Problems in which $\mathscr{F}$ cannot be chosen in this way generally arise in uncountably infinite sample spaces; for example, $\Omega =$ the reals. We shall return to this subject in Chapter 2.

We are now ready to talk about the assignment of probabilities to events. If $A \in \mathscr{F}$, the probability $P(A)$ should somehow reflect the long-run relative frequency of A in a large number of independent repetitions of the experiment. Thus $P(A)$ should be a number between 0 and 1, and $P(\Omega)$ should be 1.

Now if A and B are disjoint events, the number of occurrences of $A \cup B$ in n performances of the experiment is obtained by adding the number of occurrences of A to the number of occurrences of B. Thus we should have

$$P(A \cup B) = P(A) + P(B) \quad \text{if } A \text{ and } B \text{ are disjoint}$$

and, similarly,

$$P(A_1 \cup \cdots \cup A_n) = \sum_{i=1}^{n} P(A_i) \quad \text{if } A_1, \ldots, A_n \text{ are disjoint}$$

For mathematical convenience we require that

$$P\left(\bigcup_{n=1}^{\infty} A_n\right) = \sum_{n=1}^{\infty} P(A_n)$$

when we have a *countably infinite* family of disjoint events $A_1, A_2, \ldots$.

The assumption of countable rather than simply finite additivity has not been convincingly justified physically or philosophically; however, it leads to a much richer mathematical theory.

A function that assigns a number $P(A)$ to each set A in the sigma field $\mathscr{F}$ is called a *probability measure* on $\mathscr{F}$, provided that the following conditions are satisfied.

$$P(A) \geq 0 \qquad \text{for every } A \in \mathscr{F} \tag{1.3.4}$$

$$P(\Omega) = 1 \tag{1.3.5}$$

If $A_1, A_2, \ldots$ are disjoint sets in $\mathscr{F}$, then

$$P(A_1 \cup A_2 \cup \cdots) = P(A_1) + P(A_2) + \cdots \tag{1.3.6}$$

We may now give the underlying mathematical framework for probability theory.

DEFINITION. A *probability space* is a triple $(\Omega, \mathscr{F}, P)$, where Ω is a set, $\mathscr{F}$ a sigma field of subsets of Ω, and P a probability measure on $\mathscr{F}$.

We shall not, at this point, embark on a general study of probability measures. However, we shall establish four facts from the definition. (All sets in the arguments to follow are assumed to belong to $\mathscr{F}$.)

$$1. \quad P(\varnothing) = 0 \tag{1.3.7}$$

PROOF. $A \cup \varnothing = A$; hence $P(A \cup \varnothing) = P(A)$. But A and $\varnothing$ are disjoint and so $P(A \cup \varnothing) = P(A) + P(\varnothing)$. Thus $P(A) = P(A) + P(\varnothing)$; consequently $P(\varnothing) = 0$.

$$2. \quad P(A \cup B) = P(A) + P(B) - P(A \cap B) \tag{1.3.8}$$

PROOF. $A = (A \cap B) \cup (A \cap B^c)$, and these sets are disjoint (see Figure 1.2.4). Thus $P(A) = P(A \cap B) + P(A \cap B^c)$. Similarly $P(B) = P(A \cap B) + P(A^c \cap B)$. Thus $P(A) + P(B) - P(A \cap B) = P(A \cap B) + P(A \cap B^c) + P(A^c \cap B) = P(A \cup B)$. Intuitively, if we add the outcomes in A to those in B, we have counted those in $A \cap B$ twice; subtracting the outcomes in $A \cap B$ yields the outcomes in $A \cup B$.

$$3. \quad \text{If } B \subset A, \text{ then } P(B) \leq P(A); \text{ in fact,}$$

$$P(A - B) = P(A) - P(B) \tag{1.3.9}$$

where $A - B$ is the set of points that belong to A but not to B.

PROOF. $P(A) = P(B) + P(A - B)$, since $B \subset A$ (see Figure 1.3.1), and the result follows because $P(A - B) \geq 0$. Intuitively, if the occurrence of B

FIGURE 1.3.1

always implies the occurrence of A, A must occur at least as often as B in any sequence of performances of the experiment.

$$4. \quad P(A_1 \cup A_2 \cup \cdots) \leq P(A_1) + P(A_2) + \cdots \qquad (1.3.10)$$

That is, the probability that at least one of a finite or countably infinite collection of events will occur is less than or equal to the sum of the probabilities; note that, for the case of two events, this follows from $P(A \cup B) = P(A) + P(B) - P(A \cap B) \leq P(A) + P(B)$.

PROOF. We make use of the fact that any union may be written as a disjoint union, as follows.

$$A_1 \cup A_2 \cup \cdots = A_1 \cup (A_1^c \cap A_2) \cup (A_1^c \cap A_2^c \cap A_3) \cup \cdots \cup$$

$$(A_1^c \cap A_2^c \cap \cdots \cap A_{n-1}^c \cap A_n) \cup \cdots \qquad (1.3.11)$$

To see this, observe that if x belongs to the set on the right then $x \in A_1^c \cap \cdots \cap A_{n-1}^c \cap A_n$ for some n; hence $x \in A_n$. Thus x belongs to the set on the left. Conversely, if x belongs to the set on the left, then $x \in A_n$ for some n. Let n_0 be the smallest such n. Then $x \in A_1^c \cap \cdots \cap A_{n_0-1}^c \cap A_{n_0}$, and so x belongs to the set on the right. Thus

$$P(A_1 \cup A_2 \cup \cdots) = \sum_{n=1}^{\infty} P(A_1^c \cap \cdots \cap A_{n-1}^c \cap A_n) \leq \sum_{n=1}^{\infty} P(A_n)$$

using (1.3.9); notice that

$$A_1^c \cap \cdots \cap A_{n-1}^c \cap A_n \subset A_n.$$

REMARKS. The basic difficulty with the classical and frequency definitions of probability is that their approach is to try somehow to *prove* mathematically that, for example, the probability of picking a heart from a perfectly shuffled deck is 1/4, or that the probability of an unbiased coin coming up heads is 1/2. This cannot be done. All we can say is that if a card is picked at random and then replaced, and the process is repeated over and over again, the result that the ratio of hearts to total number of drawings will be close to 1/4 is in accord with our intuition and our physical experience. For this reason we should *assign* a probability 1/4 to the event of obtaining a heart, and similarly we should *assign* a probability 1/52 to each possible outcome of the experiment. The only reason for doing this is that the consequences agree with our experience. If you decide that some mysterious factor caused the ace of spades to be more likely than any other card, you could incorporate this factor by assigning a higher probability to the ace of spades. The mathematical development of the theory would not be affected; however, the conclusions you might draw from this assumption would be at variance with experimental results.

One can never really use mathematics to *prove* a specific physical fact. For example, we cannot prove mathematically that there is a physical quantity called "force." What we can do is postulate a mathematical entity called "force" that satisfies a certain differential equation. We can build up a collection of mathematical results that, when interpreted properly, provide a reasonable description of certain physical phenomena (reasonable until another mathematical theory is constructed that provides a better description). Similarly, in probability theory we are faced with situations in which our intuition or some physical experiments we have carried out suggest certain results. Intuition and experience lead us to an *assignment* of probabilities to events. As far as the mathematics is concerned, any assignment of probabilities will do, subject to the rules of mathematical consistency. However, our hope is to develop mathematical results that, when interpreted and related to physical experience, will help to make precise such notions as "the ratio of the number of heads to the total number of observations in a very large number of independent tosses of an unbiased coin is very likely to be very close to 1/2."

We emphasize that the insights gained by the early workers in probability are not to be discarded, but instead cast in a more precise form.

PROBLEMS

1. Write down some examples of sigma fields other than the collection of all subsets of a given set Ω.

2. Give an example to show that $P(A - B)$ need not equal $P(A) - P(B)$ if B is not a subset of A.

1.4 COMBINATORIAL PROBLEMS

We consider a class of problems in which the assignment of probabilities can be made in a natural way.

Let Ω be a *finite* or *countably infinite* set, and let $\mathscr{F}$ consist of all subsets of Ω.

For each point $\omega_i \in \Omega$, $i = 1, 2, \ldots$, assign a nonnegative number p_i, with $\sum_i p_i = 1$. If A is any subset of Ω, let $P(A) = \sum_{\omega_i \in A} p_i$. Then it may be verified that P is a probability measure; $P\{\omega_i\} = p_i$, and *the probability of any event A is found by adding the probabilities of the points of A*. An $(\Omega, \mathscr{F}, P)$ of this type is called a *discrete probability space*.

▶ **Example 1.** Throw a (biased) coin twice (see Figure 1.4.1).

Let $E_1 = \{$at least one head$\}$. Then

$$E_1 = A_1 \cup A_2 \cup A_3$$

Hence

$$P(E_1) = P(A_1) + P(A_2) + P(A_3)$$
$$= .36 + .24 + .24 = .84$$

Let $E_2 = \{$tail on first toss$\}$; then

$$E_2 = A_3 \cup A_4$$

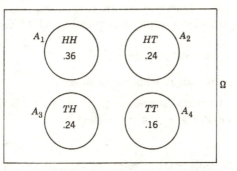

$$A_1 = \{HH\}, \quad A_2 = \{HT\}$$
$$A_3 = \{TH\}, \quad A_4 = \{TT\}$$
$$\text{Assign } P(A_1) = .36$$
$$P(A_2) = P(A_3) = .24$$
$$P(A_4) = .16$$

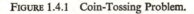

FIGURE 1.4.1 Coin-Tossing Problem.

and

$$P(E_2) = P(A_3) + P(A_4) = .4 \blacktriangleleft$$

In the special case when $\Omega = \{\omega_1, \ldots, \omega_n\}$ and $p_i = 1/n$, $i = 1, 2, \ldots, n$, we have

$$P(A) = \frac{\text{number of points of } A}{\text{total number of points in } \Omega} = \frac{\text{favorable outcomes}}{\text{total outcomes}}$$

corresponding to the classical definition of probability.

Thus, in this case, finding $P(A)$ simply involves counting the number of outcomes favorable to A. When n is large, counting by hand may not be feasible; combinatorial analysis is simply a method of counting that can often be used to avoid writing down the entire list of favorable outcomes.

There is only one basic idea in combinatorial analysis, and that is the following. Suppose that a symbol is selected from the set $\{a_1, \ldots, a_n\}$; if a_i is chosen, a symbol is selected from the set $\{b_{i1}, \ldots, b_{im}\}$. Each pair of selections (a_i, b_{ij}) is assumed to determine a "result" $f(i, j)$. If all results are distinct, the number of possible results is nm, since there is a one-to-one correspondence between results and pairs of integers (i, j), $i = 1, \ldots, n$, $j = 1, \ldots, m$.

If, after the symbol b_{ij} is chosen, a symbol is selected from the set $\{c_{ij1}, c_{ij2}, \ldots, c_{ijp}\}$, and each triple (a_i, b_{ij}, c_{ijk}) determines a distinct result $f(i, j, k)$, the number of possible results is nmp. Analogous statements may be made for any finite sequence of selections.

Certain standard selections occur frequently, and it is convenient to classify them.

Let $a_1, \ldots, a_n$ be distinct symbols.

Ordered samples of size r, with replacement

The number of ordered sequences $(a_{i_1}, \ldots, a_{i_r})$, where the a_{i_k} belong to $\{a_1, \ldots, a_n\}$, is $n \times n \times \cdots \times n$ (r times), or

$$n^r \qquad (1.4.1)$$

(The term "with replacement" refers to the fact that if the symbol a_{i_k} is selected at step k it may be selected again at any future time.)

For example, the number of possible outcomes if three dice are thrown is $6 \times 6 \times 6 = 216$.

Ordered Samples of Size r, without Replacement

The number of ordered sequences $(a_{i_1}, \ldots, a_{i_r})$, where the a_{i_k} belong to $\{a_1, \ldots, a_n\}$, but repetition is not allowed (i.e., no a_i can appear more than once in the sequence), is

$$n(n-1)\cdots(n-r+1) = \frac{n!}{(n-r)!}, \qquad r = 1, 2, \ldots, n \quad (1.4.2)$$

(The first symbol may be chosen in n ways, and the second in $n-1$ ways, since the first symbol may not be used again, and so on.) The above number is sometimes called the number of *permutations of r objects out of n*, written $(n)_r$.

For example, the number of 3-digit numbers that can be formed from $1, 2, \ldots, 9$, if no digit can be repeated, is $9(8)(7) = 504$.

Unordered Samples of Size r, without Replacement

The number of unordered sets $\{a_{i_1}, \ldots, a_{i_r}\}$, where the a_{i_k}, $k = 1, \ldots, r$, are distinct elements of $\{a_1, \ldots, a_n\}$ (i.e., the number of ways of selecting r distinct objects out of n), if order does not count, is

$$\binom{n}{r} = \frac{n!}{r!\,(n-r)!} \qquad (1.4.3)$$

To see this, consider the following process.

(a) Select r distinct objects out of n without regard to order; this can be done in $\binom{n}{r}$ ways, where $\binom{n}{r}$ is to be determined.

(b) For each set selected in (a), say $\{a_{i_1}, \ldots, a_{i_r}\}$, select an ordering of $a_{i_1}, \ldots, a_{i_r}$. This can be done in $(r)_r = r!$ ways (see Figure 1.4.2 for $n = 3$, $r = 2$).

The result of performing (a) and (b) is a permutation of r objects out of n; hence

$$\binom{n}{r}r! = (n)_r = \frac{n!}{(n-r)!}$$

or

$$\binom{n}{r} = \frac{n!}{r!\,(n-r)!}, \qquad r = 1, 2, \ldots, n$$

We define $\binom{n}{0}$ to be $n!/0!\,n! = 1$, to make the formula for $\binom{n}{r}$ valid for $r = 0, 1, \ldots, n$. Notice that $\binom{n}{k} = \binom{n}{n-k}$.

$\binom{n}{r}$ is sometimes called the number of *combinations of r objects out of n*.

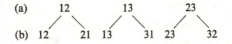

FIGURE 1.4.2   Determination of $\binom{n}{r}$.

Unordered Samples of Size r, with Replacement

We wish to find the number of unordered sets $\{a_{i_1}, \ldots, a_{i_r}\}$, where the a_{i_k} belong to $\{a_1, \ldots, a_n\}$ and repetition is allowed. As an example, let $n = 3$ and $r = 3$. Let the symbols be 1, 2, and 3. List all arrangements in a column so that a precedes b if and only if a, read as an ordinary 3-digit number, is $<b$. In an adjacent column list a new set of sequences formed from the old by adding 0 to the first digit, 1 to the second digit, and 2 to the third digit.

111	$123 = (1 + 0, 1 + 1, 1 + 2)$
112	$124 = (1 + 0, 1 + 1, 2 + 2)$
113	125
122	134
123	135
133	145
222	234
223	235
233	245
333	345

In the first column we have unordered samples of size 3 (out of 3), with replacement. In the second column we have unordered samples of size 3 (out of 5), without replacement. In this way we can set up a one-to-one correspondence between unordered samples of size r (out of n) with replacement, and unordered samples of size r (out of $n + r - 1$) without replacement. Thus the number of such samples is

$$\binom{n + r - 1}{r} \tag{1.4.4}$$

An alternative way of looking at unordered samples with replacement is to count all sequences $(a_{i_1}, \ldots, a_{i_r})$, each $a_{i_k} \in \{a_1, \ldots, a_n\}$, subject to the constraint that sequences having the same *occupancy numbers* $r_k =$ the number of occurrences of a_k, $k = 1, 2, \ldots, n$, are identified. The r_k are nonnegative integers satisfying $r_1 + r_2 + \cdots r_n = r$; hence we must count the number of nonnegative integer solutions $(r_1, \ldots, r_n)$ of the equation $r_1 + \cdots + r_n = r$. This may be done combinatorially as follows.

Consider an arrangement of r stars and $n - 1$ bars, as shown in Figure

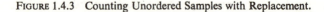

▮ * ∣ ∣ *** ▮	▮ * ∣ * ∣ ** ▮
$r_1 = 1, r_2 = 0, r_3 = 3$	$r_1 = 1, r_2 = 1, r_3 = 2$
(sample $= a_1 a_3 a_3 a_3$)	(sample $= a_1 a_2 a_3 a_3$)

FIGURE 1.4.3   Counting Unordered Samples with Replacement.

1.4.3 for $n = 3$, $r = 4$ (the thicker bars at the sides are fixed). Each arrangement corresponds to a solution of $r_1 + \cdots + r_n = r$. The number of arrangements is the number of ways of selecting r positions out of $n + r - 1$ for the stars to occur (or $n - 1$ positions for the bars); that is, $\binom{n+r-1}{r}$. For $n = 3$, $r = 4$, there are 15 solutions.

r_1	r_2	r_3	Sample
0	0	4	$a_3a_3a_3a_3$
0	4	0	$a_2a_2a_2a_2$
4	0	0	$a_1a_1a_1a_1$
0	1	3	$a_2a_3a_3a_3$
0	2	2	$a_2a_2a_3a_3$
0	3	1	$a_2a_2a_2a_3$
1	0	3	$a_1a_3a_3a_3$
2	0	2	$a_1a_1a_3a_3$
3	0	1	$a_1a_1a_1a_3$
1	3	0	$a_1a_2a_2a_2$
2	2	0	$a_1a_1a_2a_2$
3	1	0	$a_1a_1a_1a_2$
2	1	1	$a_1a_1a_2a_3$
1	2	1	$a_1a_2a_2a_3$
1	1	2	$a_1a_2a_3a_3$

▶ **Example 2.** Find the probability of obtaining four of a kind in an ordinary five-card poker hand.

There are $\binom{52}{5}$ distinct poker hands (without regard to order), and so we may take Ω to have $\binom{52}{5}$ points. To obtain the number of hands in which there are four of a kind:

(a) Choose the face value to appear four times (13 choices: A, K, Q, ..., 2)

(b) Choose the fifth card (48 ways).

Thus $p = (13)(48)/\binom{52}{5}$. Figure 1.4.4 indicates the selection process.

NOTE. The problem may also be done using ordered samples. The number of ordered poker hands is $(52)(51)(50)(49)(48) = (52)_5$ (the drawing is without replacement). The number of ordered poker hands having four of a kind is $(13)(48)\,5!$, so that $p = (13)(48)(5!)/(52)_5 = (13)(48)/\binom{52}{5}$ as before. Here we may take the space Ω' to have $(52)_5$ points; each point of Ω corresponds to 5! points of Ω'. ◀

▶ **Example 3.** Three balls are dropped into three boxes. Find the probability that exactly one box will be empty.

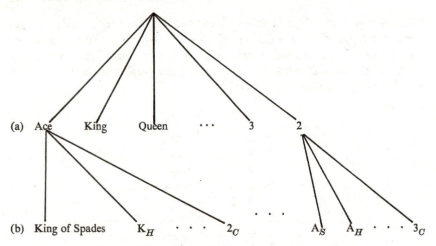

FIGURE 1.4.4 Counting Process for Selecting Four of a Kind. There is a one-to-one correspondence between paths in the diagram and favorable outcomes.

In problems of dropping r balls into n boxes, we may regard the boxes as (distinct) symbols $a_1, \ldots, a_n$; each toss of a ball corresponds to the selection of a box. Thus the sequence $a_2 a_1 a_1 a_3$ corresponds to the first ball into box 2, the second and third into box 1, and the fourth into box 3.

In general, an arrangement of r balls in n boxes corresponds to a sample of size r from the symbols $a_1, \ldots, a_n$. If we require that the sampling be with replacement, this means that a given box can contain any number of balls. Sampling without replacement means that a given box cannot contain more than one ball. If we consider ordered samples we are saying that the balls are distinguishable. For example, $a_3 a_7$ (ball 1 into box 3, ball 2 into box 7) is different from $a_7 a_3$ (ball 1 into box 7, ball 2 into box 3); in other words, we may regard the balls as being numbered $1, 2, \ldots, r$. Unordered sampling corresponds to indistinguishable balls.

If there is no restriction on the number of balls in a given box, the total number of arrangements, taking into account the order in which the balls are tossed (i.e., regarding the balls as distinct), is the number of ordered samples of size r (from $\{a_1, \ldots, a_n\}$) with replacement, or n^r. If the boxes are energy levels in physics and the balls are particles, the *Maxwell-Boltzmann* assumption is that all n^r arrangements are equally likely.

If there can be at most one ball in a given box, the number of (ordered) arrangements is $(n)_r$. If the order in which the balls are tossed is neglected, we are simply choosing r boxes out of n to be occupied; the *Fermi-Dirac* assumption takes the $\binom{n}{r}$ possible selections of boxes (or energy levels) as equally likely.

We might also mention the *Bose-Einstein* assumption; here a box may contain an unlimited number of balls, but the balls are indistinguishable; that is, the order in which the balls are tossed is neglected, so that, for example, $a_2a_1a_1a_3$ is identified with $a_1a_3a_1a_2$. Thus the number of arrangements counted in this scheme is the number of unordered samples of size r with replacement, or $\binom{n+r-1}{r}$. The Bose-Einstein assumption takes all these arrangements as equally likely.

To return to the original problem, we have boxes a_1, a_2, and a_3 and sequences of length 3 (three balls are tossed). We take all $3^3 = 27$ ordered samples with replacement as equally likely. (We shall see that this model—ordered sampling with replacement—corresponds to the tossing of the balls independently; this idea will be developed in the next section.) Now

$P\{\text{exactly 1 box empty}\} = P\{\text{box 1 empty, boxes 2 and 3 occupied}\}$

$\qquad\qquad + P\{\text{box 2 empty, boxes 1 and 3 occupied}\}$

$\qquad\qquad + P\{\text{box 3 empty, boxes 1 and 2 occupied}\}$

Furthermore

$P\{\text{box 1 empty, boxes 2 and 3 occupied}\} = P\{a_1 \text{ does not occur in the}$

$\qquad\qquad\qquad\qquad\qquad\qquad \text{sequence } a_{i_1}a_{i_2}a_{i_3}, \text{ but } a_2$

$\qquad\qquad\qquad\qquad\qquad\qquad \text{and } a_3 \text{ both occur}\}$

If a_1 does not occur, either a_2 or a_3 must occur twice, and the other symbol once. We may choose the symbol that is to occur twice in two ways; the symbol that occurs once is then determined. If, say, a_3 occurs twice and a_2 once, the position of a_2 may be any of three possibilities; the position of the two a_3's is then determined. Thus the probability that box 1 will be empty and boxes 2 and 3 occupied is $2(3)/27 = 6/27$ (in fact the six favorable outcomes are $a_2a_2a_3$, $a_2a_3a_2$, $a_3a_2a_2$, $a_3a_3a_2$, $a_3a_2a_3$, and $a_2a_3a_3$).

Thus the probability that exactly one box will be empty is, by symmetry, $3(6)/27 = 2/3$. ◀

▶ **Example 4.** In a 13-card bridge hand the probability that the hand will contain the A K Q J 10 of spades is $\binom{47}{8}/\binom{52}{13}$. (The A K Q J 10 of spades must be chosen, and afterward eight cards must be selected out of 47 that remain after the five top spades have been removed.)

Now let us find the probability of obtaining the A K Q J 10 of at least one suit. Thus, if A_S is the event that the A K Q J 10 of spades is obtained, and similarly for A_H, A_D, and A_C (hearts, diamonds, and clubs), we are looking for

$$P(A_S \cup A_H \cup A_D \cup A_C)$$

The sets are not disjoint, so that we cannot simply add probabilities. It is possible to obtain, for example, the A K Q J 10 of both spades and hearts in a 13-card hand, and this probability is easy to compute:

$$P(A_S \cap A_H) = \frac{\binom{42}{3}}{\binom{52}{13}}$$

What we need here is a way of expressing $P(A_S \cup A_H \cup A_D \cup A_C)$ in terms of the individual terms $P(A_S)$ etc., and the intersections $P(A_S \cap A_H)$ etc. We know that

$$P(A \cup B) = P(A) + P(B) - P(A \cap B)$$

If we have three events, then

$$P(A \cup B \cup C) = P(A \cup (B \cup C)) = P(A) + P(B \cup C) - P(A \cap (B \cup C))$$
$$= P(A) + P(B \cup C) - P((A \cap B) \cup (A \cap C))$$
$$= P(A) + P(B) + P(C) - P(A \cap B) - P(A \cap C)$$
$$- P(B \cap C) + P(A \cap B \cap C)$$

The general pattern is now clear and may be verified by induction.

$$P(A_1 \cup \cdots \cup A_n) = \sum_i P(A_i) - \sum_{i<j} P(A_i \cap A_j)$$
$$+ \sum_{i<j<k} P(A_i \cap A_j \cap A_k) - \cdots + (-1)^{n-1} P(A_1 \cap \cdots \cap A_n) \quad (1.4.5)$$

In the present problem the intersections taken three or four at a time are empty; hence, by symmetry,

$$P(A_S \cup A_H \cup A_D \cup A_C) = 4P(A_S) - 6P(A_S \cap A_H)$$
$$= \frac{4\binom{47}{8} - 6\binom{42}{3}}{\binom{52}{13}}$$

It is illuminating to consider an *incorrect* approach to this problem. Suppose that we first pick a suit (four choices); we then select the A K Q J 10 of that suit. The remaining eight cards can be anything (if they include the A K Q J 10 of another suit, the condition that at least one A K Q J 10 of the same suit be obtained will still be satisfied). Thus we have $\binom{47}{8}$ choices, so that the desired probability is $4\binom{47}{8}/\binom{52}{13}$.

The above procedure illustrates *multiple counting*, the nemesis of the combinatorial analyst (see Figure 1.4.5).

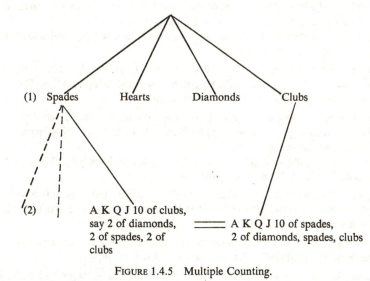

FIGURE 1.4.5 Multiple Counting.

In writing $p = 4\binom{47}{8}/\binom{52}{13}$ we are saying that there is a one-to-one correspondence between paths in Figure 1.4.5 and favorable outcomes. But this is not the case, since the two paths indicated in the diagram both lead to the same result, namely, A K Q J 10 of spades, A K Q J 10 of clubs, 2 of diamonds, 2 of spades, 2 of clubs. In fact there are $6\binom{42}{3}$ such duplications. For we can pick the two suits in $\binom{4}{2} = 6$ ways; then, after taking A K Q J 10 of each suit, we select the remaining three cards in $\binom{42}{3}$ ways. If we subtract the number of duplications, $6\binom{42}{3}$, from the original count, $4\binom{47}{8}$, we obtain the correct result.

To rephrase: the counting process we have proposed counts the number of paths in the above diagram, that is, the number of choices at Step 1 times the number of choices at Step 2. However, the paths do not in general lead to distinct "results," namely, distinct bridge hands. ◄

PROBLEMS

1. If a 3-digit number (000 to 999) is chosen at random, find the probability that exactly 1 digit will be >5.

2. Find the probability that a five-card poker hand will be:
   (a) A straight (five cards in sequence regardless of suit; ace may be high but not low).

(b) Three of a kind (three cards of the same face value x, plus two cards with face values y and z, with x, y, z distinct).

(c) Two pairs (two cards of face value x, two of face value y, and one of face value z, with x, y, z distinct).

3. An urn contains 3 red, 8 yellow, and 13 green balls; another urn contains 5 red, 7 yellow, and 6 green balls. One ball is selected from each urn. Find the probability that both balls will be of the same color.

4. An experiment consists of drawing 10 cards from an ordinary 52-card pack.
   (a) If the drawing is done with replacement, find the probability that no two cards will have the same face value.
   (b) If the drawing is done without replacement, find the probability that at least 9 cards will be of the same suit.

5. An urn contains 10 balls numbered from 1 to 10. Five balls are drawn without replacement. Find the probability that the second largest of the five numbers drawn will be 8.

6. m men and w women seat themselves at random in $m + w$ seats arranged in a row. Find the probability that all the women will be adjacent.

7. If a box contains 75 good light bulbs and 25 defective bulbs and 15 bulbs are removed, find the probability that at least one will be defective.

8. Eight cards are drawn without replacement from an ordinary deck. Find the probability of obtaining exactly three aces or exactly three kings (or both).

9. (The game of rencontre). An urn contains n tickets numbered $1, 2, \ldots, n$. The tickets are shuffled thoroughly and then drawn one by one without replacement. If the ticket numbered r appears in the rth drawing, this is denoted as a *match* (French: *rencontre*). Show that the probability of at least one match is

$$1 - \frac{1}{2!} + \frac{1}{3!} - \cdots + \frac{(-1)^{n-1}}{n!} \to 1 - e^{-1} \quad \text{as} \quad n \to \infty$$

10. A "language" consists of three "words," $W_1 = a$, $W_2 = ba$, $W_3 = bb$. Let $N(k)$ be the number of "sentences" using exactly k letters (e.g., $N(1) = 1$ (i.e., a), $N(2) = 3$ (aa, ba, bb), $N(3) = 5$ (aaa, aba, abb, baa, bba); no space is allowed between words.
    (a) Show that $N(k) = N(k-1) + 2N(k-2)$, $k = 2, 3, \ldots$ (define $N(0) = 1$).
    (b) Show that the general solution to the second-order homogeneous linear difference equation (a) [with $N(0)$ and $N(1)$ specified], is $N(k) = A2^k + B(-1)^k$, where A and B are determined by $N(0)$ and $N(1)$. Evaluate A and B in the present case.

11. (The birthday problem) Assume that a person's birthday is equally likely to fall on any of the 365 days in a year (neglect leap years). If r people are selected, find the probability that all r birthdays will be different. Equivalently, if r balls are dropped into 365 boxes, we are looking for the probability that no box will contain more than one ball. It turns out that the probability is less than 1/2

for $r \geq 23$, so that in a class of 23 or more students the odds are that two or more people will have the same birthday.

12. Fourteen balls are dropped into six boxes. Find the number of arrangements (ordered samples of size 14 with replacement, from six symbols) whose occupancy numbers coincide with 4, 4, 2, 2, 2, 0 in some order (i.e., boxes i_1 and i_2 contain four balls, boxes i_3, i_4, and i_5 two balls, and box i_6 no balls, for some $i_1, \ldots, i_6$).

13. (a) Let Ω be a set with n elements. Show that there are 2^n subsets of Ω. For example, if $\Omega = \{1, 2, 3\}$, the subsets are $\varnothing$, $\{1\}$, $\{2\}$, $\{3\}$, $\{1, 2\}$, $\{1, 3\}$, $\{2, 3\}$, and $\{1, 2, 3\} = \Omega$; $2^3 = 8$ altogether.
    (b) How many ways are there of selecting ordered pairs (A, B) of subsets of Ω such that $A \subseteq B$? For example, $A = \{1\}$, $B = \{1, 3\}$ gives such a pair, but $A = \{1, 2\}$, $B = \{1, 3\}$ does not.

14. Let Ω be a finite set. A *partition* of Ω is an (unordered) set $\{A_1, \ldots, A_n\}$, where the A_i are nonempty subsets whose union is Ω. For example, if $\Omega = \{1, 2, 3\}$, there are five partitions.

$$A_1 = \{1, 2, 3\}$$
$$A_1 = \{1, 2\}, A_2 = \{3\}$$
$$A_1 = \{1, 3\}, A_2 = \{2\}$$
$$A_1 = \{1\}, A_2 = \{2, 3\}$$
$$A_1 = \{1\}, A_2 = \{2\}, A_3 = \{3\}$$

Let $g(n)$ be the number of partitions of a set with n elements.
    (a) Show that $g(n) = \sum_{k=0}^{n-1} \binom{n-1}{k} g(k)$ [define $g(0) = 1$].
    (b) Show that $g(n) = e^{-1} \sum_{k=0}^{\infty} k^n / k!$
HINT: Show that the series satisfies the difference equation of part (a).

1.5 INDEPENDENCE

Consider the following experiment. A person is selected at random and his height is recorded. After this the last digit of the license number of the next car to pass is noted. If A is the event that the height is over 6 feet, and B is the event that the digit is ≥ 7, then, intuitively, A and B are "independent" in the sense that knowledge about the occurrence or nonoccurrence of one of the events should not influence the odds about the other. For example, say that $P(A) = .2$, $P(B) = .3$. In a long sequence of trials we would expect the following situation.

(Roughly) 20% of the time A occurs; of those cases in which A occurs:
30% B occurs
70% B does not occur

80% of the time A does not occur; of these cases:
30% B occurs
70% B does not occur

Thus, if B is independent of A, it appears that $P(A \cap B)$ should be $.2(.3) = .06 = P(A)P(B)$, and $P(A^c \cap B)$ should be $.8(.3) = .24 = P(A^c)P(B)$.

Conversely, if $P(A \cap B) = P(A)P(B) = .06$ and $P(A^c \cap B) = P(A^c)P(B) = .24$, then, if A occurs roughly 20% of the time and we look at only the cases in which A occurs, B must occur in roughly 30% of these cases in order to have $A \cap B$ occur 6% of the time. Similarly, if we look at the cases in which A does not occur (80%), then, since we are assuming that $A^c \cap B$ occurs 24% of the time, we must have B occurring in 30% of these cases. Thus the odds about B are not changed by specifying the occurrence or non-occurrence of A.

It appears that we should say that event B is independent of A iff $P(A \cap B) = P(A)P(B)$ and $P(A^c \cap B) = P(A^c)P(B)$. However, the second condition is already implied by the first. If $P(A \cap B) = P(A)P(B)$,

$$P(A^c \cap B) = P(B - A) = P(B - (A \cap B)) = P(B) - P(A \cap B)$$

since $A \cap B$ is a subset of B; hence

$$P(A^c \cap B) = P(B) - P(A)P(B) = (1 - P(A))P(B) = P(A^c)P(B)$$

Thus B is independent of A; that is, knowledge of A does not influence the odds about B, iff $P(A)P(B) = P(A \cap B)$. But this condition is perfectly symmetrical, in other words, B is independent of A iff A is independent of B. Thus we are led to the following definition.

DEFINITION. Two events A and B are *independent* iff $P(A \cap B) = P(A)P(B)$.

If we have three events A, B, C that are (intuitively) independent, knowledge of the occurrence or nonoccurrence of $A \cap B$, for example, should not change the odds about C; this leads as above to the requirement that $P(A \cap B \cap C) = P(A \cap B)P(C)$. But if A, B, and C are to be independent, we must expect that A and B are independent (as well as A and C, and B and C), so we should have all of the following conditions satisfied.

$$P(A \cap B) = P(A)P(B), \qquad P(A \cap C) = P(A)P(C),$$

$$P(B \cap C) = P(B)P(C)$$

and

$$P(A \cap B \cap C) = P(A)P(B)P(C)$$

We are led to the following definition.

DEFINITION. Let A_i, $i \in I$, where I is an arbitrary index set, possibly infinite, be an arbitrary collection of events [a fixed probability space $(\Omega, \mathscr{F}, P)$ is of course assumed].

The A_i are said to be *independent* iff for each finite set of distinct indices $i_1, \ldots, i_k \in I$ we have

$$P(A_{i_1} \cap A_{i_2} \cap \cdots \cap A_{i_k}) = P(A_{i_1})P(A_{i_2}) \cdots P(A_{i_k})$$

REMARKS

1. If the A_i, $i \in I$, are independent, it follows that

$$P(B_{i_1} \cap \cdots \cap B_{i_k}) = P(B_{i_1}) \cdots P(B_{i_k})$$

for all (distinct) $i_1, \ldots, i_k$, where each B_{i_r} may be either A_{i_r} or $A_{i_r}{}^c$. To put it simply, if the A_i are independent and we replace any event by its complement, we still have independence [see Problem 1; actually we have already done most of the work by showing that $P(A \cap B) = P(A)P(B)$ implies $P(A^c \cap B) = P(A^c)P(B)$].

2. The condition $P(A_1 \cap \cdots \cap A_n) = P(A_1) \cdots P(A_n)$ does not imply the analogous condition for any smaller family of events. For example, it is possible to have $P(A \cap B \cap C) = P(A)P(B)P(C)$, but $P(A \cap B) \neq P(A)P(B)$, $P(A \cap C) \neq P(A)P(C)$, $P(B \cap C) \neq P(B)P(C)$. In particular, A, B, and C are not independent.

Conversely it is possible to have, for example, $P(A \cap B) = P(A)P(B)$, $P(A \cap C) = P(A)P(C)$, $P(B \cap C) = P(B)P(C)$, but $P(A \cap B \cap C) \neq P(A)P(B)P(C)$. Thus A and B are independent, as are A and C, and also B and C, but A, B, and C are not independent.

▶ **Example 1.** Let two dice be tossed, and take $\Omega =$ all ordered pairs (i, j), $i, j = 1, 2, \ldots, 6$, with each point assigned probability $1/36$.
Let

$$A = \{\text{first die} = 1, 2, \text{ or } 3\}$$

$$B = \{\text{first die} = 3, 4, \text{ or } 5\}$$

$$C = \{\text{the sum of the two faces is } 9\}$$

(Thus $A \cap B = \{(3, 1), (3, 2), (3, 3), (3, 4), (3, 5), (3, 6)\}$, $A \cap C = \{(3, 6)\}$, $B \cap C = \{(3, 6), (4, 5), (5, 4)\}$, $A \cap B \cap C = \{(3, 6)\}$.)
Then

$$P(A \cap B) = \tfrac{1}{6} \neq P(A)P(B) = \tfrac{1}{2}(\tfrac{1}{2}) = \tfrac{1}{4}$$

$$P(A \cap C) = \tfrac{1}{36} \neq P(A)P(C) = \tfrac{1}{2}(\tfrac{4}{36}) = \tfrac{1}{18}$$

$$P(B \cap C) = \tfrac{1}{12} \neq P(B)P(C) = \tfrac{1}{2}(\tfrac{1}{9}) = \tfrac{1}{18}$$

But

$$P(A \cap B \cap C) = \tfrac{1}{36} = P(A)P(B)P(C)$$

Now in the same probability space let

$$A = \{\text{first die} = 1, 2, \text{ or } 3\}$$
$$B = \{\text{second die} = 4, 5, \text{ or } 6\}$$
$$C = \{\text{the sum of the two faces is } 7\}$$

(Thus $A \cap C = \{(1, 6), (2, 5), (3, 4)\} = A \cap B \cap C$, etc.) Then

$$P(A \cap B) = \tfrac{1}{4} = P(A)P(B) = \tfrac{1}{2}(\tfrac{1}{2})$$
$$P(A \cap C) = \tfrac{1}{12} = P(A)P(C) = \tfrac{1}{2}(\tfrac{1}{6})$$
$$P(B \cap C) = \tfrac{1}{12} = P(B)P(C) = \tfrac{1}{2}(\tfrac{1}{6})$$

But

$$P(A \cap B \cap C) = \tfrac{1}{12} \neq P(A)P(B)P(C) = \tfrac{1}{24} \blacktriangleleft$$

We illustrate the idea of independence by considering some problems related to the classical coin-tossing experiment.

A sequence of n *Bernoulli trials* is a sequence of n independent observations, each of which may result in exactly one of two possible situations, called "success" or "failure." At each observation the probability of success is p, and the probability of failure is $q = 1 - p$.

SPECIAL CASES

    (a) Toss a coin independently n times, with success = heads, failure = tails.

    (b) Examine components produced on an assembly line; success = acceptable, failure = defective.

    (c) Transmit binary digits through a communication channel; success = digit recieved correctly, failure = digit received incorrectly.

We take Ω = all 2^n ordered sequences of length n, with components 0 (failure) and 1 (success). To assign probabilities in accordance with the physical description given above, we reason as follows.

Consider the sample point $\omega = 11 \cdots 10 \cdots 0$ (k 1's followed by $n - k$ 0's). Let $A_i = \{\text{success on trial } i\}$ = the set of all sequences with a 1 in the ith coordinate. Because of the independence of the trials we must assign

$$P\{\omega\} = P(A_1 \cap A_2 \cap \cdots \cap A_k \cap A_{k+1}^c \cap \cdots \cap A_n^c)$$
$$= P(A_1)P(A_2) \cdots P(A_k)P(A_{k+1}^c) \cdots P(A_n^c) = p^k q^{n-k}$$

Similarly, *any point with k 1's and $n - k$ 0's is assigned probability $p^k q^{n-k}$.*

The number of such points is the number of ways of selecting k distinct positions for the 1's to occur (or selecting $n - k$ distinct positions for the 0's); that is, $\binom{n}{k}$. The sum of the probabilities assigned to all the points is

$$\sum_{k=0}^{n} \binom{n}{k} p^k q^{n-k} = (p + q)^n = 1$$

by the binomial theorem. Thus we have a legitimate assignment. Furthermore, the probability of obtaining exactly k successes is

$$p(k) = \binom{n}{k} p^k q^{n-k} \quad \cdot k = 0, 1, \ldots, n \tag{1.5.1}$$

$p(k)$, $k = 0, 1, \ldots, n$, is called the *binomial* probability function.

▶ **Example 2.** Six balls are tossed independently into three boxes A, B, C. For each ball the probability of going into a specific box is 1/3. Find the probability that box A will contain (a) exactly four balls, (b) at least two balls, (c) at least five balls.

Here we have six Bernoulli trials, with success corresponding to a ball in box A, failure to a ball in box B or C. Thus $n = 6$, $p = 1/3$, $q = 2/3$, and so the required probabilities are

(a) $\qquad\qquad p(4) = \binom{6}{4}(\frac{1}{3})^4(\frac{2}{3})^2$

(b) $\qquad\qquad 1 - p(0) - p(1) = 1 - (\frac{2}{3})^6 - \binom{6}{1}(\frac{1}{3})(\frac{2}{3})^5$

(c) $\qquad\qquad p(5) + p(6) = \binom{6}{5}(\frac{1}{3})^5(\frac{2}{3}) + (\frac{1}{3})^6$ ◀

We now consider *generalized Bernoullli trials*. Here we have a sequence of independent trials, and on each trial the result is exactly one of the k possibilities $b_1, \ldots, b_k$. On a given trial let b_i occur with probability p_i, $i = 1, 2, \ldots, k$ ($p_i \geq 0$, $\sum_{i=1}^{k} p_i = 1$).

We take Ω = all k^n ordered sequences of length n with components $b_1, \ldots, b_k$; for example, if $\omega = (b_1 b_3 b_2 b_2 \cdots)$ then b_1 occurs on trial 1, b_3 on trial 2, b_2 on trials 3 and 4, and so on. As in the previous situation, assign to the point

$$\omega = (\underbrace{b_1 b_1 \cdots b_1 b_2}_{n_1} \underbrace{\cdots b_2}_{n_2} \cdots \underbrace{b_k \cdots b_k}_{n_k})$$

the probability $p_1^{n_1} p_2^{n_2} \cdots p_k^{n_k}$. This is the probability assigned to any sequence having n_i occurrences of b_i, $i = 1, 2, \ldots, k$. To find the number of such sequences, first select n_1 positions out of n for the b_1's to occur, then n_2 positions out of the remaining $n - n_1$ for the b_2's, n_3 out of $n - n_1 - n_2$ for the b_3's, and so on. Thus the number of sequences having exactly n_1

occurrences of $b_1, \ldots, n_k$ occurrences of b_k is

$$\binom{n}{n_1}\binom{n-n_1}{n_2}\binom{n-n_1-n_2}{n_3} \cdots \binom{n-n_1-\cdots-n_{k-2}}{n_{k-1}}\binom{n_k}{n_k}$$

$$= \frac{n!}{n_1!\, n_2! \cdots n_k!}$$

The total probability assigned to all points is

$$\sum_{\substack{n_1,\ldots,n_k \text{ nonneg} \\ \text{integers, with } \sum_{i=1}^{k} n_i = n}} \frac{n!}{n_1!n_2! \cdots n_k!} p_1^{n_1} p_2^{n_2} \cdots p_k^{n_k} = (p_1 + \cdots + p_k)^n = 1$$

$$(1.5.2)$$

To see this, notice that $(p_1 + \cdots + p_k)^n = (p_1 + \cdots + p_k)(p_1 + \cdots + p_k) \cdots (p_1 + \cdots + p_k)$, n times. A typical term in the expansion is $p_1^{n_1} \cdots p_k^{n_k}$; the number of times this term appears is

$$\frac{n!}{n_1!\, n_2! \cdots n_k!}$$

since we may count the appearances by selecting p_1 from n_1 of the n factors, selecting p_2 from n_2 of the remaining $n - n_1$ factors, and so forth. Thus we have a legitimate assignment of probabilities.

The probability that b_1 will occur n_1 times, b_2 will occur n_2 times, $\ldots$, and b_k will occur n_k times is

$$p(n_1, \ldots, n_k) = \frac{n!}{n_1! \cdots n_k!} p_1^{n_1} \cdots p_k^{n_k} \qquad (1.5.3)$$

$p(n_1, \ldots, n_k)$, $n_1, \ldots, n_k$ = nonnegative integers whose sum is n, is called the *multinomial* probability function.

Note that when $k = 2$, generalized Bernoulli trials reduce to ordinary Bernoulli trials [let $b_1 =$ "success," $b_2 =$ "failure", $p_1 = p, p_2 = q = 1 - p$, $n_1 = k$, $n_2 = n - k$; then $(n!/n_1!\, n_2!)p_1^{n_1}p_2^{n_2} = \binom{n}{k}p^k q^{n-k} =$ probability of k successes in n trials].

▶ **Example 3.**   Throw four unbiased dice independently. Find the probability of exactly two 1's and one 2.

Let

$$b_1 = \text{"1 occurs" (on a } given \text{ trial)} \qquad p_1 = \tfrac{1}{6} \qquad n_1 = 2$$
$$b_2 = \text{"2 occurs"} \qquad\qquad\qquad\qquad p_2 = \tfrac{1}{6} \qquad n_2 = 1,$$
$$b_3 = \text{"3, 4, 5, or 6 occurs"} \qquad\quad p_3 = \tfrac{2}{3} \qquad n_3 = 1$$

The probability is $(4!/2!\,1!\,1!)\,(1/6)^2(1/6)^1(2/3)^1 = 1/27$. ◀

▶ **Example 4.** If 10 balls are tossed independently into five boxes, with a given ball equally likely to fall into each box, find the probability that all boxes will have the same number of balls.

Let b_i = "ball goes into box i", $p_i = 1/5$, $n_i = 2$, $i = 1, 2, 3, 4, 5, n = 10$. The probability $= (10!/2^5)(1/5)^{10}$. ◀

REMARK.   If r balls are tossed independently into n boxes, we have seen that the event {ball 1 into box i_1, ball 2 into box i_2, . . . , ball r into box i_r} must have probability $p_{i_1} p_{i_2} \cdots p_{i_r}$, where p_i is the probability that a specific ball will fall into box i. In particular, if all $p_i = 1/n$ (as is assumed in Examples 2 and 4 above), the probability of the event is $1/n^r$. In other words, all ordered samples of size r (out of n symbols), with replacement, have the same probability. This justifies the assertion we made in Example 3 of Section 1.4.

We emphasize that the independence of the tosses is an *assumption*, not a theorem. For example, if two balls are tossed into two boxes and a given box can contain at most one ball, then the events A_1 = {ball 1 goes into box 1} and A_2 = {ball 2 goes into box 1} are not independent, since $P(A_1 \cap A_2) = 0$, $P(A_1) = P(A_2) = 1/2$.

▶ **Example 5.**   An urn contains equal numbers of black, white, red, and green balls. Four balls are drawn independently, with replacement. Find the probability $p(k)$ that exactly k colors will appear in the sample, $k = 1, 2, 3, 4$.

This is a multinomial problem with $n = 4$ and $b_1 = B$ = black, $b_2 = W$ = white, $b_3 = R$ = red, $b_4 = G$ = green.

$k = 4$: The probability that all four colors will appear is given by the multinomial formula with all $n_i = 1$; that is,

$$\frac{4!}{1!\,1!\,1!\,1!}\left(\frac{1}{4}\right)^4 = 6/64 = p(4)$$

$k = 3$: The probability of obtaining two black, one white, and one red ball is given by the multinomial formula with $n_1 = 2$, $n_2 = n_3 = 1$, $n_4 = 0$; that is,

$$\frac{4!}{2!\,1!\,1!\,0!}\left(\frac{1}{4}\right)^4 = \frac{3}{64}$$

To find the total probability of obtaining exactly three colors, multiply by the number of ways of selecting three colors out of four $[\binom{4}{3} = 4]$ and the number of ways of selecting one of three colors to be repeated (3). Thus

$$p(3) = 36/64.$$

$k = 2$: The probability of obtaining two black and two white balls is

$$\frac{4!}{2!\,2!\,0!\,0!}\left(\frac{1}{4}\right)^4 = \frac{3}{128}$$

Thus the probability of obtaining two balls of one color and two of another is $3/128$ times the number of ways of selecting two colors out of $4 [(\binom{4}{2}) = 6]$, or $9/64$. Similarly, the probability of obtaining three of one color and one of another is

$$\frac{4!}{3!\,1!\,0!\,0!}\left(\frac{1}{4}\right)^4 (4)(3) = \frac{12}{64}$$

Notice that the extra factor is $(4)(3) = 12$, not $(\binom{4}{2}) = 6$, since three blacks and one white constitute a different selection from three whites and one black. Thus

$$p(2) = 9/64 + 12/64 = 21/64.$$

$k = 1$: The probability that all balls will be of the same color is

$$p(1) = \frac{4!}{4!\,0!\,0!\,0!}\left(\frac{1}{4}\right)^4 (4) = \frac{1}{64}$$

REMARK. The sample space of this problem consists of all *ordered* samples of size 4, with replacement, from the symbols B, W, R, G, with all samples assigned the same probability. The reader should resist the temptation to assign equal probability to all unordered samples of size 4, with replacement. This would imply, for example, that $\{WWWW\}$ and $\{WWWB, WWBW, WBWW, BWWW\} = \{$three whites and one black$\}$ have the same probability, and this is inconsistent with the assumption of independence. ◄

PROBLEMS

1. Show that the events A_i, $i \in I$, are independent iff $P(B_{i_1} \cap \cdots \cap B_{i_k}) = P(B_{i_1}) \cdots P(B_{i_k})$ for all (distinct) $i_1, \ldots, i_k$, where each B_{i_r} may be either A_{i_r} or $A_{i_r}{}^c$.

2. Let $p(k)$, $k = 0, 1, \ldots, n$, be the binomial probability function.
   (a) If $(n + 1)p$ is not an integer, show that $p(k)$ is strictly increasing up to $k = [(n + 1)p] = $ the largest integer $\leq (n + 1)p$, and attains a maximum at $[(n + 1)p]$. $p(k)$ is strictly decreasing for all larger values of k.
   (b) If $(n + 1)p$ is an integer, show that $p(k)$ is strictly increasing up to $k = (n + 1)p - 1$ and has a double maximum at $k = (n + 1)p - 1$ and $k = (n + 1)p$; $p(k)$ is strictly decreasing for larger values of k.

3. A single card is drawn from an ordinary deck. Give examples of events A and B associated with this experiment that are
   (a) Mutually exclusive (disjoint) but not independent
   (b) Independent but not mutually exclusive
   (c) Independent and mutually exclusive
   (d) Neither independent nor mutually exclusive

4. Of the 100 people in a certain village, 50 always tell the truth, 30 always lie, and 20 always refuse to answer. A sample of size 30 is taken with replacement.
   (a) Find the probability that the sample will contain 10 people of each category.
   (b) Find the probability that there will be exactly 12 liars.

5. Six unbiased dice are tossed independently. Find the probability that the number of 1's minus the number of 2's will be 3.

6. How many terms are there in the multinomial expansion (1.5.2)?

7. An urn contains t_1 balls of color C_1, t_2 of color C_2, . . . , t_k of color C_k.
   (a) If n balls are drawn without replacement, show that the probability of obtaining exactly n_1 of color C_1, n_2 of color C_2, . . . , n_k of color C_k is

$$\frac{\binom{t_1}{n_1}\binom{t_2}{n_2} \cdots \binom{t_k}{n_k}}{\binom{t}{n}}$$

   where $t = t_1 + t_2 + \cdots + t_k$ is the total number of balls in the urn and $\binom{t_i}{n_i}$ is defined to be 0 if $n_i > t_i$. (Notice the pattern: $t_1 + t_2 + \cdots + t_k = t$, $n_1 + n_2 + \cdots + n_k = n$.) The above expression, regarded as a function of $n_1, \ldots, n_k$, is called the *hypergeometric* probability function.
   (b) What is the probability of the event of part (a) if the balls are drawn independently, with replacement?

8. (a) If an event A is independent of itself, that is, if A and A are independent, show that $P(A) = 0$ or 1.
   (b) If $P(A) = 0$ or 1, show that A and B are independent for any event B, in particular, that A and A are independent.

1.6 CONDITIONAL PROBABILITY

If A and B are independent events, the occurrence or nonoccurrence of A does not influence the odds concerning the occurrence of B. If A and B are not independent, it would be desirable to have some way of measuring exactly how much the occurrence of one of the events changes the odds about the other.

In a long sequence of independent repetitions of the experiment, $P(A)$ measures the fraction of the trials on which A occurs. If we look only at the trials on which A occurs (say there are n_A of these) and record those trials on which B occurs also (there are n_{AB} of these, where n_{AB} is the number of trials on which both A and B occur), the ratio n_{AB}/n_A is a measure of $P(B \mid A)$, the "conditional probability of B given A," that is, the fraction of the time that B occurs, looking only at trials producing an occurrence of A. Comparing $P(B \mid A)$ with $P(B)$ will indicate the difference between the odds about B when A is known to have occurred, and the odds about B before any information about A is revealed.

The above discussion suggests that we define the *conditional probability of B given A* as

$$P(B \mid A) = \frac{P(A \cap B)}{P(A)} \tag{1.6.1}$$

This makes sense if $P(A) > 0$.

▶ **Example 1.** Throw two unbiased dice independently. Let $A = \{\text{sum of the faces} = 8\}$, $B = \{\text{faces are equal}\}$. Then

$$P(B \mid A) = \frac{P(A \cap B)}{P(A)} = \frac{P\{4 - 4\}}{P\{4 - 4, 5 - 3, 3 - 5, 6 - 2, 2 - 6\}} = \frac{1/36}{5/36} = \frac{1}{5}$$

(see Figure 1.6.1).

There is a point here that may be puzzling. In counting the outcomes favorable to A, we note that there are two ways of making an 8 using a 5 and a 3, but only one way using a 4 and a 4. The probability space consists of all 36 *ordered* pairs (i, j), $i, j = 1, 2, 3, 4, 5, 6$, each assigned probability $1/36$. The ordered pair $(4, 4)$ is the same as the ordered pair $(4, 4)$ (this is rather difficult to dispute), while $(5, 3)$ is different from $(3, 5)$. Alternatively, think of the first die to be thrown as red and the second as green. A 5 on the red die and a 3 on the green is a different outcome from a 3 on the red and a 5 on the green. However, using 4's we can make an 8 in only one way, a 4 on the red followed by a 4 on the green. ◀

The extension of the definition of conditional probability to events with probability zero will be considered in great detail later on. For now, we are

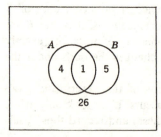

FIGURE 1.6.1 Example on Conditional Probability. Numbers Indicate Favorable Outcomes.

content to note some consequences of the above definition [whenever an expression such as $P(B \mid A)$ is written, it is assumed that $P(A) > 0$].

If A and B are independent, then $P(A \cap B) = P(A)P(B)$, so that $P(B \mid A) = P(B)$ and $P(A \mid B)$ $(= P(A \cap B)/P(B)) = P(A)$, in accordance with the intuitive notion that the occurrence of one of the events does not change the odds about the other.

The formula $P(A \cap B) = P(A)P(B \mid A)$ may be extended to more than two events.

$$P(A \cap B \cap C) = P(A \cap B)P(C \mid A \cap B)$$

Hence

$$P(A \cap B \cap C) = P(A)P(B \mid A)P(C \mid A \cap B) \qquad (1.6.2)$$

Similarly

$$P(A \cap B \cap C \cap D) = P(A)P(B \mid A)P(C \mid A \cap B)P(D \mid A \cap B \cap C)$$
$$(1.6.3)$$

and so on.

▶ **Example 2.** Three cards are drawn without replacement from an ordinary deck. Find the probability of not obtaining a heart.

Let $A_i = \{\text{card } i \text{ is not a heart}\}$. Then we are looking for

$$P(A_1 \cap A_2 \cap A_3) = P(A_1)P(A_2 \mid A_1)P(A_3 \mid A_1 \cap A_2) = \tfrac{39}{52}\tfrac{38}{51}\tfrac{37}{50}$$

[For example, to find $P(A_2 \mid A_1)$, we restrict ourselves to the outcomes favorable to A_1. If the first card is not a heart, 51 cards remain in the deck, including 13 hearts, so that the probability of not getting a heart on the second trial is 38/51.]

Notice that the above probability can be written $\binom{39}{3}/\binom{52}{3}$, which could have been derived by direct combinatorial reasoning. Furthermore, if the cards were drawn independently, with replacement, the probability would be quite different, $(3/4)^3 = 27/64$. ◀

We now prove one of the most useful theorems of the subject.

Theorem of Total Probability. *Let B_1, B_2, ... be a finite or countably infinite family of mutually exclusive and exhaustive events (i.e., the B_i are disjoint and their union is Ω). If A is any event, then*

$$P(A) = \sum_i P(A \cap B_i) \qquad (1.6.4)$$

Thus $P(A)$ is computed by finding a list of mutually exclusive, exhaustive ways in which A can happen, and then adding the individual probabilities. Also

$$P(A) = \sum_i P(B_i)P(A \mid B_i) \qquad (1.6.5)$$

where the sum is taken over those i for which $P(B_i) > 0$. Thus $P(A)$ is a weighted average of the conditional probabilities $P(A \mid B_i)$.

PROOF.

$$P(A) = P(A \cap \Omega) = P\left(A \cap \left(\bigcup_i B_i\right)\right) = P\left(\bigcup_i (A \cap B_i)\right)$$

$$= \sum_i P(A \cap B_i) = \sum_i P(B_i)P(A \mid B_i)$$

Notice that under the above assumptions we have

$$P(B_k \mid A) = \frac{P(A \cap B_k)}{P(A)} = \frac{P(B_k)P(A \mid B_k)}{\sum_i P(B_i)P(A \mid B_i)} \qquad (1.6.6)$$

This formula is sometimes referred to as *Bayes' theorem*; $P(B_k \mid A)$ is sometimes called an *a posteriori probability*. The reason for this terminology may be seen in the example below.

▶ **Example 3.**   Two coins are available, one unbiased and the other two-headed. Choose a coin at random and toss it once; assume that the unbiased coin is chosen with probability 3/4. Given that the result is heads, find the probability that the two-headed coin was chosen.

The "tree diagram" shown in Figure 1.6.2 represents the experiment.

We may take Ω to consist of the four possible paths through the tree, with each path assigned a probability equal to the product of the probabilities assigned to each branch. Notice that we are given the probabilities of the

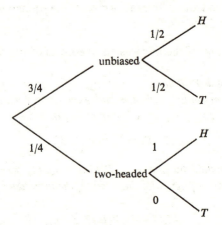

FIGURE 1.6.2   Tree Diagram.

events $B_1 = \{$unbiased coin chosen$\}$ and $B_2 = \{$two-headed coin chosen$\}$, as well as the conditional probabilities $P(A \mid B_i)$, where $A = \{$coin comes up heads$\}$. This is sufficient to determine the probabilities of all events.

Now we can compute $P(B_2 \mid A)$ using Bayes' theorem; this is facilitated if, instead of trying to identify the individual terms in (1.6.6), we simply look at the tree and write

$$P(B_2 \mid A) = \frac{P(B_2 \cap A)}{P(A)}$$

$$= \frac{P\{\text{two-headed coin chosen and coin comes up heads}\}}{P\{\text{coin comes up heads}\}}$$

$$= \frac{(1/4)(1)}{(3/4)(1/2) + (1/4)(1)} = \frac{2}{5} \blacktriangleleft$$

There are many situations in which an experiment consists of a sequence of steps, and the conditional probabilities of events happening at step $n + 1$, given outcomes at step n, are specified. In such cases a description by means of a tree diagram may be very convenient (see Problems).

▶ **Example 4.** A loaded die is tossed once; if N is the result of the toss, then $P\{N = i\} = p_i$, $i = 1, 2, 3, 4, 5, 6$. If $N = i$, an unbiased coin is tossed independently i times. Find the conditional probability that N will be odd, given that at least one head is obtained (see Figure 1.6.3).

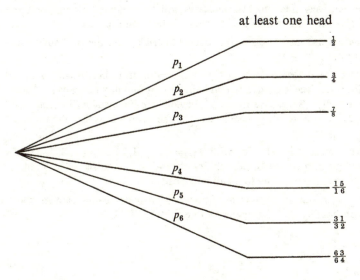

FIGURE 1.6.3

Let $A = \{$at least one head obtained$\}$, $B = \{N$ odd$\}$. Then $P(B \mid A) = P(A \cap B)/P(A)$. Now

$$P(A \cap B) = \sum_{i=1,3,5} P\{N = i \text{ and at least one head obtained}\}$$

$$= \tfrac{1}{2}p_1 + \tfrac{7}{8}p_3 + \tfrac{31}{32}p_5$$

since when an unbiased coin is tossed independently i times, the probability of at least one head is $1 - (1/2)^i$. Similarly,

$$P(A) = \sum_{i=1}^{6} P\{N = i \text{ and at least one head obtained}\}$$

$$= \sum_{i=1}^{6} p_i(1 - 2^{-i})$$

Thus

$$P(B \mid A) = \frac{\tfrac{1}{2}p_1 + \tfrac{7}{8}p_3 + \tfrac{31}{32}p_5}{\tfrac{1}{2}p_1 + \tfrac{3}{4}p_2 + \tfrac{7}{8}p_3 + \tfrac{15}{16}p_4 + \tfrac{31}{32}p_5 + \tfrac{63}{64}p_6} \qquad \blacktriangleleft$$

PROBLEMS

1. In 10 Bernoulli trials find the conditional probability that all successes will occur consecutively (i.e., no two successes will be separated by one or more failures), given that the number of successes is between four and six.

2. If X is the number of successes in n Bernoulli trials, find the probability that $X \geq 3$ given that $X \geq 1$.

3. An unbiased die is tossed once. If the face is odd, an unbiased coin is tossed repeatedly; if the face is even, a biased coin with probability of heads $p \neq 1/2$ is tossed repeatedly. (Successive tosses of the coin are independent in each case.) If the first n throws result in heads, what is the probability that the unbiased coin is being used?

4. A positive integer I is selected, with $P\{I = n\} = (1/2)^n$, $n = 1, 2, \ldots$. If I takes the value n, a coin with probability of heads e^{-n} is tossed once. Find the probability that the resulting toss will be a head.

5. A bridge player and his partner are known to have six spades between them. Find the probability that the spades will be split
   (a) 3-3
   (b) 4-2 or 2-4
   (c) 5-1 or 1-5
   (d) 6-0 or 0-6.

6. An urn contains 30 white and 15 black balls. If 10 balls are drawn with (respectively without) replacement, find the probability that the first two balls will be white, given that the sample contains exactly six white balls.

7. Let C_1 be an unbiased coin, and C_2 a biased coin with probability of heads 3/4. At time $t = 0$, C_1 is tossed. If the result is heads, then C_1 is tossed at time $t = 1$; if the result is tails, C_2 is tossed at $t = 1$. The process is repeated at $t = 2, 3, \ldots$. In general, if heads appears at $t = n$, then C_1 is tossed at $t = n + 1$; if tails appears at $t = n$, then C_2 is tossed at $t = n + 1$.

   Find y_n = the probability that the toss at $t = n$ will be a head (set up a difference equation).

8. In the switching network of Figure P.1.6.8, the switches operate independently.

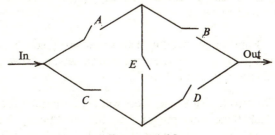

Figure P.1.6.8

Each switch closes with probability p, and remains open with probability $1 - p$.
(a) Find the probability that a signal at the input will be received at the output.
(b) Find the conditional probability that switch E is open given that a signal is received.

9. In a certain village 20% of the population has disease D. A test is administered which has the property that if a person has D, the test will be positive 90% of the time, and if he does not have D, the test will still be positive 30% of the time. All those whose test is positive are given a drug which invariably cures the disease, but produces a characteristic rash 25% of the time. Given that a person picked at random has the rash, what is the probability that he actually had D to begin with?

1.7 SOME FALLACIES IN COMBINATORIAL PROBLEMS

In this section we illustrate some common traps occurring in combinatorial problems. In the first three examples there will be a multiple count.

▶ **Example 1.** Three cards are selected from an ordinary deck, without replacement. Find the probability of not obtaining a heart.

PROPOSED SOLUTION. The total number of selections is $\binom{52}{3}$. To find the number of favorable outcomes, notice that the first card cannot be a heart; thus we have 39 choices at step 1. Having removed one card, there are 38 nonhearts left at step 2 (and then 37 at step 3). The desired probability is $(39)(38)(37)/\binom{52}{3}$.

FALLACY. In computing the number of favorable outcomes, a particular selection might be: 9 of diamonds, 8 of clubs, 7 of diamonds. Another selection is: 8 of clubs, 9 of diamonds, 7 of diamonds. In fact the $3! = 6$ possible orderings of these three cards are counted separately in the numerator (but not in the denominator). Thus the proposed answer is too high by a factor of $3!$; the actual probability is $(39)(38)(37)/3!\binom{52}{3} = \binom{39}{3}/\binom{52}{3}$ (see example 2, Section 1.6). ◄

▶ **Example 2.**   Find the probability that a five-card poker hand will result in three of a kind (three cards of the same face value x, plus two cards of face values y and z, with x, y, and z distinct).

PROPOSED SOLUTION. Pick the face value to appear three times (13 possibilities). Pick three suits out of four for the "three of a kind" ($\binom{4}{3}$ choices). Now one face value is excluded, so that 48 cards are left in the deck. Pick one of them as the fourth card; the fifth card can be chosen in 44 ways, since the fourth card excludes another face value. Thus the desired probability is $(13)\binom{4}{3}(48)(44)/\binom{52}{5}$.

FALLACY. Say the first three cards are aces. The fourth and fifth cards might be the jack of clubs and the 6 of diamonds, or equally well the 6 of diamonds and the jack of clubs. These possibilities are counted separately in the numerator but not in the denominator, so that the proposed answer is too high by a factor of 2. The actual probability is $13\binom{4}{3}(48)(44)/2\binom{52}{5} = 13\binom{4}{3}\binom{12}{2}16/\binom{52}{5}$ [see Problem 2, Section 1.4; the factor $\binom{12}{2}16$ corresponds to the selection of two distinct face values out of the remaining 12, then one card from each of these face values].

REMARK.   A more complicated approach to this problem is as follows. Pick the face value x to appear three times, then pick three suits out of four, as before. Forty-nine cards remain in the deck, and the total number of ways of selecting two remaining cards is $\binom{49}{2}$. However, if the two face values are the same, we obtain a full house; there are $12\binom{4}{2}$ selections in which this happens (select one face value out of 12, then two suits out of four). Also, if one of the two cards has face value x, we obtain four of a kind; since there is only one remaining card with face value x and 48 cards remain after this one is chosen, there are 48 possibilities. Thus the probability of obtaining three of a kind is

$$\frac{13\binom{4}{3}\left[\binom{49}{2} - 12\binom{4}{2} - 48\right]}{\binom{52}{5}}$$

(This agrees with the previous answer.) ◄

▶ **Example 3.**   Ten cards are drawn without replacement from an ordinary deck. Find the probability that at least nine will be of the same suit.

PROPOSED SOLUTION.   Pick the suit in any one of four ways, then choose nine of 13 face values. Forty-three cards now remain in the deck, so that the desired probability is $4\binom{13}{9}43/\binom{52}{10}$.

FALLACY.   Consider two possible selections.
1. Spades are chosen, then face values A K Q J 10 9 8 7 6. The last card is the 5 of spades.
2. Spades are chosen, then face values A K Q J 10 9 8 7 5. The last card is the 6 of spades (see Figure 1.7.1). Both selections yield the same 10 cards, but are counted separately in the computation. To find the number of duplications, notice that we can select 10 cards out of 13 to be involved in the duplication; each choice of one card (out of 10) for the last card yields a distinct path in Figure 1.7.1. Of the 10 possible paths corresponding to a given selection of cards, nine are redundant. Thus the actual probability is

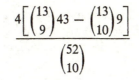

$$\frac{4\left[\binom{13}{9}43 - \binom{13}{10}9\right]}{\binom{52}{10}}$$

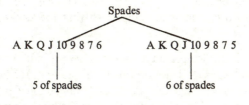

FIGURE 1.7.1   Multiple Count.

Now

$$\binom{13}{9}43 - \binom{13}{10}9 = \binom{13}{9}39 + \binom{13}{9}4 - \binom{13}{10}9$$

$$= \binom{13}{9}39 + \binom{13}{10}10 - \binom{13}{10}9$$

$$= \binom{13}{9}39 + \binom{13}{10}$$

so that the probability is

$$\frac{4\left[\binom{13}{9}39 + \binom{13}{10}\right]}{\binom{52}{10}}$$

as obtained in a straightforward manner in Problem 4, Section 1.4. ◄

▶ **Example 4.**   An urn contains 10 balls $b_1, \ldots, b_{10}$. Five balls are drawn without replacement. Find the probability that b_8 and b_9 will be included in the sample.

PROPOSED SOLUTION.   We are drawing half the balls, so that the probability that a particular ball will be included is $1/2$. Thus the probability of including both b_8 and b_9 is $(1/2)(1/2) = 1/4$.

FALLACY.   Let $A = \{b_8$ is included$\}$, $B = \{b_9$ is included$\}$. The difficulty is simply that A and B are not independent. For $P(A \cap B) = \binom{8}{3}/\binom{10}{5} = 2/9$ (after b_8 and b_9 are chosen, three balls are to be selected from the remaining eight). Also $P(A) = P(B) = \binom{9}{4}/\binom{10}{5} = 1/2$, so that $P(A \cap B) \neq P(A)P(B)$. ◄

▶ **Example 5.**   Two cards are drawn independently, with replacement, from an ordinary deck; at each selection all 52 cards are equally likely. Find the probability that the king of spades and the king of hearts will be chosen (in some order).

PROPOSED SOLUTION.   The number of unordered samples of size 2 out of 52, with replacement, is $\binom{52+2-1}{2} = \binom{53}{2}$ [see (1.4.4)]. The kings of spades and hearts constitute one such sample, so that the desired probability is $1/\binom{53}{2}$.

FALLACY.   It is not legitimate to assign equal probability to all unordered samples with replacement. If we do this we are saying, for example, that the outcomes "ace of spades, ace of spades" and "king of spades, king of hearts" have the same probability. However, this cannot be the case if independent sampling is assumed. For the probability that the ace of spades is chosen twice is $(1/52)^2$, while the probability that the spade and heart kings will be chosen (in some order) is $P\{$first card is the king of spades, second card is the king of hearts$\} + P\{$first card is the king of hearts, second card is the king of spades$\} = 2(1/52)^2$, which is the desired probability.

The main point is that we must use *ordered* samples with replacement in order to capture the idea of independence. ◀

1.8 APPENDIX: STIRLING'S FORMULA

An estimate of $n!$ that is of importance both in numerical calculations and theoretical analysis is *Stirling's formula*

$$n! \sim n^n e^{-n} \sqrt{2\pi n}$$

in the sense that

$$\lim_{n \to \infty} \frac{n!}{(n^n e^{-n} \sqrt{2\pi n})} = 1$$

PROOF. Define $(2n)!!$ (read $2n$ semifactorial) as $2n(2n - 2)(2n - 4) \cdots 6(4)(2)$, and $(2n + 1)!!$ as $(2n + 1)(2n - 1) \cdots (5)(3)(1)$. We first show that

(a) $$\frac{(2n)!!}{(2n + 1)!!} < \frac{\pi}{2} \frac{(2n - 1)!!}{(2n)!!} < \frac{(2n - 2)!!}{(2n - 1)!!}$$

Let $I_k = \int_0^{\pi/2} (\cos x)^k \, dx$, $k = 0, 1, 2, \ldots$. Then $I_0 = \pi/2$, $I_1 = 1$. Integrating by parts, we obtain $I_k = \int_0^{\pi/2} (\cos x)^{k-1} \, d(\sin x) = \int_0^{\pi/2} (k - 1)(\cos x)^{k-2} \sin^2 x \, dx$. Since $\sin^2 x = 1 - \cos^2 x$, we have $I_k = (k - 1)I_{k-2} - (k - 1)I_k$ or $I_k = [(k - 1)/k]I_{k-2}$. By iteration, we obtain $I_{2n} = (\pi/2) [(2n - 1)!!/(2n)!!]$ and $I_{2n+1} = [(2n)!!/(2n + 1)!!]$. Since $(\cos x)^k$ decreases with k, so does I_k, and hence $I_{2n+1} < I_{2n} < I_{2n-1}$, and (a) is proved.

(b) Let $Q_n = \binom{2n}{n}/2^{2n}$. Then

$$\lim_{n \to \infty} Q_n \sqrt{n\pi} = 1$$

To prove this, write

$$Q_n = \frac{(2n)!}{n! \, n! \, 2^{2n}} = \frac{(2n)!}{(2^n n!)^2}$$

$$= \frac{(2n)!}{((2n)(2n - 2) \cdots (4)(2))^2} = \frac{(2n - 1)!!}{(2n)!!}$$

Thus, by (a),

$$\frac{(2n)!!}{(2n+1)!!} < \frac{\pi}{2} Q_n < \frac{(2n-2)!!}{(2n-1)!!}$$

Multiply this inequality by

$$\frac{(2n-1)!!}{(2n-2)!!} = \frac{(2n-1)!!}{(2n)!!} \frac{(2n)!!}{(2n-2)!!} = Q_n(2n)$$

to obtain

$$\frac{2n}{2n+1} < n\pi Q_n{}^2 < 1$$

If we let $n \to \infty$, we obtain $n\pi Q_n{}^2 \to 1$, proving (b).

(c) *Proof of Stirling's formula.* Let $c_n = n!/n^n e^{-n}\sqrt{2\pi n}$. We must show that $c_n \to 1$ as $n \to \infty$. Consider $(n+1)!/n! = n+1$. We have

$$\frac{(n+1)!}{n!} = \frac{c_{n+1}(n+1)^{n+1}e^{-(n+1)}\sqrt{2\pi(n+1)}}{c_n n^n e^{-n}\sqrt{2\pi n}}$$

$$= \left(\frac{c_{n+1}}{c_n}\right)e^{-1}\left(\frac{n+1}{n}\right)^n \frac{(n+1)^{3/2}}{\sqrt{n}}$$

Thus

$$\frac{c_{n+1}}{c_n} = (n+1)(e)\left(\frac{n}{n+1}\right)^n \frac{\sqrt{n}}{(n+1)^{3/2}} = (e)\left(1 + \frac{1}{n}\right)^{-(n+1/2)}$$

Now $(1 + 1/n)^{n+1/2} > e$ for n sufficiently large (take logarithms and expand in a power series); hence $c_{n+1}/c_n < 1$ for large enough n. Since every monotone bounded sequence converges, $c_n \to$ a limit c. We must show $c = 1$. By (b),

$$\lim_{n\to\infty} \binom{2n}{n}\sqrt{n\pi}\, 2^{-2n} = 1$$

But

$$\binom{2n}{n}\sqrt{n\pi}\, 2^{-2n} = \frac{(2n)!}{n!\,n!}\frac{\sqrt{n\pi}}{2^{2n}} = \frac{c_{2n}(2n/e)^{2n}\sqrt{2\pi(2n)}}{(c_n(n/e)^n\sqrt{2\pi n})^2}\frac{\sqrt{n\pi}}{2^{2n}} = \frac{c_{2n}}{c_n{}^2}$$

Therefore $c_{2n}/c_n{}^2 \to 1$. However, $c_{2n} \to c$ and $c_n{}^2 \to c^2$, and consequently $c/c^2 = 1$, so that $c = 1$. The theorem is proved.

REMARK.   The last step requires that c be > 0. To see this, write

$$c_{n+1} = \frac{c_1}{c_0}\frac{c_2}{c_1}\ldots\frac{c_{n+1}}{c_n}$$

where c_0 is defined as 1. To show that $c_n \to a$ nonzero limit, it suffices to show that the limit of $\ln c_{n+1}$ is finite, and for this it is sufficient to show that $\sum_n \ln (c_{n+1}/c_n)$ converges to a finite limit. Now

$$\ln \frac{c_{n+1}}{c_n} = \ln \left[e \left(1 + \frac{1}{n} \right)^{-(n+1/2)} \right] = 1 - (n + \tfrac{1}{2}) \ln \left(1 + \frac{1}{n} \right)$$

$$= 1 - (n + \tfrac{1}{2}) \left(\frac{1}{n} - \frac{1}{2n^2} + \frac{\theta(n)}{n^3} \right)$$

where $\theta(n)$ is bounded by a constant independent of n. This is the order of $1/n^2$; hence $\sum_n \ln (c_{n+1}/c_n)$ converges, and the result follows.

2

Random Variables

2.1 INTRODUCTION

In Chapter 1 we mentioned that there are situations in which not all subsets of the sample space Ω can belong to the event class $\mathscr{F}$, and that difficulties of this type generally arise when Ω is uncountable. Such spaces may arise physically as approximations to discrete spaces with a very large number of points. For example, if a person is picked at random in the United States and his age recorded, a complete description of this experiment would involve a probability space with approximately 200 million points (if the data are recorded accurately enough, no two people have the same age). A more convenient way to describe the experiment is to group the data, for example, into 10-year intervals. We may define a function $q(x)$, $x = 5, 15, 25, \ldots$, so that $q(x)$ is the number of people, say in millions, between $x - 5$ and $x + 5$ years (see Figure 2.1.1).

For example, if $q(15) = 40$, there are 40 million people between the ages of 10 and 20 or, on the average, 4 million per year over that 10-year span. Now if we want the probability that a person picked at random will be between 14 and 16, we can get a reasonable figure by taking the average number of people per year $[4 = q(15)/10]$ and multiplying by the number of years (2) to obtain (roughly) 8 million people, then dividing by the total population to obtain a probability of $8/200 = .04$.

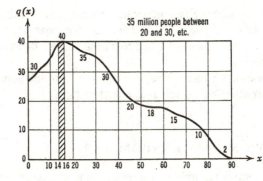

FIGURE 2.1.1 Age Statistics.

If we connect the values of $q(x)$ by a smooth curve, essentially what we are doing is evaluating $(1/200) \int_{14}^{16} [q(x)/10]\, dx$ to find the probability that a person picked at random will be between 14 and 16 years old. In general, we estimate the number of people between ages a and b by $\int_a^b [q(x)/10]\, dx$ so that $q(x)/10$ is the age density, that is, the number of people per unit age. We estimate the probability of obtaining an age between a and b by $\int_a^b [q(x)/2000]\, dx$; thus $q(x)/2000$ is the probability density, or probability per unit age. Thus we are led to the idea of assigning probabilities by means of an integral. We are taking Ω as (a subset of) the reals, and assigning $P(B) = \int_B f(x)\, dx$, where f is a real-valued function defined on the reals. There are several immediate questions, namely, what sigma field we are using, what functions f are allowed, what we mean by $\int_B f(x)\, dx$, and how we know that the resulting P is a probability.

For the moment suppose that we restrict ourselves to continuous or piecewise continuous f. Then we can certainly talk about $\int_B f(x)\, dx$, at least when B is an interval, and the integral is in the Riemann sense. Thus the appropriate sigma field $\mathscr{F}$ should contain the intervals, and hence must be at least as big as the smallest sigma field $\mathscr{B}$ containing the intervals ($\mathscr{B}$ exists; it can be described as the intersection of all sigma fields containing the intervals). The sigma field $\mathscr{B} = \mathscr{B}(E^1)$ is called the class of *Borel sets* of the reals E^1. Intuitively we may think of $\mathscr{B}$ being generated by starting with the intervals and repeatedly forming new sets by taking countable unions (and countable intersections) and complements in all possible ways (it turns out that there are subsets of E^1 that are not Borel sets).

Thus our problem will be to construct probability measures on the class of Borel sets of E^1. The reason for considering only the Borel sets rather than all subsets of E^1 is this. Suppose that we require that $P(B) = \int_B f(x)\, dx$

for all intervals B, where f is a particular nonnegative continuous function defined on E^1, and $\int_{-\infty}^{\infty} f(x)\, dx = 1$. There is no probability measure on the class of all subsets of E^1 satisfying this requirement, but there is such a measure on the Borel sets.

Before elaborating on these ideas, it is convenient to introduce the concept of a random variable; we do this in the next section.

2.2 DEFINITION OF A RANDOM VARIABLE

Intuitively, a random variable is a quantity that is measured in connection with a random experiment. If Ω is a sample space, and the outcome of the experiment is ω, a measuring process is carried out to obtain a number $R(\omega)$. Thus *a random variable is a real-valued function on a sample space*. (The formal definition, which is postponed until later in the section, is somewhat more restrictive.)

▶ **Example 1.** Throw a coin 10 times, and let R be the number of heads. We take $\Omega =$ all sequences of length 10 with components H and T; 2^{10} points altogether. A typical sample point is $\omega = HHTHTTHHTH$. For this point $R(\omega) = 6$. Another random variable, R_1, is the number of times a head is followed immediately by a tail. For the point ω above, $R_1(\omega) = 3$. ◄

▶ **Example 2.** Pick a person at random from a certain population and measure his height and weight. We may take the sample space to be the plane E^2, that is, the set of all pairs (x, y) of real numbers, with the first coordinate x representing the height and the second coordinate y the weight (we can take care of the requirement that height and weight be nonnegative by assigning probability 0 to the complement of the first quadrant). Let R_1 be the height of the person selected, and let R_2 be the weight. Then $R_1(x, y) = x$, $R_2(x, y) = y$. As another example, let R_3 be twice the height plus the cube root of the weight; that is, $R_3 = 2R_1 + \sqrt[3]{R_2}$. Then $R_3(x, y) = 2R_1(x, y) + \sqrt[3]{R_2(x, y)} = 2x + \sqrt[3]{y}$. ◄

▶ **Example 3.** Throw two dice. We may take the sample space to be the set of all pairs of integers (x, y), $x, y = 1, 2, \ldots, 6$ (36 points in all).
Let $R_1 =$ the result of the first toss. Then $R_1(x, y) = x$.
Let $R_2 =$ the sum of the two faces. Then $R_2(x, y) = x + y$.
Let $R_3 = 1$ if at least one face is an even number; $R_3 = 0$ otherwise.
Then $R_3(6, 5) = 1$; $R_3(3, 6) = 1$; $R_3(1, 3) = 0$, and so on. ◄

▶ **Example 4.** Imagine that we can observe the times at which electrons are emitted from the cathode of a vacuum tube, starting at time $t = 0$. As a sample space, we may take all infinite sequences of positive real numbers, with the components representing the emission times. Assume that the emission process never stops. Typical sample points might be $\omega_1 = (.2, 1.5, 6.3, \ldots)$, $\omega_2 = (.01, .5, .9, 1.7, \ldots)$. If R_1 is the number of electrons emitted before $t = 1$, then $R_1(\omega_1) = 1$, $R_1(\omega_2) = 3$. If R_2 is the time at which the first electron is emitted, then $R_2(\omega_1) = .2$, $R_2(\omega_2) = .01$. ◀

If we are interested in a random variable R defined on a given sample space, we generally want to know the probability of events involving R. Physical measurements of a quantity R generally lead to statements of the form $a \leq R \leq b$, and it is natural to ask for the probability that R will lie between a and b in a given performance of the experiment. Thus we are looking for $P\{\omega: a \leq R(\omega) \leq b\}$ (or, equally well, $P\{\omega: a < R(\omega) \leq b\}$, and so on). For example, if a coin is tossed independently n times, with probability p of coming up heads on a given toss, and if R is the number of heads, we have seen in Chapter 1 that

$$P\{\omega: a \leq R(\omega) \leq b\} = \sum_{k=a}^{b} \binom{n}{k} p^k (1 - p)^{n-k}$$

NOTATION. $\{\omega: a \leq R(\omega) \leq b\}$ will often be abbreviated to $\{a \leq R \leq b\}$.

As another example, if two unbiased dice are tossed independently, and R_2 is the sum of the faces (Example 3 above), then $P\{R_2 = 6\} = P\{(5, 1), (1, 5), (4, 2), (2, 4), (3, 3)\} = 5/36$.

In general an "event involving R" corresponds to a statement that the value of R lies in a set B; that is, the event is of the form $\{\omega: R(\omega) \in B\}$. Intuitively, if $P\{\omega: R(\omega) \in I\}$ is known for all intervals I, then $P\{\omega: R(\omega) \in B\}$ is determined for any "well-behaved" set B, the reason being that any such set can be built up from intervals. For example, $P\{0 \leq R < 2 \text{ or } R > 3\}$ ($= P\{R \in [0, 2) \cup (3, \infty)\}$) $= P\{0 \leq R < 2\} + P\{R > 3\}$. Thus it appears that in order to describe the nature of R completely, it is sufficient to know $P\{R \in I\}$ for each interval I. We consider in more detail the problem of characterizing a random variable in the next section; in the remainder of this section we give the formal definition of a random variable.

* For the concept of random variable to fit in with our established model for a probability space, the sets $\{a \leq R \leq b\}$ must be events; that is, they must belong to the sigma field $\mathscr{F}$. Thus a first restriction on R is that for all real a, b, the sets $\{\omega: a \leq R(\omega) \leq b\}$ are in $\mathscr{F}$. Thus we can talk intelligently about the event that R lies between a and b.

A question now comes up: Suppose that the sets $\{a \leq R \leq b\}$ are in $\mathscr{F}$

for all a, b. Can we talk about the event that R belongs to a set B of reals, for B more general than a closed interval?

For example, let $B = [a, b)$ be an interval closed on the left, open on the right. Then

$$a \leq R(\omega) < b \text{ iff } a \leq R(\omega) \leq b - \frac{1}{n} \qquad \text{for at least one } n = 1, 2, \ldots$$

Thus

$$\{\omega: a \leq R(\omega) < b\} = \bigcup_{n=1}^{\infty} \left\{\omega: a \leq R(\omega) \leq b - \frac{1}{n}\right\}$$

and this set is a countable union of sets in $\mathscr{F}$, hence belongs to $\mathscr{F}$. In a similar fashion we can handle all types of intervals. Thus $\{\omega: R(\omega) \in B\} \in \mathscr{F}$ for all intervals B.

In fact $\{\omega: R(\omega) \in B\}$ belongs to $\mathscr{F}$ for all Borel sets B. The sequence of steps by which this is proved is outlined in Problem 1.

We are now ready for the formal definition.

DEFINITION.   A *random variable* on the probability space $(\Omega, \mathscr{F}, P)$ is a real valued function R defined on Ω, such that for every Borel subset B of the reals, $\{\omega: R(\omega) \in B\}$ belongs to $\mathscr{F}$.

Notice that the probability P is not involved in the definition at all; if R is a random variable on $(\Omega, \mathscr{F}, P)$ and the probability measure is changed, R is still a random variable. Notice also that, by the above discussion, to check whether a given function R is a random variable it is sufficient to know that $\{\omega: a \leq R(\omega) \leq b\} \in \mathscr{F}$ for all real a, b. In fact (Problem 2) it is sufficient that $\{\omega: R(\omega) < b\} \in \mathscr{F}$ for all real b (or, equally well, $\{\omega: R(\omega) \leq b\} \in \mathscr{F}$ for all real b; or $\{\omega: R(\omega) > a\} \in \mathscr{F}$ for all real a; or $\{\omega: R(\omega) \geq a\} \in \mathscr{F}$ for all real a; the argument is essentially the same in all cases).

Notice that if $\mathscr{F}$ consists of all subsets of Ω, $\{\omega: R(\omega) \in B\}$ automatically belongs to $\mathscr{F}$, so that in this case any real-valued function on the sample space is a random variable. Examples 1 and 3 fall into this category.

Now let us consider Example 2. We take $\Omega =$ the plane E^2, $\mathscr{F} =$ the class of Borel subsets of E^2, that is, the smallest sigma field containing all rectangles (we shall use "rectangle" in a very broad sense, allowing open, closed, or semiclosed rectangles, as well as infinite rectangular strips).

To check that R_1 is a random variable, we have

$$\{(x, y): a \leq R_1(x, y) \leq b\} = \{(x, y): a \leq x \leq b\}$$

which is a rectangular strip and hence a set in $\mathscr{F}$. Similarly, R_2 is a random variable. For R_3, see Problem 3.

Example 2 generalizes as follows. Take $\Omega = E^n =$ all n-tuples of real numbers, $\mathscr{F}$ the smallest sigma field containing the n-dimensional "intervals." [If $a = (a_1, \ldots, a_n)$, $b = (b_1, \ldots, b_n)$, the interval (a, b) is defined as $\{x \in E^n : a_i < x_i < b_i, i = 1, \ldots, n\}$; closed and semiclosed intervals are defined similarly.] The coordinate functions, given by $R_1(x_1, \ldots, x_n) = x_1$, $R_2(x_1, \ldots, x_n) = x_2, \ldots, R_n(x_1, \ldots, x_n) = x_n$, are random variables.

Example 4 involves some serious complications, since the sample points are infinite sequences of real numbers. We postpone the discussion of situations of this type until much later (Chapter 6).

PROBLEMS

*1. Let R be a real-valued function on a sample space Ω, and let $\mathscr{C}$ be the collection of all subsets B of E^1 such that $\{\omega : R(\omega) \in B\} \in \mathscr{F}$.
   (a) Show that $\mathscr{C}$ is a sigma field.
   (b) If all intervals belong to $\mathscr{C}$, that is, if $\{\omega : R(\omega) \in B\} \in \mathscr{F}$ when B is an interval, show that all Borel sets belong to $\mathscr{C}$. Conclude that R is a random variable.

*2. Let R be a real-valued function on a sample space Ω, and assume $\{\omega : R(\omega) < b\} \in \mathscr{F}$ for all real b. Show that R is a random variable.

*3. In Example 2, show that R_3 is a random variable. Do this by showing that if R_1 and R_2 are random variables, so is $R_1 + R_2$; if R is a random variable, so is aR for any real a; if R is a random variable, so is $\sqrt[3]{R}$.

2.3 CLASSIFICATION OF RANDOM VARIABLES

If R is a random variable on the probability space $(\Omega, \mathscr{F}, P)$, we are generally interested in calculating probabilities of events involving R, that is, $P\{\omega : R(\omega) \in B\}$ for various (Borel) sets B. The way in which these probabilities are calculated will depend on the particular nature of R; in this section we examine some standard classes of random variables.

The random variable R is said to be *discrete* iff the set of possible values of R is finite or countably infinite. In this case, if $x_1, x_2, \ldots$ are the values of R that belong to B, then

$$P\{R \in B\} = P\{R = x_1 \text{ or } R = x_2 \text{ or } \cdots\}$$
$$= P\{R = x_1\} + P\{R = x_2\} + \cdots = \sum_{x \in B} p_R(x)$$

where $p_R(x)$, x real, is the *probability function* of R, defined by $p_R(x) = P\{R = x\}$. Thus the probability of an event involving R is found by summing the probability function over the set of points favorable to the event. In particular, the probability function determines the probability of all events involving R.

▶ **Example 1.**   Let R be the number of heads in two independent tosses of a coin, with the probability of heads being .6 on a given toss. Take $\Omega = \{HH, HT, TH, TT\}$ with probabilities .36, .24, .24, .16 assigned to the four points of Ω; take $\mathscr{F} =$ all subsets. Then R has three possible values, namely, 0, 1, and 2, and $P\{R = 0\} = .16$, $P\{R = 1\} = .48$, $P\{R = 2\} = .36$, by inspection or by using the binomial formula

$$P\{R = k\} = \binom{n}{k} p^k (1 - p)^{n-k} \blacktriangleleft$$

Another way of characterizing R is by means of the *distribution function*, defined by

$$F_R(x) = P\{R \le x\}, \qquad x \text{ real}$$

(see Figure 2.3.1 for a sketch of F_R and p_R in Example 1).

Observe that, for example, $P\{R \le 1\} = p_R(0) + p_R(1) = .64$, but if $0 < x < 1$, we have $P\{R \le x\} = p_R(0) = .16$. Thus F_R has a discontinuity at $x = 1$, of magnitude $.48 = p_R(1)$. In general, if R is discrete, and $P\{R = x_n\} = p_n$, $n = 1, 2, \ldots$, where the p_n are >0 and $\sum_n p_n = 1$, then F_R has a jump of magnitude p_n at $x = x_n$; F_R is constant between jumps.

In the discrete case, if we are given the probability function, we can construct the distribution function, and, conversely, given F_R, we can construct

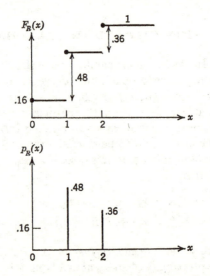

FIGURE 2.3.1   Distribution and Probability Functions of a Discrete Random Variable.

p_R. Knowledge of either function is sufficient to determine the probability of all events involving R.

We now consider the case introduced in Section 2.1, where probabilities are assigned by means of an integral.

Let f be a nonnegative Riemann integrable† function defined on E^1 with $\int_{-\infty}^{\infty} f(x)\, dx = 1$. Take $\Omega = E^1$, $\mathscr{F} =$ Borel sets. We would like to write, for each $B \in \mathscr{F}$,

$$P(B) = \int_B f(x)\, dx$$

but this makes sense only if B is an interval. However, the following result is applicable.

Theorem 1. *Let f be a nonnegative real-valued function on E^1, with $\int_{-\infty}^{\infty} f(x)\, dx = 1$. There is a unique probability measure P defined on the Borel subsets of E^1, such that $P(B) = \int_B f(x)\, dx$ for all intervals $B = (a, b]$.*

The theorem belongs to the domain of measure and integration theory, and will not be proved here.

The theorem allows us to talk about the integral of f over an arbitrary Borel set B. We simply define $\int_B f(x)\, dx$ as $P(B)$, where P is the probability measure given by the theorem.

The uniqueness part of the theorem may then be phrased as follows. If Q is a probability measure on the Borel subsets of E^1 and $Q(B) = \int_B f(x)\, dx$ for all intervals $B = (a, b]$, then $Q(B) = \int_B f(x)\, dx$ for all Borel sets B.

If R is defined on Ω by $R(\omega) = \omega$ (so that the outcome of the experiment is identified with the value of R), then

$$P\{\omega: R(\omega) \in B\} = P(B) = \int_B f(x)\, dx$$

In particular, the distribution function of R is given by

$$F_R(x) = P\{\omega: R(\omega) \le x\} = P(-\infty, x] = \int_{-\infty}^{x} f(t)\, dt$$

so that F_R is represented as an integral.

DEFINITION. The random variable R is said to be *absolutely continuous* iff there is a nonnegative function $f = f_R$ defined on E^1 such that

$$F_R(x) = \int_{-\infty}^{x} f_R(t)\, dt \qquad \text{for all real } x \qquad (2.3.1)$$

f_R is called the *density function* of R. We shall see in Section 2.5 that $F_R(x)$ must approach 1 as $x \to \infty$; hence $\int_{-\infty}^{\infty} f_R(x)\, dx = 1$.

† "Integrable" will from now on mean "Riemann integrable."

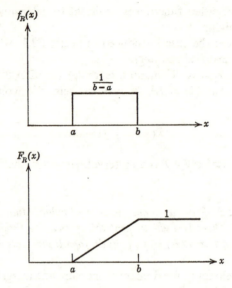

FIGURE 2.3.2 Distribution and Density Functions of a Uniformly Distributed Random Variable.

▶ **Example 2.** A number R is chosen at random between a and b; R is assumed to be *uniformly distributed*; that is, the probability that R will fall into an interval of length c depends only on c, not on the position of the interval within $[a, b]$.

We take $\Omega = E^1$, $\mathscr{F} = $ Borel sets, $R(\omega) = \omega$, $f(x) = f_R(x) = 1/(b - a)$, $a \le x \le b$; $f(x) = 0$, $x > b$ or $x < a$. Define $P(B) = \int_B f(x)\,dx$. In particular, if B is a subinterval of $[a, b]$, then $P(B) = $ (length of B)/$(b - a)$. The density and distribution function of R are shown in Figure 2.3.2. ◀

NOTE. The values of F_R are *probabilities*, but the values of f_R are not; probabilities are found by *integrating* f_R.

$$F_R(x) = P\{R \le x\} = \int_{-\infty}^{x} f_R(t)\,dt$$

If R is absolutely continuous, then

$$P\{a < R \le b\} = \int_a^b f_R(x)\,dx, \qquad a < b$$

For $\{R \le b\}$ is the disjoint union of the events $\{R \le a\}$ and $\{a < R \le b\}$; hence $P\{R \le b\} = P\{R \le a\} + P\{a < R \le b\}$. It follows that

$$P\{a < R \leq b\} = F_R(b) - F_R(a) \tag{2.3.2}$$

$$= \int_{-\infty}^{b} f_R(x)\, dx - \int_{-\infty}^{a} f_R(x)\, dx = \int_{a}^{b} f_R(x)\, dx$$

Thus, if $Q(B) = P\{R \in B\}$, we have $Q(B) = \int_B f_R(x)\, dx$ when B is an interval $(a, b]$. By Theorem 1, $Q(B) = \int_B f_R(x)\, dx$ for all Borel sets B. Therefore, *if R is absolutely continuous*,

$$P\{R \in B\} = \int_B f_R(x)\, dx \qquad \text{for all Borel sets } B$$

The basic point is that the *density function* f_R *determines the probability of all events involving R*.

If R is absolutely continuous, then

$$P\{R = c\} = P\{c \leq R \leq c\} = \int_c^c f_R(x)\, dx = 0$$

The event $\{R = c\}$ is in general *not impossible*; for example, if R is uniformly distributed between a and b, each event $\{R = x\}$, $a \leq x \leq b$, is possible; that is, the set $\{\omega: R(\omega) = x\}$ is not empty. But the event $\{R = c\}$ has probability 0. This does not contradict the axioms of probability. The definition of a probability measure requires that if the event A is impossible (i.e., $A = \varnothing$) then $P(A) = 0$; the *converse* need not be true. Intuitively, if R is uniformly distributed between a and b, it should be expected that all events $\{R = x\}$, $a \leq x \leq b$, will have the same probability. Any probability other than 0 will lead to a contradiction, since there are infinitely many points x between a and b.

As a consequence of the fact that $P\{R = x\} = 0$ in the absolutely continuous case, we have

$$P\{a \leq R \leq b\} = P\{a < R \leq b\} = P\{a \leq R < b\}$$

$$= P\{a < R < b\}$$

$$= \int_a^b f_R(x)\, dx$$

$$= F_R(b) - F_R(a) \tag{2.3.3}$$

Notice also that although in the discrete case the probability function of R determines the probability of all events involving R, in the absolutely continuous case it gives no information at all, since $p_R(x) = P\{R = x\} = 0$ for all x. However, the distribution function of R is still adequate, since F_R determines f_R. If f_R is continuous, it may be obtained from F_R by differentiation; that is,

$$\frac{d}{dx} \int_{-\infty}^{x} f_R(t)\, dt = f_R(x)$$

(the fundamental theorem of calculus). The general proof that F_R determines f_R is measure-theoretic, and we shall not pursue it here.

If f_R is continuous, we have just seen that F_R is differentiable, and its derivative is f_R. In general, if R is absolutely continuous, $F_R(x)$ will be a continuous function of x, but again we shall not pursue this.

We shall show in Section 2.5 that the distribution function of an *arbitrary* random variable must be nondecreasing [$a < b$ implies $F_R(a) \le F_R(b)$], must approach 1 as $x \to \infty$, and must approach 0 as $x \to -\infty$.

▶ **Example 3.** Let R be time of emission of the first electron from the cathode of a vacuum tube. Under certain physical assumptions, it turns out that R has the following density function:

$$f_R(x) = \lambda e^{-\lambda x}, \qquad x \ge 0$$
$$\qquad\qquad\qquad\qquad\qquad\text{(λ constant)}$$
$$= 0 \qquad\qquad x < 0$$

(see Figure 2.3.3).

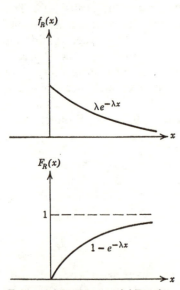

$$F_R(x) = \int_{-\infty}^{x} f_R(t)\, dt$$

so $F_R(x) = 0$, $x < 0$,

and if $x \ge 0$,

$$F_R(x) = \int_{-\infty}^{0} f_R(t)\, dt$$

$$+ \int_{0}^{x} f_R(t)\, dt = \int_{0}^{x} \lambda e^{-\lambda t}\, dt$$

$$= 1 - e^{-\lambda x}$$

FIGURE 2.3.3   Exponential Density and Distribution Functions.

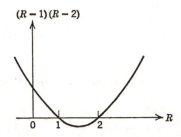

FIGURE 2.3.4 Calculation of Probabilities.

We calculate some probabilities of events involving R:

$$P\{1 \le R \le 2\} = \int_1^2 \lambda e^{-\lambda x}\, dx = e^{-\lambda} - e^{-2\lambda} = F_R(2) - F_R(1)$$

$$P\{(R-1)(R-2) \ge 0\} = P\{R \le 1 \quad \text{or} \quad R \ge 2\}$$

$$= P\{R \le 1\} + P\{R \ge 2\}$$

$$= \int_0^1 \lambda e^{-\lambda x}\, dx + \int_2^\infty \lambda e^{-\lambda x}\, dx$$

$$= 1 - e^{-\lambda} + e^{-2\lambda}$$

(see Figure 2.3.4). ◄

REMARK. You will often see the statement "Let R be an absolutely continuous random variable with density function f," with no reference made to the underlying probability space. However, we have seen that we can always supply an appropriate space, as follows. Take $\Omega = E^1$, $\mathscr{F}$ = Borel sets, $P(B) = \int_B f(x)\, dx$ for all $B \in \mathscr{F}$. If $R(\omega) = \omega$, $\omega \in \Omega$, then R is absolutely continuous and has density f.

In a sense, it does not make any difference how we arrive at Ω and P; we may equally well use a different Ω and P and a different R, as long as R is absolutely continuous with density f. No matter what construction we use, we get the same essential result, namely,

$$P\{R \in B\} = \int_B f(x)\, dx$$

Thus questions about probabilities of events involving R are answered completely by knowledge of the density f.

PROBLEMS

1. An absolutely continuous random variable R has a density function $f(x) = (1/2)e^{-|x|}$.
   (a) Sketch the distribution function of R.
   (b) Find the probability of each of the following events.

   (1) $\{|R| \leq 2\}$ (5) $\{R^3 - R^2 - R - 2 \leq 0\}$
   (2) $\{|R| \leq 2 \text{ or } R \geq 0\}$ (6) $\{e^{\sin \pi R} \geq 1\}$
   (3) $\{|R| \leq 2 \text{ and } R \leq -1\}$ (7) $\{R \text{ is irrational}\}$ ($= \{\omega : R(\omega) \text{ is}$
   (4) $\{|R| + |R - 3| \leq 3\}$ an irrational number$\})$

2. Consider a sequence of five Bernoulli trials. Let R be the number of times that a head is followed immediately by a tail. For example, if $\omega = HHTHT$ then $R(\omega) = 2$, since a head is followed directly by a tail at trials 2 and 3, and also at trials 4 and 5. Find the probability function of R.

2.4 FUNCTIONS OF A RANDOM VARIABLE

A general problem that arises in many branches of science is the following. Given a system of some sort, to which an input is applied; knowledge of some of the characteristics of the system, together with knowledge of the input, will allow some estimate of the behavior at the output. We formulate a special case of this problem. Given a random variable R_1 on a probability space, and a real-valued function g on the reals, we define a random variable R_2 by $R_2 = g(R_1)$; that is, $R_2(\omega) = g(R_1(\omega))$, $\omega \in \Omega$. R_1 plays the role of the input, and g the role of the system; the output R_2 is a random variable defined on the same space as R_1. Given the function g and the distribution or density function of the random variable R_1, the problem is to find the distribution or density function of R_2.

NOTE.   If R_1 is a random variable and we set $R_2 = g(R_1)$, the question arises
   as to whether R_2 is in fact a random variable. The answer is yes if
   g is continuous or piecewise continuous; we shall consider this problem in greater detail in Section 2.7.

▶ **Example 1.** Let R_1 be absolutely continuous, with the density f_1 given in Figure 2.4.1. Let $R_2 = R_1^2$; that is, $R_2(\omega) = R_1^2(\omega)$, $\omega \in \Omega$. Find the distribution or density function of R_2.
   We shall indicate two approaches to the problem.

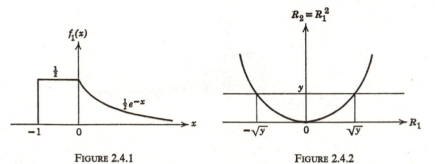

FIGURE 2.4.1   FIGURE 2.4.2

DISTRIBUTION FUNCTION METHOD.   In this method the distribution function F_2 of R_2 is found directly, by expressing the event $\{R_2 \leq y\}$ in terms of the random variable R_1. First, since $R_2 \geq 0$, we have $F_2(y) = P\{R_2 \leq y\} = 0$ for $y < 0$.

If $y \geq 0$, then $R_2 \leq y$ iff $-\sqrt{y} \leq R_1 \leq \sqrt{y}$ (see Figure 2.4.2). Thus, if $y \geq 0$,

$$P\{R_2 \leq y\} = P\{-\sqrt{y} \leq R_1 \leq \sqrt{y}\} = \int_{-\sqrt{y}}^{\sqrt{y}} f_1(x)\,dx$$

In particular, if $0 \leq y \leq 1$, then

$$F_2(y) = \int_{-\sqrt{y}}^{\sqrt{y}} f_1(x)\,dx = \int_{-\sqrt{y}}^{0} \tfrac{1}{2}\,dx + \int_{0}^{\sqrt{y}} \tfrac{1}{2}e^{-x}\,dx = \tfrac{1}{2}\sqrt{y} + \tfrac{1}{2}(1 - e^{-\sqrt{y}})$$

(see Figure 2.4.3).

If $y > 1$,

$$F_2(y) = \int_{-\sqrt{y}}^{\sqrt{y}} f_1(x)\,dx = \int_{-\sqrt{y}}^{-1} 0\,dx + \int_{-1}^{0} \tfrac{1}{2}\,dx + \int_{0}^{\sqrt{y}} \tfrac{1}{2}e^{-x}\,dx$$

$$= \tfrac{1}{2} + \tfrac{1}{2}(1 - e^{-\sqrt{y}})$$

(see Figure 2.4.4). A sketch of F_2 is given in Figure 2.4.5.

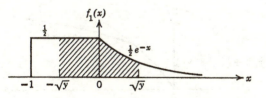

FIGURE 2.4.3

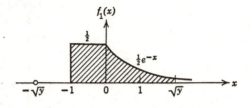

FIGURE 2.4.4

We would like to conclude, by inspection of the distribution function F_2, that the random variable R_2 is absolutely continuous. We should be able to find the density f_2 of R_2 by differentiating F_2.

$$f_2(y) = \frac{dF_2(y)}{dy} = 0, \quad y < 0$$

$$= \frac{1}{4\sqrt{y}}(1 + e^{-\sqrt{y}}), \quad 0 < y < 1$$

$$= \frac{1}{4\sqrt{y}}e^{-\sqrt{y}}, \quad y > 1$$

(see Figure 2.4.6).

It may be verified that $F_2(y)$ is given by $\int_{-\infty}^{y} f_2(t)\, dt$, so that f_2 is in fact the density of R_2. Thus, in this case, if we differentiate F_2 and then integrate the derivative, we get back to F_2.

It is reasonable to expect that a random variable R, whose distribution function is continuous everywhere and defined by an explicit formula or collection of formulas, will be absolutely continuous. The following result will cover almost all situations encountered in practice.

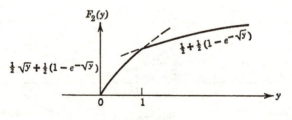

FIGURE 2.4.5

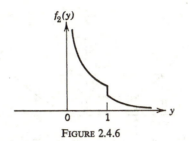

FIGURE 2.4.6

Let R be a random variable with distribution function F. Suppose that

(1) F is continuous for all x
(2) F is differentiable everywhere except possibly at a finite number of points
(3) The derivative $f(x) = F'(x)$ is continuous except possibly at a finite number of points

$$(2.4.1)$$

Then R is an absolutely continuous random variable with density function f. (The proof involves the application of the fundamental theorem of calculus; see Problem 6.)

NOTE. Density functions need not be continuous or bounded. Also, in this case there is an ambiguity in the values of $f_2(y)$ at $y = 0$ and $y = 1$, since F_2 is not differentiable at these points. However, any values may be assumed, since changing a function at a single point, or a finite or countably infinite number of points, or in fact on a set of total length (Lebesgue measure) zero, does not change the integral.

DENSITY FUNCTION METHOD. In this approach we develop an explicit formula for the density of R_2 in terms of that of R_1.

We first give an informal description. The probability that R_2 will lie in the small interval $[y, y + dy]$ is

$$\int_y^{y+dy} f_2(t) \, dt$$

which is roughly $f_2(y) \, dy$ if f_2 is well-behaved near y. But if we set $h_1(y) = \sqrt{y}$, $h_2(y) = -\sqrt{y}$, $y \geq 0$, then (see Figure 2.4.7)

$$P\{y \leq R_2 \leq y + dy\} = P\{h_1(y) \leq R_1 \leq h_1(y + dy)\}$$
$$+ P\{h_2(y + dy) \leq R_1 \leq h_2(y)\}$$

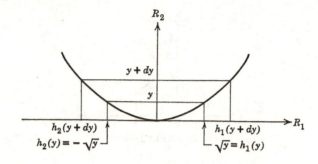

FIGURE 2.4.7

Hence

$$f_2(y)\,dy = f_1(h_1(y))\frac{[h_1(y+dy) - h_1(y)]\,dy}{dy}$$
$$+ f_1(h_2(y))\frac{[h_2(y) - h_2(y+dy)]\,dy}{dy}$$

Let $dy \to 0$ to obtain

$$f_2(y) = f_1(h_1(y))h_1'(y) + f_1(h_2(y))(-h_2'(y))$$
$$= f_1(h_1(y))\,|h_1'(y)| + f_1(h_2(y))\,|h_2'(y)|$$

In this case

$$f_2(y) = f_1(\sqrt{y})\left|\frac{d}{dy}\sqrt{y}\right| + f_1(-\sqrt{y})\left|\frac{d}{dy} - \sqrt{y}\right| = \frac{1}{2\sqrt{y}}\,[f_1(\sqrt{y}) + f_1(-\sqrt{y})]$$

Now (see Figure 2.4.1), if $0 < y < 1$,

$$f_2(y) = \frac{1}{2\sqrt{y}}\,[\tfrac{1}{2}e^{-\sqrt{y}} + \tfrac{1}{2}] = \frac{1}{4\sqrt{y}}\,(1 + e^{-\sqrt{y}})$$

If $y > 1$,

$$f_2(y) = \frac{1}{2\sqrt{y}}\,[\tfrac{1}{2}e^{-\sqrt{y}} + 0] = \frac{1}{4\sqrt{y}}\,e^{-\sqrt{y}}$$

as before.

Similar reasoning shows that in general

$$f_2(y) = f_1(h_1(y))\,|h_1'(y)| + \cdots + f_1(h_n(y))|h_n'(y)|$$

where $h_1(y), \ldots, h_n(y)$ are the values of R_1 corresponding to $R_2 = y$.

Here is the formal statement. Suppose that the domain of g can be written as the union of intervals $I_1, I_2, \ldots, I_n$. Assume that over the interval I_j, g is strictly increasing or strictly decreasing and is differentiable (except possibly at the end points), with h_j = inverse of g over I_j. Let F_1 satisfy the three conditions (2.4.1). Then

$$f_2(y) = \sum_{j=1}^{n} f_1(h_j(y)) \, |h'_j(y)|$$

where

$$f_1(h_j(y)) \, |h'_j(y)|$$

is interpreted as 0 if $y \notin$ the domain of h_j. For the proof, see Problem 7.

REMARK. If we have $R_1 = h(R_2)$, where h is a one-to-one differentiable function, and $R_2 = g(R_1)$, where g is the inverse of h, then

$$h'(y) = \frac{1}{g'(x)} \Bigg]_{x=h(y)}$$

Thus we may write

$$f_2(y) = \sum_{j=1}^{n} \frac{f_1(h_j(y))}{|g'(x)|_{x=h_j(y)}}$$

where $R_2 = g(R_1)$.

In the present example we have

$$g(x) = x^2, \qquad h_1(y) = \sqrt{y}, \qquad h_2(y) = -\sqrt{y}$$

so that

$$f_2(y) = \frac{f_1(\sqrt{y})}{|2x|_{x=\sqrt{y}}} + \frac{f_1(-\sqrt{y})}{|2x|_{x=-\sqrt{y}}} = \frac{1}{2\sqrt{y}} \, [f_1(\sqrt{y}) + f_1(-\sqrt{y})]$$

as before. ◄

▶ **Example 2.** Let R_1 be uniformly distributed between 0 and 2π; that is,

$$f_1(x) = \frac{1}{2\pi}, \qquad 0 \leq x \leq 2\pi$$

$$= 0 \qquad \text{elsewhere}$$

Let $R_2 = \sin R_1$ (see Figure 2.4.8).

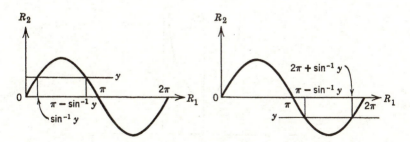

FIGURE 2.4.8

DISTRIBUTION FUNCTION METHOD.   If $0 \le y \le 1$,

$$
\begin{aligned}
F_2(y) &= P\{R_2 \le y\} \\
&= P\{0 \le R_1 \le \sin^{-1}y\} + P\{\pi - \sin^{-1}y \le R_1 \le 2\pi\} \\
&= \int_0^{\sin^{-1}y} \frac{1}{2\pi}\,dx + \int_{\pi-\sin^{-1}y}^{2\pi} \frac{1}{2\pi}\,dx \\
&= \frac{1}{2} + \frac{1}{\pi}\sin^{-1}y
\end{aligned}
$$

where the branch of the arc sin function is chosen so that $-\pi/2 \le \sin^{-1}y \le \pi/2$.

If $-1 \le y \le 0$,

$$
\begin{aligned}
F_2(y) &= P\{\pi - \sin^{-1}y \le R_1 \le 2\pi + \sin^{-1}y\} \\
&= \frac{1}{2} + \frac{1}{\pi}\sin^{-1}y
\end{aligned}
$$

as above, and

$$
f_2(y) = F_2'(y) = \frac{1}{\pi\sqrt{1-y^2}}, \quad -1 < y < 1
$$

DENSITY FUNCTION METHOD.   If $0 < y < 1$,

$$
\begin{aligned}
f_2(y) &= f_1(\sin^{-1}y)\left|\frac{d}{dy}\sin^{-1}y\right| + f_1(\pi - \sin^{-1}y)\left|\frac{d}{dy}(\pi - \sin^{-1}y)\right| \\
&= \frac{1}{\pi\sqrt{1-y^2}}
\end{aligned}
$$

Similarly,

$$
f_2(y) = \frac{1}{\pi\sqrt{1-y^2}} \quad \text{for } -1 < y < 0
$$

$[f_2(y) = 0, |y| > 1]$ (see Figure 2.4.9). ◄

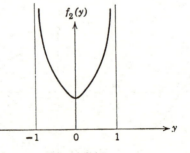

FIGURE 2.4.9

PROBLEMS

1. Let R_1 be absolutely continuous with density

$$f_1(x) = e^{-x}, \qquad x \geq 0; \qquad f_1(x) = 0, \qquad x < 0$$

Define

$$R_2 = R_1 \qquad \text{if } R_1 \leq 1$$

$$= \frac{1}{R_1} \qquad \text{if } R_1 > 1$$

Show that R_2 is absolutely continuous and find its density.

2. An absolutely continuous random variable R_1 is uniformly distributed between -1 and $+1$. Find and sketch either the density or the distribution function of the random variable R_2, where $R_2 = e^{-R_1}$.

3. Let R_1 have density $f_1(x) = 1/x^2$, $x \geq 1$; $f_1(x) = 0$, $x < 1$. Define

$$R_2 = 2R_1 \qquad \text{for } R_1 \leq 2$$

$$= R_1{}^2 \qquad \text{for } R_1 > 2$$

Find the density of R_2.

4. Let R_1 be as in Problem 3, and define

$$R_2 = 2R_1 \qquad \text{for } R_1 \leq 2$$

$$= 5 \qquad \text{for } R_1 > 2$$

Find and sketch the distribution function of R_2; is R_2 absolutely continuous?

5. (a) Let R_1 have distribution function

$$F_1(x) = 1 - e^{-x}, \qquad x \geq 0$$

$$= 0, \qquad x < 0$$

Define

$$R_2 = 1 - e^{-R_1}, \qquad R_1 \geq 0$$

$$= 0, \qquad R_1 < 0$$

Show that R_2 is uniformly distributed between 0 and 1.

(b) In general, if a random variable R_1 has a continuous distribution function $g(x) = F_1(x)$ and we define a random variable R_2 by $R_2 = g(R_1)$, show that R_2 is uniformly distributed between 0 and 1.

6. If R is a random variable with distribution function F, where F is continuous everywhere and has a continuous derivative f at all but a finite number of points, show that R is absolutely continuous with density f.

7. Establish the validity of the formula

$$f_2(y) = \sum_{j=1}^{n} f_1(h_j(y)) |h_j'(y)|$$

under the conditions given in the text.

8. Let R_1 be chosen at random between 0 and 1, with density f_1 [so that $\int_0^1 f_1(y)\, dy = 1$]. Let R_2 be the second digit in the decimal expansion of R_1. (To avoid ambiguity, write, for example, .3 as .3000 $\cdots$, not .2999 $\cdots$.)

(a) Show that $R_2 = k$ iff $i + 10^{-1}k \leq 10R_1 < i + 10^{-1}(k+1)$ for some $i = 0, 1, \ldots, 9$. Hence

$$P\{R_2 = k\} = \sum_{i=0}^{9} \int_{10^{-1}i + 10^{-2}k}^{10^{-1}i + 10^{-2}k + 10^{-2}} f_1(y)\, dy, \qquad k = 0, 1, \ldots, 9$$

(b) If R is uniformly distributed between 0 and 1, and $R_1 = \sqrt{R}$, find the probability function of $R_2 =$ the second digit in the decimal expansion of R_1.

9. A projectile is fired with initial velocity v_0 at an angle θ uniformly distributed between 0 and $\pi/2$ (see Figure P.2.4.9). If R is the distance from the launch site

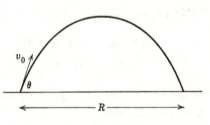

FIGURE P.2.4.9

to the point at which the projectile returns to earth, find the density of R (consider only the effect of gravity).

2.5   PROPERTIES OF DISTRIBUTION FUNCTIONS

We shall establish some general properties of the distribution function of an arbitrary random variable. We need two facts about probability measures.

Theorem 1.   *Let $(\Omega, \mathscr{F}, P)$ be a probability space.*
(a) *If $A_1, A_2, \ldots$ is an expanding sequence of sets in $\mathscr{F}$, that is, $A_n \subset A_{n+1}$ for all n, and $A = \bigcup_{n=1}^{\infty} A_n$, then $P(A) = \lim_{n \to \infty} P(A_n)$.*

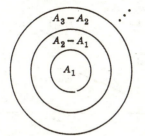

FIGURE 2.5.1 Expanding Sequence.

(b) *If A_1, A_2, ... is a contracting sequence of sets in $\mathscr{F}$, that is, $A_{n+1} \subset A_n$ for all n, and $A = \bigcap_{n=1}^{\infty} A_n$, then $P(A) = \lim_{n \to \infty} P(A_n)$.*

PROOF.
(a) We can write

$$A = A_1 \cup (A_2 - A_1) \cup (A_3 - A_2) \cup \cdots \cup (A_n - A_{n-1}) \cdots$$

(see Figure 2.5.1; note this is the expansion (1.3.11) in the special case of an expanding sequence). Since this is a disjoint union,

$$\begin{aligned}
P(A) &= P(A_1) + P(A_2 - A_1) + P(A_3 - A_2) + \cdots \\
&= P(A_1) + P(A_2) - P(A_1) + P(A_3) - P(A_2) + \cdots \quad \text{since } A_n \subset A_{n+1} \\
&= \lim_{n \to \infty} P(A_n)
\end{aligned}$$

(b) If $A = \bigcap_{n=1}^{\infty} A_n$, then, by the DeMorgan laws, $A^c = \bigcup_{n=1}^{\infty} A_n^c$. Now $A_{n+1} \subset A_n$; hence $A_n^c \subset A_{n+1}^c$. Thus the sets A_n^c form an expanding sequence, so, by (a), $P(A_n^c) \to P(A^c)$; that is; $1 - P(A_n) \to 1 - P(A)$. The result follows.

Theorem 2. *Let F be the distribution function of an arbitrary random variable R. Then*

1. *$F(x)$ is nondecreasing; that is, $a < b$ implies $F(a) \leq F(b)$*

For we have shown [see (2.3.2)] that $F(b) - F(a) = P\{a < R \leq b\} \geq 0$.

2. $\lim_{x \to \infty} F(x) = 1$

Let x_n, $n = 1, 2, \ldots$ be a sequence of real numbers increasing to $+ \infty$. Let $A_n = \{R \leq x_n\}$. Then the A_n form an expanding sequence. (Since $x_n \leq x_{n+1}$, $R \leq x_n$ implies $R \leq x_{n+1}$.) Now $\bigcup_{n=1}^{\infty} A_n = \Omega$, since, given any point

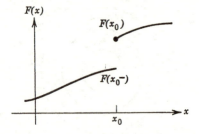

FIGURE 2.5.2   Right Continuity of Distribution Functions.

$\omega \in \Omega$, $R(\omega)$ is a real number; hence, for sufficiently large n, $R(\omega) \leq x_n$, so that $\omega \in A_n$. Thus $P(A_n) \to P(\Omega) = 1$, that is, $\lim_{n \to \infty} F(x_n) = 1$.

$$3. \quad \lim_{x \to -\infty} F(x) = 0$$

Let x_n, $n = 1, 2, \ldots$ be a sequence of real numbers decreasing to $-\infty$. Let $A_n = \{R \leq x_n\}$. Then the A_n form a contracting sequence. (Since $x_{n+1} \leq x_n$, $R \leq x_{n+1}$ implies $R \leq x_n$.) Now $\bigcap_{n=1}^{\infty} A_n = \varnothing$, since if ω is any point of Ω, $R(\omega)$ cannot always be $\leq x_n$ because $x_n \to -\infty$. Thus $P(A_n) \to P(\varnothing) = 0$; that is, $F(x_n) \to 0$.

4. *F is continuous from the right; that is,* $\lim_{x \to x_0^+} F(x) = F(x_0)$

Hence F assumes the upper value at any discontinuity; see Figure 2.5.2.

Let x_n approach x_0 from above; that is, let x_n, $n = 1, 2, \ldots$ be a (strictly) decreasing sequence whose limit is x_0. As before, let $A_n = \{R \leq x_n\}$. The A_n form a contracting sequence whose limit (intersection) is $A = \{R \leq x_0\}$. In order to show that $\bigcap_{n=1}^{\infty} A_n = \{R \leq x_0\}$, we reason as follows. If $R(\omega) \leq x_n$ for all n, then, since $x_n \to x_0$, $R(\omega) \leq x_0$. Conversely, if $R(\omega) \leq x_0$, then, since $x_0 \leq x_n$ for all n, $R(\omega) \leq x_n$ for all n. Thus $P(A_n) \to P(A)$; that is, $F(x_n) \to F(x_0)$.

$$5. \quad \lim_{x \to x_0^-} F(x) = P\{R < x_0\}$$

[We write $F(x_0^-)$ for $\lim_{x \to x_0^-} F(x)$.]

Let x_n, $n = 1, 2, \ldots$ be a (strictly) increasing sequence whose limit is x_0. Again let $A_n = \{R \leq x_n\}$. The A_n form an expanding sequence whose union is $\{R < x_0\}$. To show $\bigcup_{n=1}^{\infty} A_n = \{R < x_0\}$, we reason as follows. If $\omega \in$ some A_n, then $R(\omega) \leq x_n$, so that $R(\omega) < x_0$. Conversely, if $R(\omega) < x_0$, then, since $x_n \to x_0$, eventually $R(\omega) \leq x_n$, so that $\omega \in \bigcup_{n=1}^{\infty} A_n$. Thus $P(A_n) \to P\{R < x_0\}$, and the result follows.

$$6. \quad P\{R = x_0\} = F(x_0) - F(x_0^-)$$

Thus F is continuous at x_0 iff $P\{R = x_0\} = 0$, and if F is discontinuous at x_0, the magnitude of the jump is the probability that $R = x_0$.

For $P\{R \leq x_0\} = P\{R < x_0\} + P\{R = x_0\}$, so that

$$F(x_0) = F(x_0^-) + P\{R = x_0\}$$

REMARK. The random variable R is said to be *continuous* iff its distribution function $F_R(x)$ is a continuous function of x for all x. In any reasonable case a continuous random variable will have a density—that is, it will be absolutely continuous—but it is possible to establish the existence of random variables that are continuous but not absolutely continuous.

7. *Let F be a function from the reals to the reals, satisfying properties 1, 2, 3, and 4 above. Then F is the distribution function of some random variable.*

This is a somewhat vague statement. Let us try to clarify it, even though we omit the proof. What we are doing essentially is making the statement "Let R be a random variable with distribution function F." It is up to us to supply the underlying probability space. As we have done before, we take $\Omega = E^1$, $\mathscr{F} =$ Borel sets, $R(\omega) = \omega$. Now if F is to be the distribution function of R, we must have, for $a < b$,

$$P(a, b] = P\{a < R \leq b\} = F(b) - F(a) \qquad \text{by (2.3.2)}$$

It turns out that if F satisfies conditions 1–4, there is a unique probability measure P defined on the Borel subsets of E^1 such that $P(a, b] = F(b) - F(a)$ for all real a, b, $a < b$; thus the probabilities of all events involving R are determined by F. If we let $a \to -\infty$, we obtain $P(-\infty, b] = F(b)$, that is, $P\{R \leq b\} = F(b)$, so that in fact F is the distribution function of R. In the special case in which $F(x) = \int_{-\infty}^x f(t)\,dt$, where f is a nonnegative integrable function and $\int_{-\infty}^\infty f(x)\,dx = 1$, $P(a, b] = F(b) - F(a) = \int_a^b f(x)\,dx$. This is exactly the situation we considered in Theorem 1 of Section 2.3.

PROBLEMS

1. Let R be a random variable with the distribution function shown in Figure P.2.5.1; notice that R is neither discrete nor continuous. Find the probability

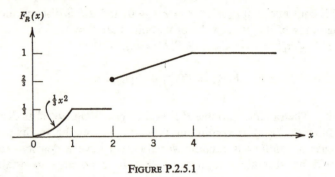

FIGURE P.2.5.1

of the following events.

(a) $\{R = 2\}$

(b) $\{R < 2\}$

(c) $\{R = 2 \text{ or } .5 \le R < 1.5\}$

(d) $\{R = 2 \text{ or } .5 \le R \le 3\}$

2. Let R be an arbitrary random variable with distribution function F. We have seen that $P\{a < R \le b\} = F(b) - F(a)$, $a < b$. Show that

$$P\{a \le R \le b\} = F(b) - F(a^-)$$
$$P\{a \le R < b\} = F(b^-) - F(a^-)$$
$$P\{a < R < b\} = F(b^-) - F(a)$$

(Of course these are all equal if F is continuous at a and b.)

2.6 JOINT DENSITY FUNCTIONS

We are going to investigate situations in which we deal simultaneously with several random variables defined on the same sample space. As an introductory example, suppose that a person is selected at random from a certain population, and his age and weight recorded. We may take as the sample space the set of all pairs (x, y) of real numbers, that is, the Euclidean plane E^2, where we interpret x as the age and y as the weight. Let R_1 be the age of the person selected, and R_2 the weight; that is, $R_1(x, y) = x$, $R_2(x, y) = y$. We wish to assign probabilities to events that involve R_1 and R_2 simultaneously. A cross-section of the available data might appear as shown in Figure 2.6.1. Thus there are 4 million people whose age is between 20 and 25 and (simultaneously) whose weight is between 150 and 160 pounds, and so on. Now suppose that we wish to estimate the number of people between 22 and 23 years, and 154 and 156 pounds. There are 4 million people spread over 5 years and 10 pounds, or 4/50 million per year-pound. We are interested in

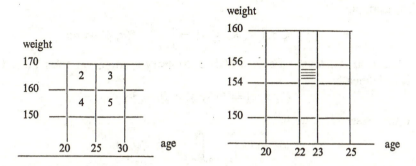

FIGURE 2.6.1 Age-Weight Data (Number of People Is in Millions).

FIGURE 2.6.2 Estimation of Probabilities.

a range of 1 year and 2 pounds, and so our estimate is $4/50 \times 1 \times 2 = 8/50$ million (see Figure 2.6.2). If the total population is 200 million, then

$$P\{22 \leq R_1 \leq 23, 154 \leq R_2 \leq 156\}$$

should be approximately

$$\frac{8/50}{200} = .0008$$

NOTATION. $\{22 \leq R_1 \leq 23, 154 \leq R_2 \leq 156\}$ means $\{22 \leq R_1 \leq 23$ *and* $154 \leq R_2 \leq 156\}$.

What we are doing is multiplying an age-weight density $4/50$ by an area 1×2 to estimate the number of people or, equally well, a probability density $4/[50(200)]$ by an area (1×2) to estimate the probability.

Thus it appears that we should assign probabilities by means of an integral over an area. Let us try to construct an appropriate probability space. We take $\Omega = E^2$, $\mathscr{F} =$ the Borel subsets of E^2. Suppose we have a nonnegative real-valued function f on E^2, with

$$\int_{-\infty}^{\infty} \int_{-\infty}^{\infty} f(x, y) \, dx \, dy = 1$$

Theorem 1 of Section 2.3 holds just as well in the two-dimensional case; there is a unique probability measure P on $\mathscr{F}$ such that $P(B) = \iint_B f(x) \, dx$ for all rectangles B.

If we define $R_1(x, y) = x$, $R_2(x, y) = y$, then

$$P\{(R_1, R_2) \in B\} = P(B) = \iint_B f(x, y) \, dx \, dy$$

For example,

$$P\{a \le R_1 \le b, c \le R_2 \le d\} = \int_{x=a}^{b} \int_{y=c}^{d} f(x, y) \, dx \, dy$$

The *joint distribution function* of two arbitrary random variables R_1 and R_2 is defined by

$$F_{12}(x, y) = P\{R_1 \le x, R_2 \le y\}$$

In the present case we have

$$F_{12}(x, y) = \int_{u=-\infty}^{x} \int_{v=-\infty}^{y} f(u, v) \, du \, dv$$

In general, if R_1 and R_2 are arbitrary random variables defined on a given probability space, the pair (R_1, R_2) is said to be *absolutely continuous* iff there is a nonnegative function $f = f_{12}$ defined on E^2 such that

$$F_{12}(x, y) = \int_{-\infty}^{x} \int_{-\infty}^{y} f_{12}(u, v) \, du \, dv \qquad \text{for all real } x, y \qquad (2.6.1)$$

f_{12} is called the *density* of (R_1, R_2) or the *joint density* of R_1 and R_2.

Just as in the one-dimensional case, if (R_1, R_2) is absolutely continuous, it follows that

$$P\{(R_1, R_2) \in B\} = \iint_{B} f_{12}(x, y) \, dx \, dy$$

for all two-dimensional Borel sets B (see Problem 1). Again, as in the one-dimensional case, if f is a nonnegative function on E^2 with

$$\int_{-\infty}^{\infty} \int_{-\infty}^{\infty} f(x, y) \, dx \, dy = 1$$

we can always find random variables R_1, R_2 such that (R_1, R_2) is absolutely continuous with density f. We take $\Omega = E^2$, $\mathcal{F} = $ Borel sets, $R_1(x, y) = x$, $R_2(x, y) = y$, $P(B) = \iint_B f(x, y) \, dx \, dy$. Even if we use a completely different construction, we get the same result, namely,

$$P\{(R_1, R_2) \in B\} = \iint_{B} f(x, y) \, dx \, dy$$

We have a similar situation in n dimensions. If the n random variables $R_1, R_2, \ldots, R_n$ are all defined on the same probability space, the *joint distribution function* of $R_1, R_2, \ldots, R_n$ is defined by

$$F_{12\ldots n}(x_1, \ldots, x_n) = P\{R_1 \le x_1, \ldots, R_n \le x_n\}$$

The *random vector* or n-tuple $(R_1, \ldots, R_n)$ is said to be absolutely continuous iff there is a nonnegative function $f_{12\ldots n}$ defined on E^n, called the *density* of $(R_1, \ldots, R_n)$ or the *joint density* of $R_1, \ldots, R_n$, such that

$$F_{12\ldots n}(x_1, \ldots, x_n) = \int_{-\infty}^{x_1} \cdots \int_{-\infty}^{x_n} f_{12\ldots n}(u_1, \ldots, u_n)\, du_1 \cdots du_n \quad (2.6.2)$$

for all real $x_1, \ldots, x_n$.

Notice that $f_{12\ldots n}$ can be recovered from $F_{12\ldots n}$ by differentiation:

$$\frac{\partial^n F_{12\ldots n}(x_1, \ldots, x_n)}{\partial x_1 \cdots \partial x_n} = f_{12\ldots n}(x_1, \ldots, x_n)$$

at least at points where $f_{12\ldots n}$ is continuous.

If $(R_1, \ldots, R_n)$ is absolutely continuous, then

$$P\{(R_1, \ldots, R_n) \in B\} = \int \cdots \int_B f_{12\ldots n}(x_1, \ldots, x_n)\, dx_1 \cdots dx_n$$

for all n-dimensional Borel sets B.

If f is a nonnegative function on E^n such that

$$\int_{-\infty}^{\infty} \cdots \int_{-\infty}^{\infty} f(x_1, \ldots, x_n)\, dx_1 \cdots dx_n = 1$$

we can always find random variables $R_1, \ldots, R_n$ such that $(R_1, \ldots, R_n)$ is absolutely continuous with density f. We take $\Omega = E^n$, $\mathscr{F} = $ Borel sets, and define $R_1(x_1, \ldots, x_n) = x_1, \ldots, R_n(x_1, \ldots, x_n) = x_n$. If B is any Borel subset of E^n, we assign

$$P(B) = \int \cdots \int_B f(x_1, \ldots, x_n)\, dx_1 \cdots dx_n$$

Then $(R_1, \ldots, R_n)$ is absolutely continuous with density f.

▶ **Example 1.** Let

$$f_{12}(x, y) = 1 \quad \text{if } 0 \leq x \leq 1 \quad \text{and} \quad 0 \leq y \leq 1$$
$$= 0 \quad \text{elsewhere}$$

(This is the *uniform density* on the unit square.) We may as well take $\Omega = E^2$, $\mathscr{F} = $ Borel sets, $R_1(x, y) = x$, $R_2(x, y) = y$,

$$P(B) = \int\int_B f_{12}(x, y)\, dx\, dy$$

Let us calculate the probability that $1/2 \leq R_1 + R_2 \leq 3/2$. Now

$$\{\tfrac{1}{2} \leq R_1 + R_2 \leq \tfrac{3}{2}\} = \{(x, y): \tfrac{1}{2} \leq R_1(x, y) + R_2(x, y) \leq \tfrac{3}{2}\}$$
$$= \{(x, y): \tfrac{1}{2} \leq x + y \leq \tfrac{3}{2}\}$$

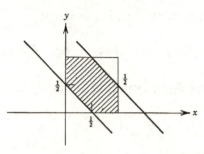

FIGURE 2.6.3

Calculation of $P\{\tfrac{1}{2} \leq R_1 + R_2 \leq \tfrac{3}{2}\}$.

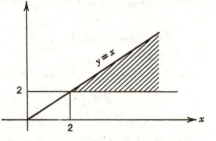

FIGURE 2.6.4

Calculation of $P\{R_1 \geq R_2 \geq 2\}$.

Thus (see Figure 2.6.3)

$$P\{\tfrac{1}{2} \leq R_1 + R_2 \leq \tfrac{3}{2}\} = \iint\limits_{1/2 \leq x+y \leq 3/2} f_{12}(x, y)\, dx\, dy$$

$$= \iint\limits_{\text{shaded area}} 1\, dx\, dy = \text{shaded area}$$

$$= 1 - 2(\tfrac{1}{8}) = \tfrac{3}{4}$$

If we want the probability that $1/2 \leq R_1 \leq 3/4$ and $0 \leq R_2 \leq 1/2$, we obtain

$$P\{\tfrac{1}{2} \leq R_1 \leq \tfrac{3}{4}, 0 \leq R_2 \leq \tfrac{1}{2}\} = \int_{x=1/2}^{3/4} \int_{y=0}^{1/2} 1\, dx\, dy = \tfrac{1}{2}(\tfrac{1}{4}) = \tfrac{1}{8} \blacktriangleleft$$

▶ **Example 2.** Let

$$f_{12}(x, y) = e^{-(x+y)}, \qquad x, y \geq 0$$

$$= 0 \qquad \text{elsewhere}$$

Let us calculate the probability that $R_1 \geq R_2 \geq 2$. We have (see Figure 2.6.4)

$$P\{R_1 \geq R_2 \geq 2\} = \iint\limits_{x \geq y \geq 2} f_{12}(x, y)\, dx\, dy$$

$$= \int_2^\infty e^{-x}\, dx \int_2^x e^{-y}\, dy = \int_2^\infty e^{-x}(e^{-2} - e^{-x})\, dx$$

$$= e^{-4} - \tfrac{1}{2}e^{-4} = \tfrac{1}{2}e^{-4} \blacktriangleleft$$

To summarize:

$$P\{(R_1, R_2) \in B\} = \iint\limits_{(x,y) \in B} f_{12}(x, y)\, dx\, dy$$

The probability of any event is found by integrating the density function over the set defined by the event. This is perhaps about as close as one can come to a one-sentence summary of the role of density functions in probability theory.

PROBLEMS

1. Let F_{12} be the joint distribution function of R_1 and R_2, where (R_1, R_2) is absolutely continuous with density f_{12}. Show that

$$P\{a_1 < R_1 \le b_1, a_2 < R_2 \le b_2\} = \int_{a_1}^{b_1} \int_{a_2}^{b_2} f_{12}(x, y) \, dx \, dy$$

The uniqueness part of Theorem 2.3.1 (generalized to two dimensions) shows that

$$P\{(R_1, R_2) \in B\} = \iint_B f_{12}(x, y) \, dx \, dy$$

for all two-dimensional Borel sets B.

HINT: If F is the joint distribution function of the random variables R_1 and R_2, show that

$$P\{a_1 < R_1 \le b_1, a_2 < R_2 \le b_2\} = F(b_1, b_2) - F(a_1, b_2) - F(b_1, a_2) + F(a_1, a_2)$$

2. If F is the joint distribution function of the random variables R_1, R_2, and R_3, express

$$P\{a_1 < R_1 \le b_1, a_2 < R_2 \le b_2, a_3 < R_3 \le b_3\}$$

in terms of F. Can you see a general pattern that will extend this result to n dimensions?

3. If

$$F(x, y) = 1 \qquad \text{for } x + y \ge 0$$
$$\qquad\quad = 0 \qquad \text{for } x + y < 0$$

(see Figure P.2.6.3.), show that F cannot possibly be the joint distribution

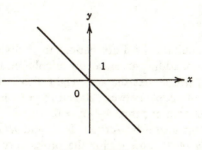

FIGURE P.2.6.3

function of a pair of random variables (see Problem 1.)

4. Let R_1 and R_2 have the following joint density:

$$f_{12}(x, y) = \tfrac{1}{4} \quad \text{if } -1 \leq x \leq 1 \quad \text{and} \quad -1 \leq y \leq 1$$
$$= 0 \quad \text{elsewhere}$$

(This corresponds to R_1 and R_2 being chosen independently, each uniformly distributed between -1 and $+1$; we elaborate on this in the next section.) Find the probability of each of the following events.

(a) $\{R_1 + R_2 \leq \tfrac{1}{2}\}$

(b) $\{R_1 - R_2 \leq \tfrac{1}{2}\}$

(c) $\{R_1 R_2 \leq \tfrac{1}{4}\}$

(d) $\left\{ \dfrac{R_2}{R_1} \leq \dfrac{1}{2} \right\}$

(e) $\left\{ \left| \dfrac{R_2}{R_1} \right| \leq \dfrac{1}{2} \right\}$

(f) $\{|R_1| + |R_2| \leq 1\}$

(g) $\{|R_2| \leq e^{R_1}\}$

2.7 RELATIONSHIP BETWEEN JOINT AND INDIVIDUAL DENSITIES; INDEPENDENCE OF RANDOM VARIABLES

If R_1 and R_2 are two random variables defined on the same probability space, we wish to investigate the relation between the characterization of the random variables individually and their characterization simultaneously. We shall consider two problems.

1. If (R_1, R_2) is absolutely continuous, are R_1 and R_2 absolutely continuous, and, if so, how can the individual densities of R_1 and R_2 be found in terms of the joint densities?

2. Given R_1, R_2 (individually) absolutely continuous, is (R_1, R_2) absolutely continuous, and, if so, can the joint density be derived from the individual density?

Problem 1

To go from simultaneous information to individual information is essentially a matter of adding across a row or column. For example, suppose that a group of 14 people has the age-weight distribution shown in Figure 2.7.1. The number of people between 20 and 25 years is found by adding the numbers in the first column; thus $4 + 2 = 6$.

Let us develop this idea a bit further. If R_1 and R_2 are discrete, the *joint probability function* of R_1 and R_2 [or the probability function of the pair

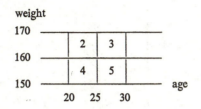

FIGURE 2.7.1 Calculation of Individual Probabilities from Joint Probabilities.

$(R_1, R_2)]$ is defined by

$$p_{12}(x, y) = P\{R_1 = x, R_2 = y\} \qquad x, y \text{ real} \qquad (2.7.1)$$

If the possible values of R_2 are $y_1, y_2, \ldots,$ then

$$\{R_1 = x\} = \{R_1 = x, R_2 = y_1\} \cup \{R_1 = x, R_2 = y_2\} \cup \cdots$$

since the events $\{R_2 = y_n\}$, $n = 1, 2, \ldots$ are mutually exclusive and exhaustive. Thus the probability function of R_1 is given by

$$p_1(x) = P\{R_1 = x\} = \sum_y p_{12}(x, y) \qquad (2.7.2)$$

Similarly,

$$p_2(y) = P\{R_2 = y\} = \sum_x p_{12}(x, y) \qquad (2.7.3)$$

There are analogous formulas in higher dimensions, for example,

$$p_{12}(x, y) = \sum_z p_{123}(x, y, z) \qquad p_2(y) = \sum_{x,z} p_{123}(x, y, z)$$

where $p_{123}(x, y, z) = P\{R_1 = x, R_2 = y, R_3 = z\}$.

Now let us return to the absolutely continuous case. If (R_1, R_2) is absolutely continuous with joint density f_{12}, we shall show that R_1 is absolutely continuous (and so is R_2) and find f_1 and f_2 in terms of f_{12}.

For any x_0 we have, intuitively,

$$P\{x_0 \leq R_1 \leq x_0 + dx_0\} \approx f_1(x_0)\, dx_0 \qquad (2.7.4)$$

But

$$P\{x_0 \leq R_1 \leq x_0 + dx_0\} = P\{x_0 \leq R_1 \leq x_0 + dx_0, -\infty < R_2 < \infty\}$$

$$= \int_{x_0}^{x_0+dx_0} dx \int_{-\infty}^{\infty} f_{12}(x, y)\, dy$$

(see Figure 2.7.2).

If f_{12} is well-behaved, this is approximately

$$dx_0 \int_{-\infty}^{\infty} f_{12}(x_0, y)\, dy \qquad (2.7.5)$$

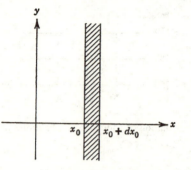

FIGURE 2.7.2   Calculation of Individual Densities from Joint Densities.

From (2.7.4) and (2.7.5) (replacing x_0 by x) we have

$$f_1(x) = \int_{-\infty}^{\infty} f_{12}(x, y)\, dy$$

To verify this formally, we work with the distribution function of R_1.

$$F_1(x_0) = P\{R_1 \leq x_0\} = P\{R_1 \leq x_0, -\infty < R_2 < \infty\}$$
$$= \int_{x=-\infty}^{x_0} \left[\int_{y=-\infty}^{\infty} f_{12}(x, y)\, dy \right] dx$$

Thus F_1 is represented as an integral, and so R_1 is absolutely continuous with density

$$f_1(x) = \int_{-\infty}^{\infty} f_{12}(x, y)\, dy \tag{2.7.6}$$

Similarly,

$$f_2(y) = \int_{-\infty}^{\infty} f_{12}(x, y)\, dx \tag{2.7.7}$$

In exactly the same way we may establish similar formulas in higher dimensions; for example,

$$f_{12}(x, y) = \int_{-\infty}^{\infty} f_{123}(x, y, z)\, dz \tag{2.7.8}$$

$$f_2(y) = \int_{-\infty}^{\infty} \int_{-\infty}^{\infty} f_{123}(x, y, z)\, dx\, dz \tag{2.7.9}$$

The process of obtaining the individual densities from the joint density is sometimes called the calculation of *marginal densities*, because of the similarity to the process of adding across a row or column.

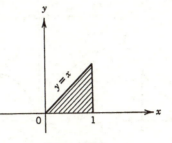

FIGURE 2.7.3

▶ **Example 1.** Let
$$f_{12}(x, y) = 8xy, \qquad 0 \leq y \leq x \leq 1$$
$$= 0 \qquad \text{elsewhere}$$
(see Figure 2.7.3).
$$f_1(x) = \int_{-\infty}^{\infty} f_{12}(x, y)\, dy$$
$$= 0 \qquad \text{if } x < 0 \quad \text{or} \quad x > 1$$
If $0 \leq x \leq 1$,
$$f_1(x) = \int_0^x 8xy\, dy = 4x^3 \qquad \text{(Figure 2.7.4a)}$$
$$f_2(y) = \int_{-\infty}^{\infty} f_{12}(x, y)\, dx$$
$$= 0 \qquad \text{if } y < 0 \quad \text{or} \quad y > 1$$
If $0 \leq y \leq 1$,
$$f_2(y) = \int_y^1 8xy\, dx = 4y(1 - y^2) \qquad \text{(Figure 2.7.4b)}$$

Sketches of f_1 and f_2 are given in Figure 2.7.5. ◀

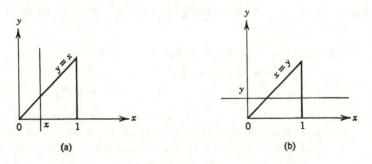

(a)   (b)

FIGURE 2.7.4

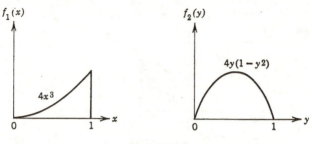

FIGURE 2.7.5

Problem 2

The second problem posed at the beginning of this section has a negative answer; that is, if R_1 and R_2 are each absolutely continuous then (R_1, R_2) is *not* necessarily absolutely continuous. Furthermore, even if (R_1, R_2) is absolutely continuous, $f_1(x)$ and $f_2(y)$ do *not* determine $f_{12}(x, y)$. We give examples later in the section.

However, there is an affirmative answer when the random variables are independent. We have considered the notion of independence of events, and this can be used to define independence of random variables. Intuitively, the random variables $R_1, \ldots, R_n$ are independent if knowledge about some of the R_i does not change the odds about the other R_i's. In other words, if A_i is an event involving R_i alone, that is, if $A_i = \{R_i \in B_i\}$, then the events $A_1, \ldots, A_n$ should be independent. Formally, we define independence as follows.

DEFINITION. Let $R_1, \ldots, R_n$ be random variables on $(\Omega, \mathscr{F}, P)$. $R_1, \ldots, R_n$ are said to be *independent* iff for all Borel subsets $B_1, \ldots, B_n$ of E^1 we have

$$P\{R_1 \in B_1, \ldots, R_n \in B_n\} = P\{R_1 \in B_1\} \cdots P\{R_n \in B_n\}$$

REMARK. If $R_1, \ldots, R_n$ are independent, so are $R_1, \ldots, R_k$ for $k < n$. For

$$P\{R_1 \in B_1, \ldots, R_k \in B_k\} = P\{R_1 \in B_1, \ldots, R_k \in B_k,$$
$$-\infty < R_{k+1} < \infty, \ldots, -\infty < R_n < \infty\}$$
$$= P\{R_1 \in B_1\} \ldots P\{R_k \in B_k\}$$

since $P\{-\infty < R_i < \infty\} = 1$. If $(R_i, i \in$ the index set $I)$, is an *arbitrary* family of random variables on the space $(\Omega, \mathscr{F}, P)$, the R_i are said to be independent iff for each finite set of distinct indices $i_1, \ldots, i_k \in I$, $R_{i_1}, \ldots, R_{i_k}$ are independent.

We may now give the solution to Problem 2 under the hypothesis of independence.

Theorem 1. *Let R_1, R_2, ..., R_n be independent random variables on a given probability space. If each R_i is absolutely continuous with density f_i, then $(R_1, R_2, ..., R_n)$ is absolutely continuous; also, for all $x_1, ..., x_n$,*

$$f_{12\cdots n}(x_1, x_2, \ldots, x_n) = f_1(x_1)f_2(x_2) \cdots f_n(x_n)$$

Thus in this sense the joint density is the product of the individual densities.

PROOF.   The joint distribution function of $R_1, \ldots, R_n$ is given by

$$F_{12\cdots n}(x_1, \ldots, x_n) = P\{R_1 \leq x_1, \ldots, R_n \leq x_n\}$$
$$= P\{R_1 \leq x_1\} \cdots P\{R_n \leq x_n\} \qquad \text{by independence}$$
$$= \int_{-\infty}^{x_1} f_1(u_1) \, du_1 \cdots \int_{-\infty}^{x_n} f_n(u_n) \, du_n$$
$$= \int_{-\infty}^{x_1} \cdots \int_{-\infty}^{x_n} f_1(u_1) \cdots f_n(u_n) \, du_1 \cdots du_n$$

It follows from the definition of absolute continuity [see (2.6.2)] that $(R_1, \ldots, R_n)$ is absolutely continuous and that the joint density is $f_{12\cdots n}(x_1, \ldots, x_n) = f_1(x_1) \cdots f_n(x_n)$.

Note that we have the following intuitive interpretation (when $n = 2$). From the independence of R_1 and R_2 we obtain

$$P\{x \leq R_1 \leq x + dx, y \leq R_2 \leq y + dy\}$$
$$= P\{x \leq R_1 \leq x + dx\}P\{y \leq R_2 \leq y + dy\}$$

If there is a joint density, we have (roughly) $f_{12}(x, y) \, dx \, dy = f_1(x) \, dx \, f_2(y) \, dy$, so that $f_{12}(x, y) = f_1(x)f_2(y)$.

As a consequence of this result, the statement "Let $R_1, \ldots, R_n$ be independent random variables, with R_i having density f_i," is unambiguous in the sense that it completely determines all probabilities of events involving the random vector $(R_1, \ldots, R_n)$; if B is an n-dimensional Borel set,

$$P\{(R_1, \ldots, R_n) \in B\} = \int \cdots \int_B f_1(x_1) \cdots f_n(x_n) \, dx_1 \cdots dx_n$$

We now show that Problem 2 has a negative answer when the hypothesis of independence is dropped. We have seen that if $(R_1, \ldots, R_n)$ is absolutely continuous then each R_i is absolutely continuous, but the converse is false in general if the R_i are not independent; that is, each of the random variables $R_1, \ldots, R_n$ can have a density without there being a density for the n-tuple $(R_1, \ldots, R_n)$.

▶ **Example 2.** Let R_1 be an absolutely continuous random variable with density f, and take $R_2 \equiv R_1$; that is, $R_2(\omega) = R_1(\omega)$, $\omega \in \Omega$. Then R_2 is absolutely continuous, but (R_1, R_2) is not. For suppose that (R_1, R_2) has a density g. Necessarily $(R_1, R_2) \in L$, where L is the line $y = x$, but

$$P\{(R_1, R_2) \in L\} = \iint_L g(x, y)\, dx\, dy$$

Since L has area 0, the integral on the right is 0. But the probability on the left is 1, a contradiction. ◀

We can also give an example to show that if R_1 and R_2 are each absolutely continuous (but not necessarily independent), then even if (R_1, R_2) is absolutely continuous, the joint density is not determined by the individual densities.

▶ **Example 3.** Let

$$f_{12}(x, y) = \tfrac{1}{4}(1 + xy), \qquad \begin{matrix} -1 \le x \le 1 \\ -1 \le y \le 1 \end{matrix}$$

$$= 0 \qquad \text{elsewhere}$$

Since

$$\int_{-1}^{1} x\, dx = \int_{-1}^{1} y\, dy = 0,$$

$$f_1(x) = \int_{-\infty}^{\infty} f_{12}(x, y)\, dy = \tfrac{1}{2}, \qquad -1 \le x \le 1$$

$$= 0 \qquad \text{elsewhere}$$

$$f_2(y) = \tfrac{1}{2}, \qquad -1 \le y \le 1$$

$$= 0 \qquad \text{elsewhere}$$

But if

$$f_{12}(x, y) = \tfrac{1}{4}, \qquad \begin{matrix} -1 \le x \le 1 \\ -1 \le y \le 1 \end{matrix}$$

$$= 0 \qquad \text{elsewhere}$$

we get the same individual densities. ◀

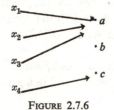

FIGURE 2.7.6

Now intuitively, if R_1 and R_2 are independent, then, say, e^{R_1} and $\sin R_2$ should be independent, since information about e^{R_1} should not change the odds concerning R_2 and hence should not affect $\sin R_2$ either. We shall prove a theorem of this type, but first we need some additional terminology.

If g is a function that maps points in the set D into points in the set E,† and $T \subset E$, we define the *preimage* of T under g as

$$g^{-1}(T) = \{x \in D : g(x) \in T\}$$

For example, let $D = \{x_1, x_2, x_3, x_4\}$, $E = \{a, b, c\}$, $g(x_1) = g(x_2) = g(x_3) = a$, $g(x_4) = c$ (see Figure 2.7.6). We then have

$$g^{-1}\{a\} = \{x_1, x_2, x_3\}$$
$$g^{-1}\{a, b\} = \{x_1, x_2, x_3\}$$
$$g^{-1}\{a, c\} = \{x_1, x_2, x_3, x_4\}$$
$$g^{-1}\{b\} = \varnothing$$

Note that, by definition of preimage, $x \in g^{-1}(T)$ iff $g(x) \in T$.

Now let $R_1, \ldots, R_n$ be random variables on a given probability space, and let $g_1, \ldots, g_n$ be functions of *one* variable, that is, functions from the reals to the reals. Let $R_1' = g_1(R_1), \ldots, R_n' = g_n(R_n)$; that is, $R_i'(\omega) = g_i(R_i(\omega))$, $\omega \in \Omega$. We assume that the R_i' are also random variables; this will be the case if the g_i are continuous or piecewise continuous. Specifically, we have the following result, which we shall use without proof.

If g is a real-valued function defined on the reals, and g is piecewise continuous, then for each Borel set $B \subset E^1$, $g^{-1}(B)$ is also a Borel subset of E^1. (A function with this property is said to be *Borel measurable*.)

Now we show that if g_i is piecewise continuous or, more generally, Borel measurable, R_i' is a random variable. Let B_i' be a Borel subset of E^1. Then

$$R_i'^{-1}(B_i') = \{\omega : R_i'(\omega) \in B_i'\}$$
$$= \{\omega : g_i(R_i(\omega)) \in B_i'\}$$
$$= \{\omega : R_i(\omega) \in g_i^{-1}(B_i')\} \in \mathscr{F}$$

since $g_i^{-1}(B_i')$ is a Borel set.

† A common notation for such a function is $g : D \to E$. It means simply that $g(x)$ is defined and belongs to E for each x in D.

Similarly, if g is a continuous real-valued function defined on E^n, then, for each Borel set $B \subset E^1$, $g^{-1}(B)$ is a Borel subset of E^n. It follows that if $R_1, \ldots, R_n$ are random variables, so is $g(R_1, \ldots, R_n)$.

Theorem 2. *If* $R_1, \ldots, R_n$ *are independent, then* $R'_1, \ldots, R'_n$ *are also independent.* (For short, "functions of independent random variables are independent.")

PROOF. If $B'_1, \ldots, B'_n$ are Borel subsets of E^1, then

$$P\{R'_1 \in B'_1, \ldots, R'_n \in B'_n\} = P\{g_1(R_1) \in B'_1 \ldots, g_n(R_n) \in B'_n\}$$

$$= P\{R_1 \in g_1^{-1}(B'_1), \ldots, R_n \in g_n^{-1}(B'_n)\}$$

$$= \prod_{i=1}^{n} P\{R_i \in g_i^{-1}(B'_i)\} \quad \text{by independence of the } R_i$$

$$= \prod_{i=1}^{n} P\{g_i(R_i) \in B'_i\} = \prod_{i=1}^{n} P\{R'_i \in B'_i\}$$

PROBLEMS

1. Let (R_1, R_2) have the following density function.

$$
\begin{aligned}
f_{12}(x, y) &= 4xy && \text{if } 0 \le x \le 1, 0 \le y \le 1, x \ge y \\
&= 6x^2 && \text{if } 0 \le x \le 1, 0 \le y \le 1, x < y \\
&= 0 && \text{elsewhere}
\end{aligned}
$$

(a) Find the individual density functions f_1 and f_2.
(b) If $A = \{R_1 \le \frac{1}{2}\}$, $B = \{R_2 \le \frac{1}{2}\}$, find $P(A \cup B)$.

2. If (R_1, R_2) is absolutely continuous with

$$
\begin{aligned}
f_{12}(x, y) &= 2e^{-(x+y)}, && 0 \le y \le x \\
&= 0 && \text{elsewhere}
\end{aligned}
$$

find $f_1(x)$ and $f_2(y)$.

3. Let (R_1, R_2) be uniformly distributed over the parallelogram with vertices $(-1, 0)$, $(1, 0)$, $(2, 1)$, and $(0, 1)$.
(a) Find and sketch the density functions of R_1 and R_2.
(b) A new random variable R_3 is defined by $R_3 = R_1 + R_2$. Show that R_3 is absolutely continuous, and find and sketch its density.

4. If $R_1, R_2, \ldots, R_n$ are independent, show that the joint distribution function is the product of the individual distribution functions; that is,

$$F_{12\cdots n}(x_1, x_2, \ldots, x_n) = F_1(x_1)F_2(x_2) \cdots F_n(x_n) \quad \text{for all real } x_1, \ldots, x_n$$

[Conversely, it can be shown that if $F_{12\cdots n}(x_1, \ldots, x_n) = F_1(x_1) \cdots F_n(x_n)$ for all real $x_1, \ldots, x_n$, then $R_1, \ldots, R_n$ are independent.]

5. Show that a random variable R is independent of itself—in other words, R and R are independent—if and only if R is *degenerate*, that is, essentially constant ($P\{R = c\} = 1$ for some c).

6. Under what conditions will R and $\sin R$ be independent? (Use Problem 5 and the result that functions of independent random variables are independent.)

7. If $(R_1, \ldots, R_n)$ is absolutely continuous and $f_{12\cdots n}(x_1, \ldots, x_n) = f_1(x_1) \cdots f_n(x_n)$ for all $x_1, \ldots, x_n$, show that $R_1, \ldots, R_n$ are independent.

8. Let (R_1, R_2) be absolutely continuous with density $f_{12}(x, y) = (x + y)/8$, $0 \leq x \leq 2$, $0 \leq y \leq 2$; $f_{12}(x, y) = 0$ elsewhere.

   (a) Find the probability that $R_1^2 + R_2 \leq 1$.

   (b) Find the conditional probability that exactly one of the random variables R_1, R_2 is ≤ 1, given that at least one of the random variables is ≤ 1.

   (c) Determine whether or not R_1 and R_2 are independent.

2.8 FUNCTIONS OF MORE THAN ONE RANDOM VARIABLE

We are now equipped to consider a wide variety of problems of the following sort. If $R_1, \ldots, R_n$ are random variables with a given joint density, and we define $R = g(R_1, \ldots, R_n)$, we ask for the distribution or density function of R. We shall use a distribution function approach to these problems; that is, we shall find the distribution function of R directly. There is also a density function method, but it is usually not as convenient; the density function approach is outlined in Problem 12. The distribution function method can be described as follows.

$$F_R(z) = P\{R \leq z\} = P\{g(R_1, \ldots, R_n) \leq z\}$$

$$= \underset{g(x_1, \ldots, x_n) \leq z}{\int \cdots \int} f_{12\cdots n}(x_1, \ldots, x_n)\, dx_1 \cdots dx_n$$

▶ **Example 1.** Let R_1 and R_2 be uniformly distributed between 0 and 1, and independent.

(a) Let $R_3 = R_1 + R_2$. Then, since $f_{12}(x, y) = f_1(x)f_2(y)$ by independence,

$$F_3(z) = P\{R_1 + R_2 \leq z\} = \underset{x+y \leq z}{\iint} f_1(x)f_2(y)\, dx\, dy$$

If $0 \leq z \leq 1$ (see Figure 2.8.1a),

$$F_3(z) = \underset{\text{shaded area}}{\iint} 1\, dx\, dy = \text{shaded area} = \frac{z^2}{2}$$

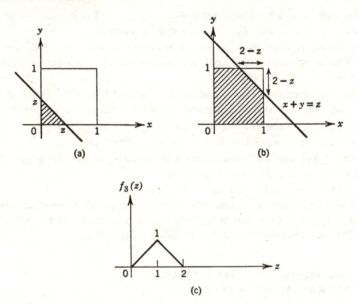

FIGURE 2.8.1   (a) Calculation of $F_3(z)$, $0 \le z \le 1$. (b) Calculation of $F_3(z)$, $1 \le z \le 2$.
(c) $f_3(z)$.

If $1 \le z \le 2$ (see Figure 2.8.1b),

$$F_3(z) = \text{shaded area} = 1 - \frac{(2 - z)^2}{2}$$

Thus $f_3(z) = z$, $0 \le z \le 1$; $f_3(z) = 2 - z$, $1 \le z \le 2$; $f_3(z) = 0$ elsewhere (see Figure 2.8.1c).

(b) Let $R_3 = R_1 R_2$. (Notice that $0 \le R_3 \le 1$.) If $0 \le z \le 1$ (see Figure 2.8.2),

$$F_3(z) = P\{R_1 R_2 \le z\} = \iint\limits_{xy \le z} f_{12}(x, y) \, dx \, dy$$

$$= \text{shaded area} = z + \int_z^1 \frac{z}{x} \, dx = z - z \ln z$$

$$f_3(z) = - \ln z \qquad 0 < z \le 1$$

$$= 0 \qquad \text{elsewhere}$$

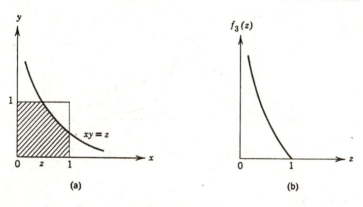

FIGURE 2.8.2

(c) Let $R_3 = \max (R_1, R_2)$. If $0 \leq z \leq 1$ (see Figure 2.8.3),

$$P\{R_3 \leq z\} = P\{R_1 \leq z, R_2 \leq z\} = \text{shaded area} = z^2$$

[Alternatively, $F_3(z) = P\{R_1 \leq z\}P\{R_2 \leq z\} = z^2$ by independence.]

$$f_3(z) = 2z, \qquad 0 \leq z \leq 1 \blacktriangleleft$$

Before the next example, we introduce the *Gaussian* or *normal* density function.

$$f(x) = \frac{1}{\sqrt{2\pi b}}\, e^{-(x-a)^2/2b^2}, \qquad x \text{ real } (b > 0,\ a \text{ any real number}) \quad (2.8.1)$$

This is the familiar bell-shaped curve centered at a (Figure 2.8.4); the smaller the value of b, the higher the peak and the more f is concentrated close to $x = a$. To check that this is a legitimate density, we must show that the area

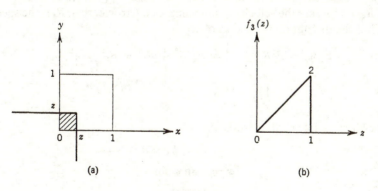

FIGURE 2.8.3

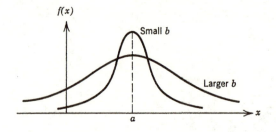

FIGURE 2.8.4   Normal Density.

under f is 1. Let

$$I = \int_{-\infty}^{\infty} e^{-x^2}\, dx$$

Then

$$I^2 = \int_{-\infty}^{\infty} e^{-x^2}\, dx \int_{-\infty}^{\infty} e^{-y^2}\, dy = \int_{-\infty}^{\infty}\int_{-\infty}^{\infty} e^{-(x^2+y^2)}\, dx\, dy = \text{(in polar coordinates)}$$

$$\int_0^{2\pi} d\theta \int_0^{\infty} r e^{-r^2}\, dr = \pi$$

so that

$$\int_{-\infty}^{\infty} e^{-x^2}\, dx = \sqrt{\pi} \tag{2.8.2}$$

Thus

$$\int_{-\infty}^{\infty} \frac{1}{\sqrt{2\pi b}}\, e^{-(x-a)^2/2b^2}\, dx = \left(\text{with } y = \frac{x-a}{\sqrt{2}\,b}\right) \int_{-\infty}^{\infty} \frac{1}{\sqrt{\pi}}\, e^{-y^2}\, dy = 1$$

▶ **Example 2.** Let R_1, R_2, and R_3 be independent, each normally distributed (i.e., having the normal density), with $a = 0$, $b = 1$. Let $R_4 = (R_1^2 + R_2^2 + R_3^2)^{1/2}$; take the positive square root so that $R_4 \geq 0$. (For example, if R_1, R_2, and R_3 are the velocity components of a particle, then R_4 is the speed.) Find the distribution function of R_4.

$$F_4(w) = P\{R_4 \leq w\} = P\{R_1^2 + R_2^2 + R_3^2 \leq w^2\}$$

$$= \iiint_{x^2+y^2+z^2 \leq w^2} (2\pi)^{-3/2} e^{-(x^2+y^2+z^2)/2}\, dx\, dy\, dz$$

We switch to spherical coordinates:

$$x = r \sin \phi \cos \theta$$
$$y = r \sin \phi \sin \theta$$
$$z = r \cos \phi$$

(ϕ is the "cone angle," and θ the "polar coordinate angle.") Then

$$F_4(w) = \int_0^{2\pi} d\theta \int_0^{\pi} d\phi \int_0^{w} (2\pi)^{-3/2} e^{-r^2/2} r^2 \sin \phi \, dr$$

$$= (2\pi)^{-3/2}(2\pi)(2) \int_0^{w} r^2 e^{-r^2/2} \, dr$$

Thus R_4 has a density given by

$$f_4(w) = \frac{2}{\sqrt{2\pi}} w^2 e^{-w^2/2} \qquad w \geq 0$$

$$= 0, \qquad w < 0 \blacktriangleleft$$

▶ **Example 3.** There are certain situations in which it is possible to avoid all integration in an n-dimensional problem. Suppose that $R_1, \ldots, R_n$ are independent and F_i is the distribution function of R_i, $i = 1, 2, \ldots, n$.

Let T_k be the kth smallest of the R_i. [For example, if $n = 4$ and $R_1(\omega) = 3$, $R_2(\omega) = 1.5$, $R_3(\omega) = -10$, $R_4(\omega) = 7$, then

$$T_1(\omega) = \min_i R_i(\omega) = R_3(\omega) = -10, \qquad T_2(\omega) = R_2(\omega) = 1.5$$

$$T_3(\omega) = R_1(\omega) = 3, \qquad T_4(\omega) = \max_i R_i(\omega) = R_4(\omega) = 7]$$

(Ties may be broken, for example, by favoring the random variable with the smaller subscript.)

We wish to find the distribution function of T_k. When $k = 1$ or $k = n$, the calculation is brief.

$$P\{T_n \leq x\} = P\{\max (R_1, \ldots, R_n) \leq x\} = P\{R_1 \leq x, \ldots, R_n \leq x\}$$

$$= \prod_{i=1}^{n} P\{R_i \leq x\} \qquad \text{by independence}$$

Thus

$$F_{T_n}(x) = \prod_{i=1}^{n} F_i(x)$$

$$P\{T_1 \leq x\} = 1 - P\{T_1 > x\} = 1 - P\{\min (R_1, \ldots, R_n) > x\}$$

$$= 1 - P\{R_1 > x, \ldots, R_n > x\} = 1 - \prod_{i=1}^{n} P\{R_i > x\}$$

Thus

$$F_{T_1}(x) = 1 - \prod_{i=1}^{n} (1 - F_i(x))$$

REMARK.   We may also calculate $F_{T_1}(x)$ as follows.

$$P\{T_1 \leq x\} = P\{\text{at least one } R_i \text{ is } \leq x\}$$
$$= P(A_1 \cup A_2 \cup \cdots \cup A_n) \quad \text{where } A_i = \{R_i \leq x\}$$
$$= P(A_1) + P(A_1^c \cap A_2) + \cdots$$
$$+ P(A_1^c \cap \cdots \cap A_{n-1}^c \cap A_n) \quad \text{by (1.3.11)}$$

But

$$P(A_1^c \cap \cdots \cap A_{i-1}^c \cap A_i) = P\{R_1 > x, \ldots, R_{i-1} > x, R_i \leq x\}$$
$$= (1 - F_1(x)) \cdots (1 - F_{i-1}(x))F_i(x)$$

Thus

$$F_{T_1}(x) = F_1(x) + (1 - F_1(x))F_2(x) + (1 - F_1(x))(1 - F_2(x))F_3(x)$$
$$+ \cdots + (1 - F_1(x)) \cdots (1 - F_{n-1}(x))F_n(x)$$

Hence

$$1 - F_{T_1} = (1 - F_1)[1 - F_2 - (1 - F_2)F_3 - \cdots$$
$$- (1 - F_2) \cdots (1 - F_{n-1})F_n]$$
$$= (1 - F_1)(1 - F_2)[1 - F_3 - (1 - F_3)F_4$$
$$- \cdots - (1 - F_3) \cdots (1 - F_{n-1})F_n]$$
$$= \prod_{i=1}^{n} (1 - F_i)$$

as above.

We now make the simplifying assumption that the R_i are absolutely continuous (as well as independent), each with the same density f. [Note that

$$P\{R_i = R_j\} = \iint_{x_i = x_j} f(x_i)f(x_j) \, dx_i \, dx_j = 0 \quad \text{(if } i \neq j)$$

Hence

$$P\{R_i = R_j \text{ for at least one } i \neq j\} \leq \sum_{i \neq j} P\{R_i = R_j\} = 0$$

Thus ties occur with probability zero and can be ignored.]

We shall show that the T_k are absolutely continuous, and find the density explicitly. We do this intuitively first. We have

$$P\{x < T_k < x + dx\} = P\{x < T_k < x + dx, T_k = R_1\}$$
$$+ P\{x < T_k < x + dx, T_k = R_2\} + \cdots + P\{x < T_k < x + dx, T_k = R_n\}$$

by the theorem of total probability. (The events $\{T_k = R_i\}$, $i = 1, \ldots, n$, are mutually exclusive and exhaustive.) Thus

$$P\{x < T_k < x + dx\} = nP\{x < T_k < x + dx, T_k = R_1\} \quad \text{by symmetry}$$
$$= nP\{T_k = R_1, x < R_1 < x + dx\}$$

Now for R_1 to be the kth smallest and fall between x and $x + dx$, exactly $k - 1$ of the random variables $R_2, \ldots, R_n$ must be $< R_1$, and the remaining $n - k$ must be $> R_1$ [and R_1 must lie in $(x, x + dx)$].

Since there are $\binom{n-1}{k-1}$ ways of selecting $k - 1$ distinct objects out of $n - 1$, we have

$$P\{x < T_k < x + dx\} = n\binom{n-1}{k-1} P\{x < R_1 < x + dx,$$

$$R_2 < R_1, \ldots, R_k < R_1, R_{k+1} > R_1, \ldots, R_n > R_1\}$$

But if R_1 falls in $(x, x + dx)$, $R_i < R_1$ is essentially the same thing as $R_i < x$, so that

$$P\{x < T_k < x + dx\} = n\binom{n-1}{k-1} f(x)\, dx (P\{R_i < x\})^{k-1}(P\{R_j > x\})^{n-k}$$

$$= n\binom{n-1}{k-1} f(x)(F(x))^{k-1}(1 - F(x))^{n-k}\, dx$$

Since $P\{x < T_k < x + dx\} = f_k(x)\, dx$, where f_k is the density of T_k (assumed to exist), we have

$$f_k(x) = n\binom{n-1}{k-1} f(x)(F(x))^{k-1}(1 - F(x))^{n-k}$$

[When $k = n$ we get $nf(x)(F(x))^{n-1} = (d/dx)F(x)^n$, and when $k = 1$ we get $nf(x)(1 - F(x))^{n-1} = (d/dx)(1 - (1 - F(x))^n)$, in agreement with the previous results if all R_i have distribution function F and the density f can be obtained by differentiating F.]

To obtain the result formally, we reason as follows.

$$P\{T_k \leq x\} = \sum_{i=1}^{n} P\{T_k \leq x, T_k = R_i\} = nP\{T_k \leq x, T_k = R_1\}$$

$$= nP\{R_1 \leq x, \text{ exactly } k - 1 \text{ of the variables } R_2, \ldots, R_n \text{ are} < R_1, \text{ and the remaining } n - k \text{ variables are} > R_1\}$$

$$= n\binom{n-1}{k-1} P\{R_1 \leq x, R_2 < R_1, \ldots, R_k < R_1, R_{k+1} > R_1, \ldots,$$

$$R_n > R_1\} \quad \text{by symmetry}$$

$$= n\binom{n-1}{k-1} \int_{x_1=-\infty}^{x} \int_{x_2=-\infty}^{x_1} \cdots \int_{x_k=-\infty}^{x_1} \int_{x_{k+1}=x_1}^{\infty}$$

$$\cdots \int_{x_n=x_1}^{\infty} f(x_1) \cdots f(x_n)\, dx_1 \cdots dx_n$$

$$= \int_{-\infty}^{x} n\binom{n-1}{k-1} f(x_1)(F(x_1))^{k-1}(1 - F(x_1))^{n-k}\, dx_1$$

The integrand is the density of T_k, in agreement with the intuitive approach. $T_1, \ldots, T_n$ are called the *order statistics* of $R_1, \ldots, R_n$.

REMARK.   All events

$$\{R_{i_1} \leq x, R_{i_2} < R_{i_1}, \ldots, R_{i_k} < R_{i_1}, R_{i_{k+1}} > R_{i_1}, \ldots, R_{i_n} > R_{i_1}\}$$

have the same probability, namely,

$$\int_{-\infty}^{x} f(x_{i_1}) \, dx_{i_1} \int_{-\infty}^{x_{i_1}} f(x_{i_2}) \, dx_{i_2} \cdots \int_{-\infty}^{x_{i_1}} f(x_{i_k}) \, dx_{i_k} \int_{x_{i_1}}^{\infty} f(x_{i_{k+1}}) \, dx_{i_{k+1}}$$

$$\cdots \int_{x_{i_1}}^{\infty} f(x_{i_n}) \, dx_{i_n}$$

This justifies the appeal to symmetry in the above argument. ◄

PROBLEMS

1. Let R_1 and R_2 be independent and uniformly distributed between 0 and 1. Find and sketch the distribution or density function of the random variable $R_3 = R_2/R_1^2$.

2. If R_1 and R_2 are independent random variables, each with the density function $f(x) = e^{-x}, x \geq 0; f(x) = 0, x < 0$, find and sketch the distribution or density function of the random variable R_3, where
(a) $R_3 = R_1 + R_2$
(b) $R_3 = R_2/R_1$

3. Let R_1 and R_2 be independent, absolutely continuous random variables, each normally distributed with parameters $a = 0$ and $b = 1$; that is,

$$f_1(x) = f_2(x) = \frac{1}{\sqrt{2\pi}} e^{-x^2/2}$$

Find and sketch the density or distribution function of the random variable $R_3 = R_2/R_1$.

4. Let R_1 and R_2 be independent, absolutely continuous random variables, each uniformly distributed between 0 and 1. Find and sketch the distribution or density function of the random variable R_3, where

$$R_3 = \frac{\max(R_1, R_2)}{\min(R_1, R_2)}$$

REMARK.   The example in which $R_3 = \max(R_1, R_2) + \min(R_1, R_2)$ may occur to the reader. However, this yields nothing new, since

$\max(R_1, R_2) + \min(R_1, R_2) = R_1 + R_2$ (the sum of two numbers is the larger plus the smaller).

5. A point-size worm is inside an apple in the form of the sphere $x^2 + y^2 + z^2 = 4a^2$. (Its position is uniformly distributed.) If the apple is eaten down to a core determined by the intersection of the sphere and the cylinder $x^2 + y^2 = a^2$, find the probability that the worm will be eaten.

6. A point (R_1, R_2, R_3) is uniformly distributed over the region in E^3 described by $x^2 + y^2 \leq 4$, $0 \leq z \leq 3x$. Find the probability that $R_3 \leq 2R_1$.

7. Solve Problem 6 under the assumption that (R_1, R_2, R_3) has density $f(x, y, z) = kz^2$ over the given region and $f(x, y, z) = 0$ outside the region.

8. Let $T_1, \ldots, T_n$ be the order statistics of $R_1, \ldots, R_n$, where $R_1, \ldots, R_n$ are independent, each with density f. Show that the joint density of $T_1, \ldots, T_n$ is given by

$$g(x_1, \ldots, x_n) = n!\, f(x_1) \cdots f(x_n), \qquad x_1 < x_2 < \cdots < x_n$$
$$= 0 \qquad \text{elsewhere}$$

HINT: Find $P\{T_1 \leq b_1, \ldots, T_n \leq b_n, R_1 < R_2 < \cdots < R_n\}$.

9. Let R_1, R_2, and R_3 be independent, each with density

$$f(x) = e^{-x}, \qquad x \geq 0$$
$$= 0, \qquad x < 0$$

Find the probability that $R_1 \geq 2R_2 \geq 3R_3$.

10. A man and a woman agree to meet at a certain place some time between 11 and 12 o'clock. They agree that the one arriving first will wait z hours, $0 \leq z \leq 1$, for the other to arrive. Assuming that the arrival times are independent and uniformly distributed, find the probability that they will meet.

11. If n points $R_1, \ldots, R_n$ are picked independently and with uniform density on a straight line of length L, find the probability that no two points will be less than distance d apart; that is, find

$$P\{\min_{i \neq j} |R_i - R_j| \geq d\}$$

HINT: First find $P\{\min_{i \neq j} |R_i - R_j| \geq d, R_1 < R_2 < \cdots < R_n\}$; show that the region of integration defined by this event is

$$x_{n-1} + d \leq x_n \leq L$$
$$x_{n-2} + d \leq x_{n-1} \leq L - d$$
$$x_{n-3} + d \leq x_{n-2} \leq L - 2d$$
$$\cdot$$
$$\cdot$$
$$\cdot$$
$$x_1 + d \leq x_2 \leq L - (n-2)d$$
$$0 \leq x_1 \leq L - (n-1)d$$

12. (The density function method for functions of more than one random variable.) Let $(R_1, \ldots, R_n)$ be absolutely continuous with density $f_{12 \cdots n}(x_1, \ldots, x_n)$. Define random variables $W_1, \ldots, W_n$ by $W_i = g_i(R_1, \ldots, R_n)$, $i = 1, 2, \ldots,$ n; thus $(W_1, \ldots, W_n) = g(R_1, \ldots, R_n)$. Assume that g is one-to-one, continuously differentiable with a nonzero Jacobian J_g (hence g has a continuously differentiable inverse h). Show that $(W_1, \ldots, W_n)$ is absolutely continuous with density

$$f^*_{12 \cdots n}(y) = f_{12 \cdots n}(h(y)) \, |J_h(y)|, \qquad y = (y_1, \ldots, y_n)$$
$$= \frac{f_{12 \cdots n}(h(y))}{|J_g(x)|_{x = h(y)}}$$

[The result is the same if g is defined only on some open subset D of E^n and $P\{(R_1, \ldots, R_n) \in D\} = 1$.]

13. Let R_1 and R_2 be independent random variables, each normally distributed with $a = 0$ and the same b. Define random variables R_0 and θ_0 by

$$R_1 = R_0 \cos \theta_0 \qquad (\text{taking } R_0 \geq 0)$$
$$R_2 = R_0 \sin \theta_0$$

Show that R_0 and θ_0 are independent, and find their density functions.

14. Let R_1 and R_2 be independent, absolutely continuous, positive random variables and let $R_3 = R_1 R_2$. Show that the density function of R_3 is given by

$$f_3(z) = \int_0^\infty \frac{1}{w} f_1\left(\frac{z}{w}\right) f_2(w) \, dw, \qquad z > 0$$
$$= 0, \qquad z < 0$$

Note: This problem may be done by the distribution function method or by applying Problem 12 as follows.

$$R_3 = R_1 R_2$$
$$R_4 = R_2$$

Use the results of Problem 12 to obtain $f_{34}(z, w)$ and from this find $f_3(z)$.

15. Because of inefficiency of production, the resistances R_1 and R_2 in Figure P.2.8.15 may be regarded as independent random variables, each uniformly

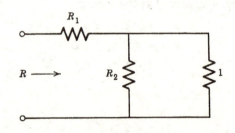

FIGURE P.2.8.15

distributed between 0 and 1 ohm. Find the probability that the total resistance R of the network is $\leq \frac{1}{2}$ ohm.

16. A chamber consists of the inside of the cylinder $x^2 + y^2 = 1$. A particle at the origin is given initial velocity components $v_x = R_1$ and $v_y = R_2$, where R_1 and R_2 are independent random variables, each with normal density $f(x) = (2\pi)^{-1/2} e^{-x^2/2}$. (There is no motion in the z-direction, and no force acting on the particle after the initial "push" at time $t = 0$.) If T is the time at which the particle strikes the wall of the chamber, find the distribution and density functions of T.

2.9 SOME DISCRETE EXAMPLES

In this section we examine some typical problems involving one or more discrete random variables. We first introduce the Poisson distribution, which may be regarded as an approximation to the binomial when the number n of trials is large and the probability p of success on a given trial is small.

Let R_n be the number of successes in n Bernoulli trials, with probability p_n of success on a given trial. We have seen (Section 1.5) that R_n has the *binomial distribution*;† that is, the probability function of R_n is

$$p_{R_n}(k) = \binom{n}{k} p_n^{\ k}(1 - p_n)^{n-k}, \qquad k = 0, 1, \ldots, n$$

We now let $n \to \infty$, $p_n \to 0$ in such a way that $np_n \to \lambda = $ constant. We shall show that

$$p_{R_n}(k) \to \frac{e^{-\lambda}\lambda^k}{k!}, \qquad k = 0, 1, \ldots$$

To see this, write

$$p_{R_n}(k) = \frac{n(n-1)\cdots(n-k+1)}{k!} p_n^{\ k}(1 - p_n)^{n-k}$$

$$= \frac{(1 - 1/n)(1 - 2/n)\cdots(1 - (k-1)/n)}{k!} (np_n)^k \left(1 - \frac{np_n}{n}\right)^{n-k}$$

Now $(1 - np_n/n)^{-k} \to 1$ and $(1 - np_n/n)^n \to e^{-\lambda}$ (Problem 1), and the result follows.

We call

$$p(k) = \frac{e^{-\lambda}\lambda^k}{k!}, \qquad k = 0, 1, 2, \ldots \tag{2.9.1}$$

---

† When a probabilist says he knows the *distribution* of a random variable R, he generally means that he has some way of calculating $P\{R \in B\}$ for all Borel sets B. For example, he might know the distribution function of R, or the probability function if R is discrete, or the density function if R is absolutely continuous. Thus to say that R has the normal distribution means that R has a density given by the formula (2.8.1).

the *Poisson* probability function; a random variable which has this probability function is said to have the *Poisson distribution*. (To check that it is a legitimate probability function:

$$\sum_{k=0}^{\infty} \frac{e^{-\lambda}\lambda^k}{k!} = e^{-\lambda}\left[1 + \lambda + \frac{\lambda^2}{2!} + \cdots\right] = e^{-\lambda}e^{\lambda} = 1)$$

We shall show that if R_1 and R_2 are independent, each having the Poisson distribution, then $R_1 + R_2$ also has the Poisson distribution. We first need a characterization of independence in the discrete case.

Theorem 1. Let $R_1, \ldots, R_n$ be discrete random variables on a given probability space, with probability functions $p_1, \ldots, p_n$. Let $p_{12\cdots n}$ be the joint probability function of $R_1, \ldots, R_n$, defined by

$$p_{12\cdots n}(x_1, \ldots, x_n) = P\{R_1 = x_1, \ldots, R_n = x_n\} \qquad (2.9.2)$$

Then $R_1, \ldots, R_n$ are independent if and only if

$$p_{12\cdots n}(x_1, \ldots, x_n) = p_1(x_1) \cdots p_n(x_n) \qquad \text{for all } x_1, \ldots, x_n$$

PROOF. If $R_1, \ldots, R_n$ are independent, then

$$\begin{aligned}
p_{12\cdots n}(x_1, \ldots, x_n) &= P\{R_1 = x_1, \ldots, R_n = x_n\} \\
&= P\{R_1 = x_1\} \cdots P\{R_n = x_n\} \qquad \text{by independence} \\
&= p_1(x_1) \cdots p_n(x_n)
\end{aligned}$$

Conversely, if $p_{12\cdots n}(x_1, \ldots, x_n) = p_1(x_1) \cdots p_n(x_n)$, then for all one-dimensional Borel sets $B_1, \ldots, B_n$,

$$\begin{aligned}
P\{R_1 \in B_1, \ldots, R_n \in B_n\} &= \sum_{x_1 \in B_1, \ldots, x_n \in B_n} P\{R_1 = x_1, \ldots, R_n = x_n\} \\
&= \sum_{x_1 \in B_1, \ldots, x_n \in B_n} p_1(x_1) \cdots p_n(x_n) \\
&= \sum_{x_1 \in B_1} p_1(x_1) \cdots \sum_{x_n \in B_n} p_n(x_n) \\
&= P\{R_1 \in B_1\} \cdots P\{R_n \in B_n\}
\end{aligned}$$

Hence $R_1, \ldots, R_n$ are independent.

REMARKS. If R_1 and R_2 are not independent, the joint probability function of R_1 and R_2 is not determined by the individual probability functions. For example, if $P\{R_1 = 1, R_2 = 1\} = P\{R_1 = 2, R_2 = 2\} = a$,

$P\{R_1 = 1, R_2 = 2\} = P\{R_1 = 2, R_2 = 1\} = \frac{1}{2} - a, 0 \leq a \leq \frac{1}{2}$, then $P\{R_1 = 1\} = P\{R_1 = 2\} = \frac{1}{2}$ and $P\{R_2 = 1\} = P\{R_2 = 2\} = \frac{1}{2}$. Thus we have uncountably many joint probability functions giving rise to the same individual probability functions.

If we wish to define discrete random variables $R_1, \ldots, R_n$ having a specified joint probability function $p_{12 \ldots n}$, there is no difficulty in constructing an appropriate probability space. Take $\Omega = E^n$, $\mathscr{F} = $ all subsets of Ω (since the random variables are discrete, there is no need to restrict to Borel sets), $P(B) = \sum_{(x_1, \ldots, x_n) \in B} p_{12 \ldots n}(x_1, \ldots, x_n), B \in \mathscr{F}$.

Now let R_1 and R_2 be independent, with R_i having the Poisson distribution with parameter λ_i, $i = 1, 2$. By Theorem 1, the joint probability function of R_1 and R_2 is

$$p_{12}(j, k) = P\{R_1 = j, R_2 = k\} = e^{-(\lambda_1 + \lambda_2)} \frac{\lambda_1^{\,j} \lambda_2^{\,k}}{j!\,k!}$$

We find the probability function of $R_1 + R_2$.

$$P\{R_1 + R_2 = m\} = \sum_{j+k=m} p_{12}(j, k)$$

$$= e^{-(\lambda_1 + \lambda_2)} \sum_{j=0}^{m} \frac{\lambda_1^{\,j} \lambda_2^{\,m-j}}{j!\,(m-j)!}$$

But $(\lambda_1 + \lambda_2)^m = \sum_{j=0}^{m} \binom{m}{j} \lambda_1^j \lambda_2^{m-j}$ by the binomial theorem, so that

$$P\{R_1 + R_2 = m\} = \frac{e^{-(\lambda_1 + \lambda_2)} (\lambda_1 + \lambda_2)^m}{m!}, \qquad m = 0, 1, \ldots$$

Thus $R_1 + R_2$ has the Poisson distribution with parameter $\lambda_1 + \lambda_2$.

By induction, it follows that the sum of n independent random variables $R_1, \ldots, R_n$, where R_i is Poisson with parameter λ_i, has the Poisson distribution with parameter $\lambda_1 + \cdots + \lambda_n$.

The use of the Poisson distribution as an approximation to the binomial is illustrated in the problems.

▶ **Example 1.**   Six unbiased dice are tossed independently. Let R_1 be the number of ones, R_2 the number of twos; R_1 and R_2 have the binomial distribution with $n = 6$, $p = 1/6$; that is,

$$p_1(k) = p_2(k) = \binom{6}{k} \left(\frac{1}{6}\right)^k \left(\frac{5}{6}\right)^{6-k}, \qquad k = 0, 1, 2, 3, 4, 5, 6$$

Let us find the joint probability function $p_{12}(j, k)$ of R_1 and R_2. This is a multinomial problem.

$b_1 =$ "1 occurs" on a given toss $p_1 = \frac{1}{6}$ $n_1 = j$

$b_2 =$ "2 occurs" $p_2 = \frac{1}{6}$ $n_2 = k$ $n = 6$

$b_3 =$ "3, 4, 5, or 6 occurs" $p_3 = \frac{2}{3}$ $n_3 = 6 - j - k$

Thus

$$p_{12}(j, k) = \frac{6!}{j!\, k!\, (6 - j - k)!}\left(\frac{1}{6}\right)^{j+k}\left(\frac{2}{3}\right)^{6-j-k},$$

$$j, k = 0, 1, 2, 3, 4, 5, 6; j + k \leq 6$$

Thus the multinomial formula appears as the joint probability function of a number of random variables, each of which is individually binomial.

Now let us find the conditional probability function of R_1 given R_2; that is,

$$p_1(j \mid k) = P\{R_1 = j \mid R_2 = k\}$$

$$= \frac{p_{12}(j, k)}{p_2(k)} = \frac{[6!/j!\, k!\, (6 - j - k)!](1/6)^{j+k}(4/6)^{6-j-k}}{[6!/k!\, (6 - k)!](1/6)^{k}(5/6)^{6-k}}$$

$$= \frac{(6 - k)!}{j!\, (6 - j - k)!}\,\frac{4^{6-j-k}}{5^{6-k}}\left(\frac{5^{-j}}{5^{-j}}\right) = \binom{6 - k}{j}\left(\frac{1}{5}\right)^{j}\left(\frac{4}{5}\right)^{6-k-j}$$

Intuitively, given $R_2 = k$, there are $6 - k$ remaining tosses. The possible outcomes are 1, 3, 4, 5, or 6 (2 is not permitted), all equally likely. Thus, given $R_2 = k$, R_1 should be binomial with $n = 6 - k$, $p = 1/5$. This is verified by the formal calculation above.

REMARK.   Since the discrete random variables R_1 and R_2 are independent iff $p_{12}(j, k) = p_1(j)p_2(k)$ for all j, k, it follows that independence is equivalent to $p_1(j \mid k) = p_1(j)$ for all j, k [such that $p_2(k) > 0$].

In the present case $p_1(j \mid k)$ is the binomial probability function with $n = 6 - k$, $p = 1/5$, and $p_1(j)$ is the binomial probability function with $n = 6$, $p = 1/6$. Thus R_1 and R_2 are not independent. This is clear intuitively; for example, if we know that $R_1 = 6$, the odds about R_2 are certainly affected; in fact, R_2 must be 0. ◄

PROBLEMS

1. (a) If $|x| \leq 1/2$, show that
$$\ln (1 + x) = x + \theta x^2$$
where $|\theta| \leq 1$, θ depending on x.

(b) Show that if $x_n \to \lambda$, then

$$\left(1 - \frac{x_n}{n}\right)^n \to e^{-\lambda}$$

2. If R has the binomial distribution with n large and p small, the Poisson approximation with $\lambda = np$ may be used (a rule of thumb that has been given is that the approximation will be good to several decimal places if $n \geq 100$ and $p \leq .01$). Feller (An Introduction to Probability Theory and Its Applications, vol. 1, John Wiley and Sons, 1950) gives several examples of such random variables:

   (i) The number of color-blind people in a large group (or the number of people possessing some other rare characteristic).
   (ii) The number of misprints on a page.
   (iii) The number of radioactive particles (or particles with some other distinguishing characteristic) passing through a counting device in a given time interval.
   (iv) The number of flying bomb hits on a particular area of London during World War II (n is the number of bombs in a given period of time, p the probability that a single bomb will hit the area).
   (v) The number of raisins in a cookie.

[Here the assumptions are not entirely clear. Perhaps what is envisioned is that the dough is bombarded by a raisin gun at some stage in the cookie-making process. It would seem that this is simply a peaceful version of example (iv).]

   In the following exercises, use the Poisson approximation to calculate the probabilities.

   (a) If $p = .001$, how large must n be if $P\{R \geq 1\} \geq .99$?
   (b) If $np = 2$, find $P\{R \geq 3\}$.

3. The joint probability function of two discrete random variables R_1 and R_2 is as follows:

   $p_{12}(1, 1) = .4$
   $p_{12}(1, 2) = .3$
   $p_{12}(2, 1) = .2$
   $p_{12}(2, 2) = .1$
   $p_{12}(j, k) = 0$ elsewhere

   (a) Determine whether or not R_1 and R_2 are independent.
   (b) Find the probability that $R_1 R_2 \leq 2$.

4. Let R_1 and R_2 be independent; assume that R_1 has the binomial distribution with parameters n and p, and R_2 has the binomial distribution with parameters m and p. Find $P\{R_1 = j \mid R_1 + R_2 = k\}$, and interpret the result intuitively. [Note: one approach involves establishing the formula $\binom{n+m}{k} = \sum_{i=0}^{k} \binom{n}{i}\binom{m}{k-i}$.]

3

Expectation

3.1 INTRODUCTION

We begin here the study of the long-run convergence properties of situations involving a very large number of independent repetitions of a random experiment. As an introductory example, suppose that we observe the length of a telephone call made from a specific phone booth at a given time of the day, say, the first call after 12 o'clock noon. Suppose that we repeat the experiment independently n times, where n is very large, and record the cost of each call (which is determined by its length). If we take the arithmetic average of the costs, that is, add the total cost of all n calls and then divide by n, we expect physically that the arithmetic average will converge in some sense to a number that we should interpret as the long-run average cost of a call. We shall try first to pin down the notion of average more precisely.

Assume that the cost R_2 of a call in terms of its length R_1 is as follows.

$$\text{If } 0 \leq R_1 \leq 3 \text{ (minutes)} \qquad R_2 = 10 \text{ (cents)}$$
$$\text{If } 3 < R_1 \leq 6 \qquad\qquad R_2 = 20$$
$$\text{If } 6 < R_1 \leq 9 \qquad\qquad R_2 = 30$$

(Assume for simplicity that the telephone is automatically disconnected after 9 minutes.)

Thus R_2 takes on three possible values, 10, 20, and 30; say $P\{R_2 = 10\} = .6$, $P\{R_2 = 20\} = .25$, $P\{R_2 = 30\} = .15$. If we observe N calls, where N is very large, then, roughly, $\{R_2 = 10\}$ will occur $.6N$ times; the total cost of calls of this type is $10(.6N) = 6N$. $\{R_2 = 20\}$ will occur approximately $.25N$ times, giving rise to a total cost of $20(.25N) = 5N$. $\{R_2 = 30\}$ will occur approximately $.15N$ times, producing a total cost of $30 (.15N) = 4.5N$. The total cost of all calls is $6N + 5N + 4.5N = 15.5N$, or 15.5 cents per call on the average.

Observe how we have computed the average.

$$\frac{10(.6N) + 20(.25N) + 30(.15N)}{N} = 10(.6) + 20(.25) + 30(.15)$$
$$= \sum_y yP\{R_2 = y\}$$

Thus we are taking a *weighted average* of the possible values of R_2, where the weights are the probabilities of R_2 assuming those values. This suggests the following definition.

Let R be a *simple* random variable, that is, a discrete random variable taking on only *finitely* many possible values. Define the *expectation* [also called the *expected value*, *average value*, *mean value*, or *mean*] of R as

$$E(R) = \sum_x xP\{R = x] \tag{3.1.1}$$

Since R is simple, this is a finite sum and there are no convergence problems. In particular, if R is identically constant, say $R = c$, then $E(R) = cP\{R = c\} = c$. For short,

$$E(c) = c \tag{3.1.2}$$

Note that if R takes the values $x_1, \ldots, x_n$, each with probability $1/n$, then $E(R) = (x_1 + \cdots + x_n)/n$, as we would expect intuitively. In this case each x_i is given the same weight, namely, $1/n$.

We now have the problem of extending the definition to more general random variables. If R is an arbitrary discrete random variable, the natural choice for $E(R)$ is again $\sum_x xP\{R = x\}$, provided that the sum makes sense. (Theorem 1 will make this precise.)

Similarly, let R_1 be discrete and $R_2 = g(R_1)$. Since R_2 is also discrete, we have $E(R_2) = \sum_y yP\{R_2 = y\}$. However, if $x_1, x_2, \ldots$ are the values of R_1, then with probability $p_{R_1}(x_i)$ we have $R_1 = x_i$, hence $R_2 = g(x_i)$. Thus if our definition of expectation is sound, we should have the following alternate expression for $E(R_2)$:

$$E(R_2) = E[g(R_1)] = \sum_i g(x_i)p_{R_1}(x_i) \tag{3.1.3}$$

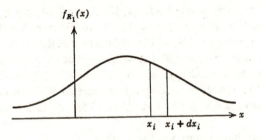

FIGURE 3.1.1

again a weighted average of possible values of R_2, but expressed in terms of the probability function of R_1.

If R_1 is absolutely continuous, this approach breaks down completely, since $P\{R_1 = x\} = 0$ for all x. However, we may get some idea as to how to compute $E(R_2) = E[g(R_1)]$ when R_1 is absolutely continuous, by making a discrete approximation. If we split the real line into intervals $(x_i, x_i + dx_i]$, then, roughly, the probability that $x_i < R_1 \leq x_i + dx_i$ is $f_{R_1}(x_i) \, dx_i$ (see Figure 3.1.1). If R_1 falls into this interval, $g(R_1)$ is approximately $g(x_i)$, at least if g is continuous. Thus an approximation to $E(R_2)$ should be

$$\sum_i g(x_i) f_R(x_i) \, dx_i$$

which suggests that if a general definition of expectation is formulated properly, and R_1 is absolutely continuous,

$$E[g(R_1)] = \int_{-\infty}^{\infty} g(x) f_{R_1}(x) \, dx \qquad (3.1.4)$$

In the telephone call example above, if R_1 has density f_1, we obtain

$$E(R_2) = \int_{-\infty}^{\infty} g(x) f_1(x) \, dx$$

where

$$R_2 = g(R_1) = 10 \qquad \text{if } 0 \leq R_1 \leq 3$$
$$= 20 \qquad \text{if } 3 < R_1 \leq 6$$
$$= 30 \qquad \text{if } 6 < R_1 \leq 9$$

Thus

$$E(R_2) = 10 \int_0^3 f_1(x) \, dx + 20 \int_3^6 f_1(x) \, dx + 30 \int_6^9 f_1(x) \, dx$$

$$= 10(.6) + 20(.25) + 30(.15) \qquad \text{as before}$$

If we have an n-dimensional situation, for instance $R_0 = g(R_1, \ldots, R_n)$, the preceding formulas generalize in a natural way. If $R_1, \ldots, R_n$ are discrete,

$$E[g(R_1, \ldots, R_n)] = \sum_{x_1, \ldots, x_n} g(x_1, \ldots, x_n) P\{R_1 = x_1, \ldots, R_n = x_n\} \quad (3.1.5)$$

If $(R_1, \ldots, R_n)$ is absolutely continuous with density $f_{12 \ldots n}$,

$$E[g(R_1, \ldots, R_n)] = \int_{-\infty}^{\infty} \cdots \int_{-\infty}^{\infty} g(x_1, \ldots, x_n) f_{12\ldots n}(x_1, \ldots, x_n) \, dx_1 \cdots dx_n$$

$$(3.1.6)$$

We shall outline very briefly a general definition of expectation that includes all the previous special cases.

If R is a simple random variable on $(\Omega, \mathscr{F}, P)$, we define

$$E(R) = \sum_x x P\{R = x\}$$

just as above. Now let R be a nonnegative random variable. We approximate R by simple random variables as follows.

Define

$$R_n(\omega) = \frac{k-1}{2^n} \quad \text{if } \frac{k-1}{2^n} \leq R(\omega) < \frac{k}{2^n}, \quad k = 1, 2, \ldots, n2^n$$

and let

$$R_n(\omega) = n \quad \text{if } R(\omega) \geq n$$

(see Figure 3.1.2 for an illustration with $n = 2$).

For any fixed ω, eventually $R(\omega) < n$, so that $0 \leq R(\omega) - R_n(\omega) < 2^{-n}$. Thus $R_n(\omega) \to R(\omega)$. In fact $R_n(\omega) \leq R_{n+1}(\omega)$ for all n, ω. For example, if $3/4 \leq R(\omega) < 7/8$, then $R_2(\omega) = R_3(\omega) = 3/4$; if $7/8 \leq R(\omega) < 1$, then $R_2(\omega) = 3/4$, $R_3(\omega) = 7/8$. In general, if $R(\omega)$ lies in the lower half of the interval $[(k-1)/2^n, k/2^n)$, then $R_n(\omega) = R_{n+1}(\omega)$; if $R(\omega)$ lies in the upper half, $R_n(\omega) < R_{n+1}(\omega)$.

Thus we have constructed a sequence of nonnegative simple functions R_n converging monotonically up to R. We have already defined $E(R_n)$, and since $R_n \leq R_{n+1}$ we have $E(R_n) \leq E(R_{n+1})$. We define

$$E(R) = \lim_{n \to \infty} E(R_n) \quad \text{(this may be } +\infty)$$

It is possible to show that if $\{R_n'\}$ is any other sequence of nonnegative simple functions converging monotonically up to R,

$$\lim_{n \to \infty} E(R_n') = \lim_{n \to \infty} E(R_n)$$

and thus $E(R)$ is well defined.

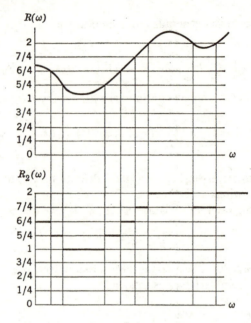

FIGURE 3.1.2   Approximation of a Nonnegative Random Variable by Simple Random Variables.

Finally, if R is an arbitrary random variable, let $R^+ = \max(R, 0)$, $R^- = \max(-R, 0)$; that is,

$$R^+(\omega) = R(\omega) \quad \text{if } R(\omega) \geq 0; \qquad R^+(\omega) = 0 \quad \text{if } R(\omega) < 0$$

$$R^-(\omega) = -R(\omega) \quad \text{if } R(\omega) \leq 0; \qquad R^-(\omega) = 0 \quad \text{if } R(\omega) > 0$$

R^+ and R^- are called the *positive* and *negative* parts of R (see Figure 3.1.3). It follows that $R = R^+ - R^-$ (and $|R| = R^+ + R^-$), and we define $E(R) = E(R^+) - E(R^-)$ if this is not of the form $+\infty - \infty$; if it is, we say that the expectation does not exist. Note that $E(R)$ is finite if and only if $E(R^+)$ and $E(R^-)$ are both finite. Since it can be shown that $E(|R|) = E(R^+) + E(R^-)$, it follows that

$$E(R) \text{ is finite if and only if } E(|R|) \text{ is finite} \qquad (3.1.7)$$

The expectation of a nonnegative random variable always exists; it may be $+\infty$.

The following results may be proved.

Let $R_1, R_2, \ldots, R_n$ be random variables on $(\Omega, \mathscr{F}, P)$, and let $R_0 = g(R_1, \ldots, R_n)$, where g is a function from E^n to E^1. Assume that g has the property that $g^{-1}(B)$ is a Borel subset of E^n whenever B is a Borel subset of E^1. Then, as we indicated in Section 2.7, R_0 is a random variable.

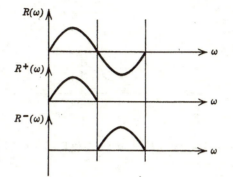

FIGURE 3.1.3 Positive and Negative Parts of a Random Variable.

Theorem 1. If $R_1, \ldots, R_n$ are discrete, then

$$E[g(R_1, \ldots, R_n)] = \sum_{x_1, \ldots, x_n} g(x_1, \ldots, x_n) P\{R_1 = x_1, \ldots, R_n = x_n\}$$

if $g(x_1, \ldots, x_n) \geq 0$ for all $x_1, \ldots, x_n$, or if the series on the right is absolutely convergent.

Theorem 2. If $(R_1, \ldots, R_n)$ is absolutely continuous with density $f_{12\ldots n}$, then

$$E[g(R_1, \ldots, R_n)] = \int_{-\infty}^{\infty} \cdots \int_{-\infty}^{\infty} g(x_1, \ldots, x_n) f_{12\ldots n}(x_1, \ldots, x_n) \, dx_1, \ldots, dx_n$$

if $g(x_1, \ldots, x_n) \geq 0$ for all $x_1, \ldots, x_n$, or if the integral on the right is absolutely convergent.

We shall look at examples that are neither discrete nor absolutely continuous in Chapter 4.

Notice that it is quite possible for the expectation to exist and be infinite, or not to exist at all. For example, let

$$f_R(x) = \frac{1}{x^2}, \qquad x \geq 1; \qquad f_R(x) = 0, \qquad x < 1$$

(see Figure 3.1.4a). Then

$$E(R) = \int_{-\infty}^{\infty} x f_R(x) \, dx = \int_1^{\infty} x \frac{1}{x^2} \, dx = \infty$$

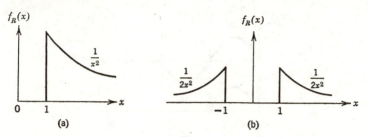

FIGURE 3.1.4 (a) $E(R) = \infty$. (b) $E(R)$ Does Not Exist.

As another example, let $f_R(x) = 1/2x^2$, $|x| \geq 1$; $f_R(x) = 0$, $|x| < 1$ (Figure 3.1.4b). Then (see Figure 3.1.5)

$$E(R^+) = \int_{-\infty}^{\infty} x^+ f_R(x)\, dx = \int_0^{\infty} x f_R(x)\, dx = \tfrac{1}{2} \int_0^{\infty} x \frac{1}{x^2}\, dx = \infty$$

$$E(R^-) = \int_{-\infty}^{\infty} x^- f_R(x)\, dx = \int_{-\infty}^0 -x f_R(x)\, dx = \tfrac{1}{2} \int_{-\infty}^{-1} -x \frac{1}{x^2}\, dx$$

$$= \tfrac{1}{2} \int_1^{\infty} \frac{1}{x}\, dx = \infty$$

Thus $E(R)$ does not exist.

Finally it can be shown that if two random variables R_1 and R_2 are "essentially" equal, that is, if $P\{\omega : R_1(\omega) \neq R_2(\omega)\} = 0$, then $E(R_1) = E(R_2)$ if the expectations exist.

REMARK. Theorem 1 fails if the series on the right is conditionally but not absolutely convergent. For example, let $P\{R_1 = n\} = (1/2)^n$, $n = 1$, $2, \ldots$, and $R_2 = g(R_1)$, where $g(n) = (-1)^{n+1} 2^n/n$. If $R_1(\omega) = n$, n odd, then $R_2(\omega) = 2^n/n$; hence $R_2^+(\omega) = g(n) = 2^n/n$, $R_2^-(\omega) = 0$. If $R_1(\omega) = n$, n even, then $R_2(\omega) = -2^n/n$; hence $R_2^+(\omega) = 0$, $R_2^-(\omega) = -g(n) = 2^n/n$. Therefore, by the nonnegative case of Theorem 1,

$$E(R_2^+) = \sum_{n \text{ odd}} g(n) P\{R_1 = n\} = 1 + \tfrac{1}{3} + \tfrac{1}{5} + \cdots = \infty$$

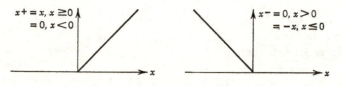

FIGURE 3.1.5

and

$$E(R_2^-) = \sum_{n \text{ even}} -g(n)P\{R_1 = n\} = \tfrac{1}{2} + \tfrac{1}{4} + \tfrac{1}{6} + \cdots = \infty$$

Hence $E(R_2)$ does not exist, although

$$\sum_{n=1}^{\infty} g(n)P\{R_1 = n\} = 1 - \tfrac{1}{2} + \tfrac{1}{3} - \tfrac{1}{4} + \cdots$$

is conditionally convergent. From an intuitive standpoint, the expectation should not change if the series is rearranged; a conditionally but not absolutely convergent series will not have this property.

3.2 TERMINOLOGY AND EXAMPLES

If R is a random variable on a given probability space, the kth *moment* of R ($k > 0$, not necessarily an integer) is defined by

$$\alpha_k = E(R^k) \qquad \text{if the expectation exists}$$

Thus

$$\alpha_k = \sum_x x^k p_R(x) \qquad \text{if } R \text{ is discrete}$$

$$= \int_{-\infty}^{\infty} x^k f_R(x)\, dx \qquad \text{if } R \text{ is absolutely continuous}$$

α_1 is simply $E(R)$, the expectation of R, often written as m and called the *mean* of R. If R has density f_R, m may be regarded as the abscissa of the centroid of the region in the plane between the x-axis and the graph of f_R (see Figure 3.2.1). To see this, notice that the total moment of the region

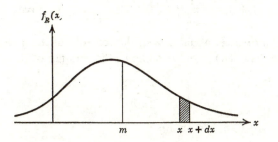

FIGURE 3.2.1 Geometric Interpretation of $E(R)$. The "Strip" Between x and $x + dx$ Contributes $(x - m)f_R(x)\, dx$ to the Moment about $x = m$.

about the line $x = m$ is

$$\int_{-\infty}^{\infty} (x - m)f_R(x)\, dx = m - m = 0$$

The expectation of R is a *measure of central tendency* in the sense that the arithmetic average of n independent observations of R converges (in a sense yet to be made precise) to $E(R)$.

The kth *central moment* of R ($k > 0$) is defined by

$$\beta_k = E[(R - m)^k] \qquad \text{if } m \text{ is finite and the expectation exists}$$

$$= \sum_x (x - m)^k p_R(x) \qquad \text{if } R \text{ is discrete}$$

$$= \int_{-\infty}^{\infty} (x - m)^k f_R(x)\, dx \qquad \text{if } R \text{ is absolutely continuous}$$

Notice that $\beta_1 = E(R - m) = m - m = 0$.

$\beta_2 = E[(R - m)^2]$ is called the *variance* of R, written σ^2, $\sigma^2(R)$, or Var R. σ (the positive square root of β_2) is called the *standard deviation* of R. Note that if R has finite mean, then, since $(R - m)^2 \geq 0$, Var R always exists; it may be infinite.

If R has density f_R, the variance of R may be regarded as the moment of inertia of the region in the plane between the x-axis and the graph of f_R, about the axis $x = m$.

The variance may be interpreted as a *measure of dispersion*. A large variance corresponds to a high probability that R will fall far from its mean, while a small variance indicates that R is likely to be close to its mean (see Figure 3.2.2). We shall make a quantititative statement to this effect (Chebyshev's inequality) in Section 3.7.

▶ **Example 1.** Consider the *normal density function*

$$f_R(x) = \frac{1}{\sqrt{2\pi}\, b}\, e^{-(x-a)^2/2b^2}, \qquad b > 0,\, a \text{ real}$$

Since f_R is symmetrical about $x = a$, the centroid of the area under f_R has abscissa a, so that $E(R) = a$. We compute the variance of R.

$$\sigma^2 = \int_{-\infty}^{\infty} \frac{(x - a)^2}{\sqrt{2\pi}\, b}\, e^{-(x-a)^2/2b^2}\, dx$$

Let $y = (x - a)/\sqrt{2}\, b$. We obtain

$$\sigma^2 = \int_{-\infty}^{\infty} \frac{2b^2}{\sqrt{2\pi}\, b}\, y^2 e^{-y^2} \sqrt{2}\, b\, dy = \frac{2b^2}{\sqrt{\pi}} \int_{-\infty}^{\infty} y^2 e^{-y^2}\, dy$$

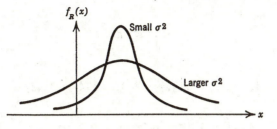

FIGURE 3.2.2

Now, by (2.8.2),

$$\sqrt{\pi} = \int_{-\infty}^{\infty} e^{-y^2} d(y)$$

Integrate by parts to obtain

$$\sqrt{\pi} = ye^{-y^2}]_{-\infty}^{\infty} - \int_{-\infty}^{\infty} -2y^2 e^{-y^2} dy$$

It follows that

$$\int_{-\infty}^{\infty} y^2 e^{-y^2} dy = \tfrac{1}{2}\sqrt{\pi}$$

Hence $\sigma^2 = b^2$. Thus we may write

$$f_R(x) = \frac{1}{\sqrt{2\pi}\,\sigma} e^{-(x-m)^2/2\sigma^2} \qquad (3.2.1)$$

In this case the mean and variance determine the density completely.

If R has the normal density with mean m and variance σ^2, we sometimes write "R is normal (m, σ^2)" for short. ◄

Before looking at the next example, it will be convenient to introduce the *gamma function*, defined by

$$\Gamma(r) = \int_0^\infty x^{r-1} e^{-x} dx, \qquad r > 0 \qquad (3.2.2)$$

Integrating by parts, we have

$$\Gamma(r) = \int_0^\infty e^{-x} d\left(\frac{x^r}{r}\right) = \frac{x^r e^{-x}}{r}\bigg]_0^\infty + \int_0^\infty \frac{x^r}{r} e^{-x} dx$$

$$= \int_0^\infty \frac{x^r}{r} e^{-x} dx = \frac{\Gamma(r+1)}{r}$$

Thus

$$\Gamma(r+1) = r\Gamma(r) \qquad (3.2.3)$$

Since

$$\Gamma(1) = \int_0^\infty e^{-x}\,dx = 1$$

we have

$$\Gamma(2) = 1\Gamma(1) = 1, \quad \Gamma(3) = 2\Gamma(2) = 2\cdot 1, \quad \Gamma(4) = 3\Gamma(3) = 3\cdot 2\cdot 1 = 3!$$

and

$$\Gamma(n+1) = n!, \qquad n = 0, 1, \ldots \tag{3.2.4}$$

We also need $\Gamma(1/2)$.

$$\Gamma(\tfrac{1}{2}) = \int_0^\infty x^{-1/2}e^{-x}\,dx$$

Let $x = y^2$ to obtain

$$\Gamma(\tfrac{1}{2}) = \int_0^\infty \frac{1}{y}e^{-y^2}2y\,dy = 2\int_0^\infty e^{-y^2}\,dy = \int_{-\infty}^\infty e^{-y^2}\,dy$$

By (2.8.2), we have

$$\Gamma(\tfrac{1}{2}) = \sqrt{\pi} \tag{3.2.5}$$

▶ **Example 2.** Let R_1 be absolutely continuous with density $f_1(x) = e^{-x}$, $x \geq 0; f_1(x) = 0, x < 0$. Let $R_2 = R_1{}^2$. We may compute $E(R_2)$ in two ways.

1. $E(R_2) = E(R_1{}^2) = \displaystyle\int_{-\infty}^\infty x^2 f_1(x)\,dx = \int_0^\infty x^2 e^{-x}\,dx = \Gamma(3) = 2$ by (3.2.4)

2. We may find the density of R_2 by the technique of Section 2.4 (see Figure 3.2.3). We have

$$f_2(y) = f_1(\sqrt{y})\frac{d}{dy}\sqrt{y} = \frac{e^{-\sqrt{y}}}{2\sqrt{y}}, \quad y > 0$$

$$= 0, \quad y < 0$$

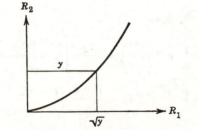

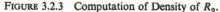

FIGURE 3.2.3 Computation of Density of R_2.

Then

$$E(R_2) = \int_{-\infty}^{\infty} yf_2(y)\,dy$$

$$= \int_0^{\infty} y\,\frac{e^{-\sqrt{y}}}{2\sqrt{y}}\,dy$$

$$= (\text{with } y = x^2) \int_0^{\infty} x^2 e^{-x}\,dx = 2$$

as before.

Notice that both methods must give the same answer by Theorem 2 of Section 3.1. For $R_2(\omega) = (R_1(\omega))^2$; applying the theorem with $g(R_1) = R_1^2$, we obtain

$$E(R_1^2) = \int_{-\infty}^{\infty} x^2 f_1(x)\,dx$$

Applying the theorem with $g(R_2) = R_2$, we have

$$E(R_2) = \int_{-\infty}^{\infty} yf_2(y)\,dy$$

Generally the first method is easier, since the computation of the density of R_2 is avoided. ◄

▶ **Example 3.** Let R_1 and R_2 be independent, each with density $f(x) = e^{-x}$, $x \geq 0$; $f(x) = 0$, $x < 0$. Let $R_3 = \max(R_1, R_2)$. We compute $E(R_3)$.

$$E(R_3) = E[g(R_1, R_2)] = \int_{-\infty}^{\infty}\int_{-\infty}^{\infty} g(x, y) f_{12}(x, y)\,dx\,dy$$

$$= \int_0^{\infty}\int_0^{\infty} \max(x, y) e^{-x} e^{-y}\,dx\,dy$$

Now $\max(x, y) = x$ if $x \geq y$; $\max(x, y) = y$ if $x \leq y$ (see Figure 3.2.4). Thus

$$E(R_3) = \iint_A xe^{-x}e^{-y}\,dx\,dy + \iint_B ye^{-x}e^{-y}\,dx\,dy$$

$$= \int_{x=0}^{\infty} xe^{-x}\int_{y=0}^{x} e^{-y}\,dy\,dx + \int_{y=0}^{\infty} ye^{-y}\int_{x=0}^{y} e^{-x}\,dx\,dy$$

on A, $\max(x, y) = x$
on B, $\max(x, y) = y$

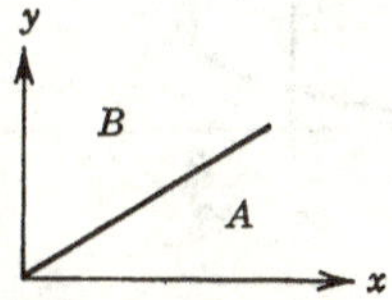

FIGURE 3.2.4

The two integrals are equal, since one may be obtained from the other by interchanging x and y. Thus

$$E(R_3) = 2\int_0^\infty xe^{-x}\int_0^x e^{-y}\,dy\,dx = 2\int_0^\infty xe^{-x}(1 - e^{-x})\,dx$$

$$= 2\int_0^\infty xe^{-x}\,dx - 2\int_0^\infty \frac{z}{2}e^{-z}\frac{dz}{2} = \frac{3}{2}\Gamma(2) = \frac{3}{2} \;\blacktriangleleft$$

The moments and central moments of a random variable R, especially the mean and variance, give some information about the behavior of R. In many situations it may be difficult to compute the distribution function of R explicitly, but the calculation of some of the moments may be easier. We shall examine some problems of this type in Section 3.5.

Another parameter that gives some information about a random variable R is the *median* of R, defined when F_R is continuous as a number μ (not necessarily unique) such that $F_R(\mu) = 1/2$ (see Figure 3.2.5a and b).

In general the median of a random variable R is a number μ such that

$$F_R(\mu) = P\{R \leq \mu\} \geq \tfrac{1}{2}$$

$$F_R(\mu^-) = P\{R < \mu\} \leq \tfrac{1}{2}$$

(see Figure 3.2.5c).

Loosely speaking, μ is the halfway point of the distribution function of R.

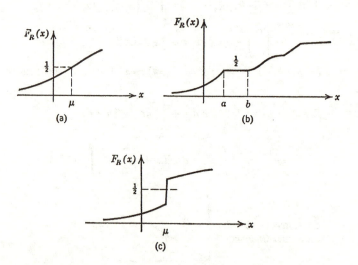

FIGURE 3.2.5   (a) μ is the Unique Median. (b) Any Number Between a and b is a Median. (c) μ is the Unique Median.

PROBLEMS

1. Let R be normally distributed with mean 0 and variance 1. Show that

$$E(R^n) = 0, \quad n \text{ odd}$$
$$= (n-1)(n-3)\cdots(5)(3)(1), \quad n \text{ even}$$

2. Let R_1 have the exponential density $f_1(x) = e^{-x}$, $x \geq 0$; $f_1(x) = 0$, $x < 0$. Let $R_2 = g(R_1)$ be the largest integer $\leq R_1$ (if $0 \leq R_1 < 1$, $R_2 = 0$; if $1 \leq R_1 < 2$, $R_2 = 1$, and so on).
   (a) Find $E(R_2)$ by computing $\int_{-\infty}^{\infty} g(x)f_1(x)\, dx$.
   (b) Find $E(R_2)$ by evaluating the probability function of R_2 and then computing $\sum_y y p_{R_2}(y)$.

3. Let R_1 and R_2 be independent random variables, each with the exponential density $f(x) = e^{-x}$, $x \geq 0$; $f(x) = 0$, $x < 0$. Find the expectation of
   (a) $R_1 R_2$
   (b) $R_1 - R_2$
   (c) $|R_1 - R_2|$

4. Let R_1 and R_2 be independent, each uniformly distributed between -1 and $+1$. Find $E[\max(R_1, R_2)]$.

5. Suppose that the density function for the length R of a telephone call is

$$f(x) = xe^{-x}, \quad x \geq 0$$
$$= 0, \quad x < 0$$

The cost of a call is
$$C(R) = 2, \quad 0 < R \leq 3$$
$$= 2 + 6(R-3), \quad R > 3$$

Find the average cost of a call.

6. Two machines are put into service at $t = 0$, processing the same data. Let R_i ($i = 1, 2$) be the time (in hours) at which machine i breaks down. Assume that R_1 and R_2 are independent random variables, each having the exponential density function $f(x) = \lambda e^{-\lambda x}$, $x \geq 0$; $f(x) = 0$, $x < 0$. Suppose that we start counting down time if and only if *both* machines are out of service. No repairs are allowed during the working day (which is T hours long), but any machine that has failed during the day is assumed to be completely repaired by the time the next day begins. For example, if $T = 8$ and the machines fail at $t = 2$ and $t = 6$, the down time is 2 hours.
   (a) Find the probability that at least one machine will fail during a working day.
   (b) Find the average down time per day. (Leave the answer in the form of an integral.)

7. Show that if R has the binomial distribution with parameters n and p, that is, R is the number of successes in n Bernoulli trials with probability of success p on

a given trial, then $E(R) = np$, as one should expect intuitively. HINT: in $E(R) = \sum_{k=0}^{n} k \binom{n}{k} p^k (1 - p)^{n-k}$, factor out np and use the binomial theorem.

REMARK. In Section 3.5 we shall calculate the mean and variance of R in an indirect but much more efficient way.

8. If R has the Poisson distribution with parameter λ, show that

$$E[R(R - 1)(R - 2) \cdots (R - r + 1)] = \lambda^r$$

Conclude that $E(R) = \operatorname{Var} R = \lambda$.

3.3 PROPERTIES OF EXPECTATION

In this section we list several basic properties of the expectation of a random variable. A precise justification of these properties would require a detailed analysis of the general definition of $E(R)$ that we gave in Section 3.1; what we actually did there was to outline the construction of the abstract Lebesgue integral. Instead we shall give plausibility arguments or proofs in special cases.

1. Let $R_1, \ldots, R_n$ be random variables on a given probability space. Then

$$E(R_1 + \cdots + R_n) = E(R_1) + \cdots + E(R_n)$$

CAUTION. Recall that $E(R)$ can be $\pm \infty$, or not exist at all. The complete statement of property 1 is: If $E(R_i)$ exists for all $i = 1, 2, \ldots, n$, and $+\infty$ and $-\infty$ do not *both* appear in the sum $E(R_1) + \cdots + E(R_n)$ ($+\infty$ alone or $-\infty$ alone is allowed), then $E(R_1 + \cdots + R_n)$ exists and equals $E(R_1) + \cdots + E(R_n)$.

For example, suppose that (R_1, R_2) has density f_{12}, and $R' = g(R_1, R_2)$, $R'' = h(R_1, R_2)$. Then

$$E(R' + R'') = E[g(R_1, R_2) + h(R_1, R_2)]$$

$$= \int_{-\infty}^{\infty} \int_{-\infty}^{\infty} [g(x, y) + h(x, y)] f_{12}(x, y) \, dx \, dy$$

$$= \int_{-\infty}^{\infty} \int_{-\infty}^{\infty} g(x, y) f_{12}(x, y) \, dx \, dy + \int_{-\infty}^{\infty} \int_{-\infty}^{\infty} h(x, y) f_{12}(x, y) \, dx \, dy$$

$$= E(R') + E(R'')$$

2. If R is a random variable whose expectation exists, and a is any real number, then $E(aR)$ exists and

$$E(aR) = aE(R)\dagger$$

For example, if R_1 has density f_1 and $R_2 = aR_1$, then

$$E(R_2) = \int_{-\infty}^{\infty} axf_1(x)\,dx = aE(R_1)$$

Basically, properties 1 and 2 say that the expectation is linear.

3. If $R_1 \leq R_2$, then $E(R_1) \leq E(R_2)$, assuming that both expectations exist. For example, if R has density f, and $R_1 = g(R)$, $R_2 = h(R)$, and $g \leq h$, we have

$$E(R_1) = \int_{-\infty}^{\infty} g(x)f(x)\,dx \leq \int_{-\infty}^{\infty} h(x)f(x)\,dx = E(R_2)$$

4. If $R \geq 0$ and $E(R) = 0$, then R is *essentially* 0; that is, $P\{R = 0\} = 1$. This we can actually prove, from the previous properties. Define $R_n = 0$ if $0 \leq R < 1/n$; $R_n = 1/n$ if $R \geq 1/n$. Then $0 \leq R_n \leq R$, so that, by property 3, $E(R_n) = 0$. But R_n has only two possible values, 0 and $1/n$, and so

$$E(R_n) = \sum_y yp_{R_n}(y) = 0P\{R_n = 0\} + \frac{1}{n}P\left\{R_n = \frac{1}{n}\right\}$$

Thus

$$P\left\{R_n = \frac{1}{n}\right\} = P\left\{R \geq \frac{1}{n}\right\} = 0 \quad \text{for all } n$$

But

$$P\{R > 0\} = P\left[\bigcup_{n=1}^{\infty}\left\{R \geq \frac{1}{n}\right\}\right] \leq \sum_{n=1}^{\infty} P\left\{R \geq \frac{1}{n}\right\} = 0$$

Hence

$$P\{R = 0\} = 1$$

Notice that if R is discrete, the argument is much faster: if $\sum_{x\geq 0} xp_R(x) = 0$, then $xp_R(x) = 0$ for all $x \geq 0$; hence $p_R(x) = 0$ for $x > 0$, and therefore $p_R(0) = 1$.

COROLLARY. If Var $R = 0$, then R is essentially constant.

PROOF. If $m = E(R)$, then $E[(R - m)^2] = 0$, hence $P\{R = m\} = 1$.

† Since $E(R)$ is allowed to be infinite, expressions of the form $0 \cdot \infty$ will occur. The most convenient way to handle this is simply to define $0 \cdot \infty = 0$; no inconsistency will result.

5. Let $R_1, \ldots, R_n$ be *independent* random variables.

(a) If all the R_i are nonnegative, then

$$E(R_1 R_2 \cdots R_n) = E(R_1)E(R_2) \cdots E(R_n)$$

(b) If $E(R_i)$ is finite for all i (whether or not the $R_i \geq 0$), then $E(R_1 R_2 \cdots R_n)$ is finite and

$$E(R_1 R_2 \cdots R_n) = E(R_1)E(R_2) \cdots E(R_n)$$

We can prove this when all the R_i are discrete, if we accept certain facts about infinite series. For

$$E(R_1 R_2 \cdots R_n) = \sum_{x_1, \ldots, x_n} x_1 x_2 \cdots x_n p_{12 \ldots n}(x_1, \ldots, x_n)$$

$$= \sum_{x_1, \ldots, x_n} x_1 \cdots x_n p_1(x_1) \cdots p_n(x_n)$$

Under hypothesis (a) we may restrict the x_i's to be ≥ 0. Under hypothesis (b) the above series is absolutely convergent. Since a nonnegative or absolutely convergent series can be summed in any order, we have

$$E(R_1 R_2 \cdots R_n) = \sum_{x_1} x_1 p_1(x_1) \cdots \sum_{x_n} x_n p_n(x_n) = E(R_1)E(R_2) \cdots E(R_n)$$

If $(R_1, \ldots, R_n)$ is absolutely continuous, the argument is similar, with sums replaced by integrals.

$$E(R_1 R_2 \cdots R_n) = \int_{-\infty}^{\infty} \cdots \int_{-\infty}^{\infty} x_1 \cdots x_n f_{12 \ldots n}(x_1, \ldots, x_n) \, dx_1 \cdots dx_n$$

$$= \int_{-\infty}^{\infty} \cdots \int_{-\infty}^{\infty} x_1 \cdots x_n f_1(x_1) \cdots f_n(x_n) \, dx_1 \cdots dx_n$$

$$= \int_{-\infty}^{\infty} x_1 f_1(x_1) \, dx_1 \cdots \int_{-\infty}^{\infty} x_n f_n(x_n) \, dx_n$$

$$= E(R_1) \cdots E(R_n)$$

6. Let R be a random variable with finite mean m and variance σ^2 (possibly infinite). If a and b are real numbers, then

$$\text{Var}\,(aR + b) = a^2 \sigma^2$$

PROOF. Since $E(aR + b) = am + b$ by properties 1 and 2 [and (3.1.2)], we have

$$\text{Var}\,(aR + b) = E[(aR + b - (am + b))^2]$$
$$= E[a^2(R - m)^2]$$
$$= a^2 E[(R - m)^2] \quad \text{by property 2}$$
$$= a^2 \sigma^2.$$

7. Let $R_1, \ldots, R_n$ be independent random variables, each with finite mean. Then

$$\text{Var}\,(R_1 + \cdots + R_n) = \text{Var}\,R_1 + \cdots + \text{Var}\,R_n$$

PROOF.   Let $m_i = E(R_i)$. Then

$$\text{Var}\,(R_1 + \cdots + R_n) = E\!\left[\left(\sum_{i=1}^{n} R_i - \sum_{i=1}^{n} m_i\right)^{\!2}\right] = E\!\left[\left(\sum_{i=1}^{n} (R_i - m_i)\right)^{\!2}\right]$$

If this is expanded, the "cross terms" are 0, since, if $i \neq j$,

$$E[(R_i - m_i)(R_j - m_j)] = E(R_i R_j - m_i R_j - m_j R_i + m_i m_j)$$
$$= E(R_i)E(R_j) - m_i E(R_j) - m_j E(R_i) + m_i m_j$$

$$\text{by properties 5, 1, and 2}$$

$$= 0 \qquad \text{since } E(R_i) = m_i,\; E(R_j) = m_j$$

Thus

$$\text{Var}\,(R_1 + \cdots + R_n) = \sum_{i=1}^{n} E(R_i - m_i)^2 = \sum_{i=1}^{n} \text{Var}\,R_i$$

COROLLARY.   If $R_1, \ldots, R_n$ are independent, each with finite mean, and $a_1, \ldots, a_n, b$ are real numbers, then

$$\text{Var}\,(a_1 R_1 + \cdots + a_n R_n + b) = a_1^2 \,\text{Var}\,R_1 + \cdots + a_n^2 \,\text{Var}\,R_n$$

PROOF.   This follows from properties 6 and 7. (Notice that $a_1 R_1, \ldots, a_n R_n$ are still independent; see Problem 1.)

8. The central moments $\beta_1, \ldots, \beta_n$ $(n \geq 2)$ can be obtained from the moments $\alpha_1, \ldots, \alpha_n$, provided that $\alpha_1, \ldots, \alpha_{n-1}$ are finite and α_n exists. To see this, expand $(R - m)^n$ by the binomial theorem.

$$(R - m)^n = \sum_{k=0}^{n} \binom{n}{k} R^k (-m)^{n-k}$$

Thus

$$\beta_n = E[(R - m)^n] = \sum_{k=0}^{n} \binom{n}{k} (-m)^{n-k} \alpha_k$$

Notice that since $\alpha_1, \ldots, \alpha_{n-1}$ are finite, no terms of the form $+\infty - \infty$ can appear in the summation, and thus we may take the expectation term by term, by property 1.

This result is applied most often when $n = 2$. If R has finite mean $[E(R^2)$ always exists since $R^2 \geq 0]$, then $(R - m)^2 = R^2 - 2mR + m^2$; hence

$$\text{Var}\,R = E(R^2) - 2mE(R) + m^2$$

That is,

$$\sigma^2 = E(R^2) - [E(R)]^2 \tag{3.3.1}$$

which is the "mean of the square" minus the "square of the mean."

9. If $E(R^k)$ is finite and $0 < j < k$, then $E(R_j)$ is also finite.

PROOF

$$|R(\omega)|^j \leq |R(\omega)|^k \quad \text{if } |R(\omega)| \geq 1$$
$$\leq 1 \quad \text{if } |R(\omega)| < 1$$

Thus

$$|R(\omega)|^j \leq 1 + |R(\omega)|^k \quad \text{for all } \omega$$

Hence

$$E(|R|^j) \leq 1 + E(|R|^k) < \infty$$

and the result follows. Notice that the expectation of a random variable is finite if and only if the expectation of its absolute value is finite; see (3.1.7).

Thus in property 8, if α_{n-1} is finite, automatically $\alpha_1, \ldots, \alpha_{n-2}$ are finite as well.

REMARK. Properties 5 and 7 fail without the hypothesis of independence. For example, let $R_1 = R_2 = R$, where R has finite mean. Then $E(R_1 R_2) \neq E(R_1)E(R_2)$ since $E(R^2) - [E(R)]^2 = \text{Var } R$, which is >0 unless R is essentially constant, by the corollary to property 4. Also, $\text{Var } (R_1 + R_2) = \text{Var } (2R) = 4 \text{ Var } R$, which is not the same as $\text{Var } R_1 + \text{Var } R_2 = 2 \text{ Var } R$ unless R is essentially constant.

PROBLEMS

1. If $R_1, \ldots, R_n$ are independent random variables, show that $a_1 R_1 + b_1, \ldots, a_n R_n + b_n$ are independent for all possible choices of the constants a_i and b_i.

2. If R is normally distributed with mean m and variance σ^2, evaluate the central moments of R (see Problem 1, Section 3.2).

3. Let θ be uniformly distributed between 0 and 2π. Define $R_1 = \cos \theta$, $R_2 = \sin \theta$. Show that $E(R_1 R_2) = E(R_1)E(R_2)$, and also $\text{Var } (R_1 + R_2) = \text{Var } R_1 + \text{Var } R_2$, but R_1 and R_2 are not independent. Thus, in properties 5 and 7, the converse assertion is false.

4. If $E(R)$ exists, show that $|E(R)| \leq E(|R|)$.

5. Let R be a random variable with finite mean. Indicate how and under what conditions the moments of R can be obtained from the central moments. In

particular show that $E(R^2) < \infty$ if and only if Var $R < \infty$. More generally, α_n is finite if and only if β_n is finite.

3.4 CORRELATION

If R_1 and R_2 are random variables on a given probability space, we may define *joint moments* associated with R_1 and R_2

$$\alpha_{jk} = E(R_1^{\,j} R_2^{\,k}), \qquad j, k > 0$$

and *joint central moments*

$$\beta_{jk} = E[(R_1 - m_1)^j (R_2 - m_2)^k], \qquad m_1 = E(R_1), m_2 = E(R_2)$$

We shall study $\beta_{11} = E[(R_1 - m_1)(R_2 - m_2)] = E(R_1 R_2) - E(R_1)E(R_2)$, which is called the *covariance* of R_1 and R_2, written Cov (R_1, R_2).

In this section we assume that $E(R_1)$ and $E(R_2)$ are finite, and $E(R_1 R_2)$ exists; then the covariance of R_1 and R_2 is well defined.

Theorem 1. *If R_1 and R_2 are independent, then* Cov $(R_1, R_2) = 0$, *but not conversely.*

PROOF. By property 5 of Section 3.3, independence of R_1 and R_2 implies that $E(R_1 R_2) = E(R_1)E(R_2)$; hence Cov $(R_1, R_2) = 0$. An example in which Cov $(R_1, R_2) = 0$ but R_1 and R_2 are not independent is given in Problem 3 of Section 3.3.

We shall try to find out what the knowledge of the covariance of R_1 and R_2 tells us about the random variables themselves. We first establish a very useful inequality.

Theorem 2 (Schwarz Inequality). *Assume that $E(R_1^2)$ and $E(R_2^2)$ are finite (R_1 and R_2 then automatically have finite mean, by property 9 of Section 3.3, and finite variance, by property 8). Then $E(R_1 R_2)$ is finite, and*

$$|E(R_1 R_2)|^2 \leq E(R_1^2)E(R_2^2)$$

PROOF. If R_1 is essentially 0, the inequality is immediate, so assume R_1 not essentially 0; then $E(R_1^2) > 0$. For any real number x let

$$h(x) = E[(x\,|R_1| + |R_2|)^2] = E(R_1^2)x^2 + 2E(|R_1 R_2|)x + E(R_2^2)$$

Since $h(x)$ is the expectation of a nonnegative random variable, it must be ≥ 0 for all x. The quadratic equation $h(x) = 0$ has either no real roots or, at

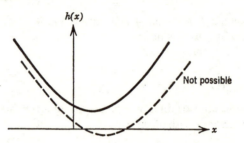

FIGURE 3.4.1   Proof of the Schwarz Inequality.

worst, one real repeated root (see Figure 3.4.1). Thus the discriminant must be ≤ 0; hence

$$(E(|R_1 R_2|))^2 \leq E(R_1{}^2) E(R_2{}^2) < \infty$$

Since $E(|R_1 R_2|)$ is finite, so is $E(R_1 R_2)$, by (3.1.7). Furthermore, $|E(R_1 R_2)| \leq E(|R_1 R_2|)$ (Problem 4, Section 3.3), and the result follows.

Now assume that $E(R_1{}^2)$ and $E(R_2{}^2)$ are finite and, in addition, that the variances $\sigma_1{}^2$ and $\sigma_2{}^2$ of R_1 and R_2 are >0. Define the *correlation coefficient* of R_1 and R_2 as

$$\rho(R_1, R_2) = \frac{\mathrm{Cov}\,(R_1, R_2)}{\sigma_1 \sigma_2}$$

By Theorem 1, if R_1 and R_2 are independent, they are uncorrelated; that is, $\rho(R_1, R_2) = 0$, but not conversely.

Theorem 3. $-1 \leq \rho(R_1, R_2) \leq 1$.

PROOF.   Apply the Schwarz inequality to $R_1 - E(R_1)$ and $R_2 - E(R_2)$.

$$|E[(R_1 - ER_1)(R_2 - ER_2)]|^2 \leq E[(R_1 - ER_1)^2] E[(R_2 - ER_2)^2]$$

Thus $|\mathrm{Cov}\,(R_1, R_2)|^2 \leq \sigma_1{}^2 \sigma_2{}^2$, and the result follows.

We shall show that ρ is a *measure of linear dependence* between R_1 and R_2 [more precisely, between $R_1 - E(R_1)$ and $R_2 - E(R_2)$], in the following sense.

Let us try to estimate $R_2 - ER_2$ by a linear combination $c(R_1 - ER_1) + d$, that is, find the c and d that minimize

$$
\begin{aligned}
E\{[(R_2 - ER_2) - (c(R_1 - ER_1) + d)]^2\} \\
= \sigma_2{}^2 - 2c\,\mathrm{Cov}\,(R_1, R_2) + c^2 \sigma_1{}^2 + d^2 \\
= \sigma_2{}^2 - 2c\,\rho \sigma_1 \sigma_2 + c^2 \sigma_1{}^2 + d^2
\end{aligned}
$$

Clearly we can do no better than to take $d = 0$. Now the minimum of $Ax^2 + 2Bx + D$ occurs for $x = -B/A$; hence $\sigma_1^2 c^2 - 2\rho\sigma_1\sigma_2 c + \sigma_2^2$ is minimized when

$$c = \frac{\rho\sigma_1\sigma_2}{\sigma_1^2} = \rho\,\frac{\sigma_2}{\sigma_1}$$

Thus the minimum expectation is $\sigma_2^2 - 2\rho^2\sigma_2^2 + \rho^2\sigma_2^2 = \sigma_2^2(1 - \rho^2)$. For a given σ_2^2, the closer $|\rho|$ is to 1, the better R_2 is approximated (in the mean square sense) by a linear combination $aR_1 + b$. In particular, if $|\rho| = 1$, then

$$E\left[\left(R_2 - ER_2 - \frac{\rho\sigma_2}{\sigma_1}(R_1 - ER_1)\right)^2\right] = 0$$

so that

$$R_2 - ER_2 = \frac{\rho\sigma_2}{\sigma_1}(R_1 - ER_1)$$

with probability 1.

Thus, if $|\rho| = 1$, then $R_1 - E(R_1)$ and $R_2 - E(R_2)$ are linearly dependent. (The random variables $R_1, \ldots, R_n$ are said to be linearly dependent iff there are real numbers $a_1, \ldots, a_n$, not all 0, such that $P\{a_1 R_1 + \cdots + a_n R_n = 0\} = 1$.) Conversely, if $R_1 - E(R_1)$ and $R_2 - E(R_2)$ are linearly dependent, that is, if $a(R_1 - ER_1) + b(R_2 - ER_2) = 0$ with probability 1 for some constants a and b, not both 0, then $|\rho| = 1$ (Problem 1).

PROBLEMS

1. If $R_1 - E(R_1)$ and $R_2 - E(R_2)$ are linearly dependent, show that $|\rho(R_1, R_2)| = 1$.

2. If $aR_1 + bR_2 = c$ for some constants a, b, c, where a and b are not both 0, show that $R_1 - E(R_1)$ and $R_2 - E(R_2)$ are linearly dependent. Thus $|\rho(R_1, R_2)| = 1$ if and only if there is a line L in the plane such that $(R_1(\omega), R_2(\omega))$ lies on L for "almost" all ω, that is, for all ω outside a set of probability 0.

3. Show that equality occurs in the Schwarz inequality, $|E(R_1 R_2)|^2 = E(R_1^2)E(R_2^2)$, if and only if R_1 and R_2 are linearly dependent.

4. Prove the following results.
   (a) *Schwarz inequality for sums:* For any real numbers $a_1, \ldots, a_n, b_1, \ldots, b_n$, $(\sum_{i=1}^n a_i b_i)^2 \le \sum_{i=1}^n a_i^2 \sum_{i=1}^n b_i^2$.
   (b) *Schwarz inequality for integrals:* If $\int_a^b g^2(x)\,dx$ and $\int_a^b h^2(x)\,dx$ are finite, so is $\int_a^b g(x)h(x)\,dx$, and furthermore $(\int_a^b g(x)h(x)\,dx)^2 \le \int_a^b g^2(x)\,dx \int_a^b h^2(x)\,dx$. HINT: show that both (a) and (b) are special cases of Theorem 2.

5. Show that if $R_1, \ldots, R_n$ are arbitrary random variables with $E(R_i^2)$ finite for all i, then

$$\text{Var}\,(R_1 + \cdots + R_n) = \sum_{i=1}^{n} \text{Var}\, R_i + 2 \sum_{\substack{i,j=1 \\ i<j}}^{n} \text{Cov}\,(R_i, R_j)$$

3.5 THE METHOD OF INDICATORS

In this section we introduce a technique that in certain cases allows the expectation of a random variable to be computed quickly, without any knowledge of the distribution function. This is especially useful in situations when the distribution function is difficult to calculate.

The *indicator* of an event A is a random variable I_A defined as follows.

$$I_A(\omega) = 1 \qquad \text{if } \omega \in A$$
$$ = 0 \qquad \text{if } \omega \notin A$$

Thus $I_A = 1$ if A occurs and 0 if A does not occur. (Sometimes I_A is called the "characteristic function" of A, but we do not use this terminology since we reserve the term "characteristic function" for something quite different.)

The expectation of I_A is given by

$$E(I_A) = 0P\{I_A = 0\} + 1P\{I_A = 1\} = P\{I_A = 1\} =: P(A)$$

The "method of indicators" simply involves expressing, if possible, a given random variable R as a sum of indicators, say, $R = I_{A_1} + \cdots + I_{A_n}$. Then

$$E(R) = \sum_{j=1}^{n} E(I_{A_j}) = \sum_{j=1}^{n} P(A_j)$$

Hopefully, it will be easier to compute the $P(A_j)$ than to evaluate $E(R)$ directly.

▶ **Example 1.** Let R be the number of successes in n Bernoulli trials, with probability of success p on a given trial; then R has the binomial distribution with parameters n and p; that is,

$$P\{R = k\} = \binom{n}{k} p^k (1 - p)^{n-k}, \qquad k = 0, 1, \ldots, n$$

We have found by a direct evaluation that $E(R) = np$ (Problem 7, Section 3.2), but the method of indicators does the job more smoothly. Let A_i be the event that there is a success on trial i, $i = 1, 2, \ldots, n$. Then $R = I_{A_1} + \cdots + I_{A_n}$ (note that I_{A_i} may be regarded as the number of successes on trial i). Thus

$$E(R) = \sum_{i=1}^{n} E(I_{A_i}) = \sum_{i=1}^{n} P(A_i) = np$$

Now, since $A_1, \dots, A_n$ are independent, the indicators $I_{A_1}, \dots, I_{A_n}$ are independent (Problem 1), and so there is a bonus, namely (by property 7, Section 3.3),

$$\text{Var } R = \sum_{i=1}^{n} \text{Var } I_{A_i}$$

But $I_{A_i}{}^2 = I_{A_i}$; hence

$$E(I_{A_i}{}^2) = E(I_{A_i}) = P(A_i) = p$$

Therefore

$$\text{Var } I_{A_i} = E(I_{A_i}{}^2) - [E(I_{A_i})]^2 \quad \text{by (3.3.1)}$$
$$= p - p^2 = p(1 - p)$$

Thus

$$\text{Var } R = np(1 - p) \blacktriangleleft$$

▶ **Example 2.** A single unbiased die is tossed independently n times. Let R_1 be the number of 1's obtained, and R_2 the number of 2's. Find $E(R_1 R_2)$.

If A_i is the event that the ith toss results in a 1, and B_i the event that the ith toss results in a 2, then

$$R_1 = I_{A_1} + \cdots + I_{A_n}$$
$$R_2 = I_{B_1} + \cdots + I_{B_n}$$

Hence

$$E(R_1 R_2) = \sum_{i,j=1}^{n} E(I_{A_i} I_{B_j})$$

Now if $i \neq j$, I_{A_i} and I_{B_j} are independent (see Problem 1); hence

$$E(I_{A_i} I_{B_j}) = E(I_{A_i}) E(I_{B_j}) = P(A_i) P(B_j) = \tfrac{1}{36}$$

If $i = j$, A_i and B_i are disjoint, since the ith toss cannot simultaneously result in a 1 and a 2. Thus $I_{A_i} I_{B_i} = I_{A_i \cap B_i} = 0$ (see Problem 2). Thus

$$E(R_1 R_2) = \frac{n(n-1)}{36}$$

since there are $n(n-1)$ ordered pairs (i,j) of integers $\in \{1, 2, \dots, n\}$ such that $i \neq j$.

Note that the $I_{A_i} I_{B_j}$, $i, j = 1, \dots, n$, are not independent [for instance, if $I_{A_1}(\omega) I_{B_2}(\omega) = 1$, then $I_{A_2}(\omega) I_{B_3}(\omega)$ must be 0], so that we cannot compute the variance of $R_1 R_2$ in the same way as in Example 1. ◀

PROBLEMS

1. If the events $A_1, \ldots, A_n$ are independent, show that the indicators $I_{A_1}, \ldots, I_{A_n}$ are independent random variables, and conversely.

2. Establish the following properties of indicators:
   (a) $I_\Omega = 1$, $\quad I_\emptyset = 0$
   (b) $I_{A \cap B} = I_A I_B$, $\quad I_{A \cup B} = I_A + I_B - I_{A \cap B}$
   (c) $I_{\bigcup_{i=1}^{\infty} A_i} = \sum_{i=1}^{\infty} I_{A_i}$ $\quad$ if the A_i are disjoint
   (d) If $A_1, A_2, \ldots$ is an expanding sequence of events ($A_n \subset A_{n+1}$ for all n) and $\bigcup_{n=1}^{\infty} A_n = A$, or if $A_1, A_2, \ldots$ is a contracting sequence ($A_{n+1} \subset A_n$ for all n) and $\bigcap_{n=1}^{\infty} A_n = A$, then $I_{A_n} \to I_A$; that is, $\lim_{n \to \infty} I_{A_n}(\omega) = I_A(\omega)$ for all ω.

3. In Example 2, find the joint probability function of R_1 and R_2. Notice how unwieldy is the direct expression for $E(R_1 R_2)$.

$$E(R_1 R_2) = \sum_{j,k=0}^{n} jk P\{R_1 = j, R_2 = k\}$$

4. In a sequence of n Bernoulli trials, let R_0 be the number of times a success is followed immediately by a failure. For example, if $n = 7$ and $\omega = (S\overline{SF}F\overline{SF}S)$, then $R_0(\omega) = 2$, as indicated. Find $E(R_0)$.

5. Find Var R_0 in Problem 4.

6. 100 balls are tossed independently and at random into 50 boxes. Let R be the number of empty boxes. Find $E(R)$.

3.6 SOME PROPERTIES OF THE NORMAL DISTRIBUTION

Let R_1 be normally distributed with mean m and variance σ^2.

$$f_1(x) = \frac{1}{\sqrt{2\pi}\,\sigma} e^{-(x-m)^2/2\sigma^2}$$

If $R_2 = aR_1 + b$, $a \neq 0$, we shall show that R_2 is also normally distributed [necessarily $E(R_2) = am + b$, Var $R_2 = a^2\sigma^2$ by properties 1, 2, and 6 of Section 3.3].

We may use the technique of Section 2.4 to find the density of R_2. $R_2 = y$ corresponds to $R_1 = h(y) = (y - b)/a$. Thus

$$f_2(y) = f_1(h(y)) |h'(y)| = \frac{1}{|a|} f_1\left(\frac{y-b}{a}\right)$$

$$= \frac{1}{\sqrt{2\pi}\,|a|\,\sigma} \exp\left[-\frac{(y - (am + b))^2}{2a^2\sigma^2}\right]$$

so that R_2 has the normal density with mean $am + b$ and variance $a^2\sigma^2$. We may use this result in the calculation of probabilities of events involving a normally distributed random variable. If R has the normal density with $E(R) = m$, Var $R = \sigma^2$, then

$$P\{a \leq R \leq b\} = \int_a^b \frac{1}{\sqrt{2\pi}\,\sigma}\, e^{-(x-m)^2/2\sigma^2}\, dx$$

One must resort to tables to evaluate this. The point we wish to bring out is that, regardless of m and σ^2, only one table is needed, namely, that of the normal distribution function when $m = 0$, $\sigma^2 = 1$; that is,

$$F^*(x) = \int_{-\infty}^x \frac{1}{\sqrt{2\pi}}\, e^{-t^2/2}\, dt$$

For if R is normally distributed with $E(R) = m$, Var $R = \sigma^2$, then $R^* = (R - m)/\sigma$ is normally distributed with $E(R^*) = 0$ and Var $R^* = 1$. Thus

$$P\{a \leq R \leq b\} = P\left\{\frac{a - m}{\sigma} \leq R^* \leq \frac{b - m}{\sigma}\right\}$$

$$= F^*\left(\frac{b - m}{\sigma}\right) - F^*\left(\frac{a - m}{\sigma}\right)$$

A brief table of values of F^* is given at the end of the book.

REMARK.   If a random variable has a density function f that is symmetrical about 0 [i.e., an even function: $f(-x) = f(x)$], then the distribution function has the property that $F(-x) = 1 - F(x)$. For (see Figure 3.6.1)

$$F(-x) = P\{R \leq -x\} = \int_{-\infty}^{-x} f(t)\, dt = \int_x^{\infty} f(t)\, dt$$

$$= P\{R > x\} = 1 - F(x)$$

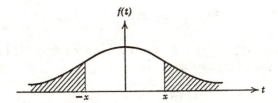

FIGURE 3.6.1   Symmetrical Density.

In particular, the distribution function F^* has this property, and thus once the values of $F^*(x)$ for positive x are known, the values of $F^*(x)$ for negative x are determined.

PROBLEMS

1. Let R be normally distributed with $m = 1$, $\sigma^2 = 9$.
   (a) Find $P\{-.5 \leq R \leq 4\}$
   (b) If $P\{R \geq c\} = .9$, find c.

2. If R is normally distributed and k is a positive real number, show that $P\{|R - m| \geq k\sigma\}$ does not depend on m or σ; thus one can speak unambiguously of the "probability that a normally distributed random variable lies at least k standard deviations from its mean." Show that when $k = 1.96$, the probability is .05.

3.7 CHEBYSHEV'S INEQUALITY AND THE WEAK LAW OF LARGE NUMBERS

In this section we are going to prove a result that corresponds to the physical statement that the arithmetic average of a very large number of independent observations of a random variable R is very likely to be very close to $E(R)$. We first establish a quantitative result about the variance as a measure of dispersion.

Theorem 1.

(a) *Let R be a nonnegative random variable, and b a positive real number. Then*

$$P\{R \geq b\} \leq \frac{E(R)}{b}$$

PROOF. We first consider the absolutely continuous case. We have

$$E(R) = \int_{-\infty}^{\infty} x f_R(x)\, dx = \int_{0}^{\infty} x f_R(x)\, dx$$

since $R \geq 0$, so that $f_R(x) = 0$, $x < 0$. Now if we drop the integral from 0 to b, we get something smaller.

$$E(R) \geq \int_{b}^{\infty} x f_R(x)\, dx$$

Since $x \geq b$,

$$\int_b^\infty x f_R(x)\, dx \geq \int_b^\infty b f_R(x)\, dx = bP\{R \geq b\}$$

This is the desired result.

The general proof is based on the same idea. Let $A_b = \{R \geq b\}$; then $R \geq RI_{A_b}$. For if $\omega \notin A_b$, this says simply that $R(\omega) \geq 0$; if $\omega \in A_b$, it says that $R(\omega) \geq R(\omega)$. Thus $E(R) \geq E(RI_{A_b})$. But $RI_{A_b} \geq bI_{A_b}$, since $\omega \in A_b$ implies that $R(\omega) \geq b$. Thus

$$E(R) \geq E(RI_{A_b}) \geq E(bI_{A_b}) = bE(I_{A_b}) = bP(A_b)$$

Consequently $P(A_b) \leq E(R)/b$, as desired.

(b) *Let R be an arbitrary random variable, c any real number, and ε and m positive real numbers. Then*

$$P\{|R - c| \geq \varepsilon\} \leq \frac{E[|R - c|^m]}{\varepsilon^m}$$

PROOF.

$$P\{|R - c| \geq \varepsilon\} = P\{|R - c|^m \geq \varepsilon^m\} \leq \frac{E[|R - c|^m]}{\varepsilon^m} \qquad \text{by (a)}$$

(c) *If R has finite mean m and finite variance $\sigma^2 > 0$, and k is a positive real number, then*

$$P\{|R - m| \geq k\sigma\} \leq \frac{1}{k^2}$$

PROOF. This follows from (b) with $c = m$, $\varepsilon = k\sigma$, $m = 2$.

All three parts of Theorem 1 go under the name of *Chebyshev's inequality*. Part (c) says that the probability that a random variable will fall k or more standard deviations from its mean is $\leq 1/k^2$. Notice that nothing at all is said about the distribution function of R; Chebyshev's inequality is therefore quite a general statement. When applied to a particular case, however, it may be quite weak. For example, let R be normally distributed with mean m and variance σ^2. Then (Problem 2, Section 3.6) $P\{|R - m| \geq 1.96\sigma\} = .05$. In this case Chebyshev's inequality says only that

$$P\{|R - m| \geq 1.96\sigma\} \leq \frac{1}{(1.96)^2} = .26$$

which is a much weaker statement. The strength of Chebyshev's inequality lies in its universality.

We are now ready for the main result.

Theorem 2. (*Weak Law of Large Numbers*). *For each* $n = 1, 2, \ldots,$ *suppose that* $R_1, R_2, \ldots, R_n$ *are independent random variables on a given probability space, each having finite mean and variance. Assume that the variances are uniformly bounded; that is, assume that there is some finite positive number* M *such that* $\sigma_i^2 \leq M$ *for all* i. *Let* $S_n = \sum_{i=1}^{n} R_i$. *Then, for any* $\varepsilon > 0$,

$$P\left\{\left|\frac{S_n - E(S_n)}{n}\right| \geq \varepsilon\right\} \to 0 \qquad \text{as } n \to \infty$$

Before proving the theorem, we consider two cases of interest.

SPECIAL CASES

1. Suppose that $E(R_i) = m$ for all i, and $\text{Var } R_i = \sigma^2$ for all i. Then

$$E(S_n) = \sum_{i=1}^{n} E(R_i) = nm$$

$$\frac{S_n - E(S_n)}{n} = \frac{S_n}{n} - m = \frac{R_1 + \cdots + R_n}{n} - m$$

Therefore, for any arbitrary $\varepsilon > 0$, there is for large n a high probability that the arithmetic average and the expectation m will differ by $< \varepsilon$.

This case covers the situation when $R_1, R_2, \ldots, R_n$ are independent observations of a given random variable R. All this means is that $R_1, \ldots, R_n$ are independent, and the R_i all have the same distribution function, namely, F_R. In particular, $E(R_i) = E(R)$, so that for large n there is a high probability that $(R_1 + \cdots + R_n)/n$ and $E(R)$ will differ by $< \varepsilon$.

2. Consider a sequence of Bernoulli trials, and let R_i be the number of successes on trial i; that is, $R_i = I_{A_i}$, where $A_i = \{\text{success on trial } i\}$. Then $(R_1 + \cdots + R_n)/n$ is the relative frequency of successes in n trials. Now $E(R_i) = P(A_i) = p$, the probability of success on a given trial, so that for large n there is a high probability that the relative frequency will differ from p by $< \varepsilon$.

PROOF OF THEOREM 2. By the second form of Chebyshev's inequality [part (b) of Theorem 1], with $R = (S_n - E(S_n))/n$, $c = 0$, and $m = 2$, we have

$$P\left\{\left|\frac{S_n - E(S_n)}{n}\right| \geq \varepsilon\right\} \leq \frac{1}{\varepsilon^2} E\left[\left(\frac{S_n - E(S_n)}{n}\right)^2\right] = \frac{1}{n^2\varepsilon^2} \text{Var } S_n$$

But since $R_1, \ldots, R_n$ are independent,

$$\text{Var } S_n = \sum_{i=1}^{n} \text{Var } R_i \qquad \text{by property 7 of Section 3.3}$$

$$\leq nM \qquad \text{by hypothesis}$$

Thus

$$P\left\{\left|\frac{S_n - E(S_n)}{n}\right| \geq \varepsilon\right\} \leq \frac{nM}{n^2\varepsilon^2} = \frac{M}{n\varepsilon} \to 0$$

REMARK. If a coin with probability p of heads is tossed indefinitely, the successive tosses being independent, we expect that as a practical matter the relative frequency of heads will converge, in the ordinary sense of convergence of a sequence of real numbers, to p. This is a somewhat stronger statement than the weak law of large numbers, which says that for large n the relative frequency of heads in n trials is very likely to be very close to p. The first statement, when properly formulated, becomes the *strong law of large numbers*, which we shall examine in detail later.

PROBLEMS

1. Let R have the exponential density $f(x) = e^{-x}, x \geq 0; f(x) = 0, x < 0$. Evaluate $P\{|R - m| \geq k\sigma\}$ and compare with the Chebyshev bound.

2. Suppose that we have a sequence of random variables R_n such that $P\{R_n = e^n\} = 1/n, P\{R_n = 0\} = 1 - 1/n, n = 1, 2, \ldots$.
   (a) State and prove a theorem that expresses the fact that for large n, R_n is very likely to be 0.
   (b) Show that $E(R_n{}^k) \to \infty$ as $n \to \infty$ for any $k > 0$.

3. Suppose that R_n is the amount you win on trial n in a game of chance. Assume that the R_i are independent random variables, each with finite mean m and finite variance σ^2. Make the realistic assumption that $m < 0$. Show that $P\{(R_1 + \cdots + R_n)/n < m/2\} \to 1$ as $n \to \infty$. What is the moral of this result?

4

Conditional Probability
and Expectation

4.1 INTRODUCTION

We have thus far defined the conditional probability $P(B \mid A)$ only when $P(A) > 0$. However, there are many situations when it is natural to talk about a conditional probability given an event of probability 0. For example, suppose that a real number R is selected at random, with density f. If R takes the value x, a coin with probability of heads $g(x)$ is tossed ($0 \leq g(x) \leq 1$). It is natural to assert that the conditional probability of obtaining a head, given $R = x$, is $g(x)$. But since R is absolutely continuous, the event $\{R = x\}$ has probability 0, and thus conditional probabilities given $R = x$ are not as yet defined.

If we ignore this problem for the moment, we can find the over-all probability of obtaining a head by the following intuitive argument. The probability that R will fall into the interval $(x, x + dx]$ is roughly $f(x) \, dx$; given that R falls into this interval, the probability of a head is roughly $g(x)$. Thus we should expect, from the theorem of total probability, that the probability of a head will be $\sum_x g(x) f(x) \, dx$, which approximates $\int_{-\infty}^{\infty} g(x) f(x) \, dx$. Thus the probability in question is a weighted average of conditional probabilities, the weights being assigned in accordance with the density f.

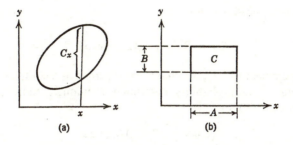

FIGURE 4.1.1

Let us examine what is happening here. We have two random variables R_1 and R_2 [$R_1 = R$, $R_2 = $ (say) the number of heads obtained]. We are specifying the density of R_1, and for each x and each Borel set B we are specifying a quantity $P_x(B)$ that is to be interpreted intuitively as the conditional probability that $R_2 \in B$ given that $R_1 = x$. (We shall often write $P\{R_2 \in B \mid R_1 = x\}$ for $P_x(B)$.)

We would like to conclude that the probabilities of all events involving R_1 and R_2 are now determined. Suppose that C is a two-dimensional Borel set. What is a reasonable figure for $P\{(R_1, R_2) \in C\}$? Intuitively, the probability that R_1 falls into $(x, x + dx]$ is $f_1(x)\,dx$. Given that this happens, that is, (roughly) given $R_1 = x$, the only way (R_1, R_2) can lie in C is if R_2 belongs to the "section" $C_x = \{y: (x, y) \in C\}$ (see Figure 4.1.1a). This happens with probability $P_x(C_x)$. Thus we expect that the total probability that (R_1, R_2) will belong to C is

$$\int_{-\infty}^{\infty} P_x(C_x) f_1(x)\, dx$$

In particular, if $C = A \times B = \{(x, y): x \in A, y \in B\}$ (see Figure 4.1.1b),

$$C_x = \varnothing \quad \text{if } x \notin A; \qquad C_x = B \quad \text{if } x \in A$$

Thus

$$P\{(R_1, R_2) \in C\} = P\{R_1 \in A, R_2 \in B\} = \int_A P_x(B) f_1(x)\, dx$$

The above reasoning may be formalized as follows. Let $\Omega = E^2$, $\mathscr{F} = $ Borel subsets, $R_1(x, y) = x$, $R_2(x, y) = y$. Let f_1 be a density function on E^1, that is, a nonnegative function such that $\int_{-\infty}^{\infty} f_1(x)\, dx = 1$. Suppose that for each real x we are given a probability measure P_x on the Borel subsets of E^1. Assume also that $P_x(B)$ is a piecewise continuous function of x for each fixed B.

Then it turns out that there is a unique probability measure P on $\mathscr{F}$ such that for all Borel subsets A, B of E^1

$$P(A \times B) = \int_A P_x(B) f_1(x) \, dx \qquad (4.1.1)$$

Thus the requirement (4.1.1), which may be regarded as a continuous version of the theorem of total probability, determines P uniquely. In fact, if $C \in \mathscr{F}$, $P(C)$ is given explicitly by

$$P(C) = \int_{-\infty}^{\infty} P_x(C_x) f_1(x) \, dx \qquad (4.1.2)$$

Notice that if $R_1(x, y) = x$, $R_2(x, y) = y$, then

$$P(A \times B) = P\{R_1 \in A, R_2 \in B\}$$

and

$$P(C) = P\{(R_1, R_2) \in C\}$$

Furthermore, the distribution function of R_1 is given by

$$F_1(x_0) = P\{R_1 \leq x_0\} = P\{R_1 \in A, R_2 \in B\}$$

$$\text{where } A = (-\infty, x_0], \ B = (-\infty, \infty)$$

$$= \int_A P_x(B) f_1(x) \, dx = \int_{-\infty}^{x_0} f_1(x) \, dx$$

Thus f_1 is in fact the density of R_1. Notice also that

$$P\{R_2 \in B\} = P\{R_1 \in A, R_2 \in B\}$$

where $A = (-\infty, \infty)$; hence

$$P\{R_2 \in B\} = \int_{-\infty}^{\infty} P_x(B) f_1(x) \, dx \qquad (4.1.3)$$

To summarize: If we start with a density for R_1 and a set of probabilities $P_x(B)$ that we interpret as $P\{R_2 \in B \mid R_1 = x\}$, the probabilities of events of the form $\{(R_1, R_2) \in C\}$ are determined in a natural way, if you believe that there should be a continuous version of the theorem of total probability; $P\{(R_1, R_2) \in C\}$ is given explicitly by (4.1.2), which reduces to (4.1.1) in the special case when $C = A \times B$.

We have not yet answered the question of how to define $P\{R_2 \in B \mid R_1 = x\}$ for arbitrarily specified random variables R_1 and R_2; we attack this problem later in the chapter. Instead we have approached the problem in a somewhat oblique way. However, there are many situations in which one specifies the density of R_1, and then the conditional probability of events involving R_2, given $R_1 = x$. We now know how to formulate such problems precisely. Consider again the problem at the beginning of the section. If R_1 has density f, and a coin with probability of heads $g(x)$ is tossed whenever $R_1 = x$ (and a head corresponds to $R_2 = 1$, a tail to $R_2 = 0$), then the probability of obtaining a head is

$$P\{R_2 = 1\} = \int_{-\infty}^{\infty} P\{R_2 = 1 \mid R_1 = x\} f_1(x)\, dx \qquad \text{by (4.1.3)}$$

$$= \int_{-\infty}^{\infty} g(x) f_1(x)\, dx$$

in agreement with the previous intuitive argument.

4.2 EXAMPLES

We apply the general results of this section to some typical special cases.

▶ **Example 1.** A point is chosen with uniform density between 0 and 1. If the number R_1 selected is x, then a coin with probability x of heads is tossed independently n times. If R_2 is the resulting number of heads, find $p_2(k) = P\{R_2 = k\}$, $k = 0, 1, \ldots, n$.

Here we have $f_1(x) = 1$, $0 \le x \le 1$; $f_1(x) = 0$ elsewhere. Also

$$P_x\{k\} = P\{R_2 = k \mid R_1 = x\} = \binom{n}{k} x^k (1 - x)^{n-k}$$

By (4.1.3),

$$P\{R_2 = k\} = \int_0^1 \binom{n}{k} x^k (1 - x)^{n-k}\, dx$$

This is an instance of the *beta function*, defined by

$$\beta(r, s) = \int_0^1 x^{r-1}(1 - x)^{s-1}\, dx, \qquad r, s > 0$$

It can be shown that the beta function can be expressed in terms of the gamma function [see (3.2.2)] by

$$\beta(r, s) = \frac{\Gamma(r)\Gamma(s)}{\Gamma(r + s)} \tag{4.2.1}$$

(see Problem 1). Thus

$$p_2(k) = \binom{n}{k} \beta(k + 1, n - k + 1)$$

$$= \binom{n}{k} \frac{\Gamma(k + 1)\Gamma(n - k + 1)}{\Gamma(n + 2)}$$

$$= \binom{n}{k} \frac{k!\,(n - k)!}{(n + 1)!} = \frac{1}{n + 1}, \qquad k = 0, 1, \ldots, n \ ◀$$

▶ **Example 2.**   A nonnegative number R_1 is chosen with the density $f_1(x) = xe^{-x}, x \geq 0; f_1(x) = 0, x < 0$. If $R_1 = x$, a number R_2 is chosen with uniform density between 0 and x. Find $P\{R_1 + R_2 \leq 2\}$.

Now we must have $0 \leq R_2 \leq R_1$; hence, if $0 \leq R_1 \leq 1$, then necessarily $R_1 + R_2 \leq 2$. If $1 < R_1 \leq 2$, then $R_1 + R_2 \leq 2$ provided that $R_2 \leq 2 - R_1$. If $R_1 > 2$, then $R_1 + R_2$ cannot be ≤ 2. By (4.1.2),

$$P\{R_1 + R_2 \leq 2\}$$

$$= \int_0^\infty xe^{-x}P\{R_1 + R_2 \leq 2 \mid R_1 = x\}\, dx$$

$$= \int_0^1 xe^{-x}(1)\, dx + \int_1^2 xe^{-x}P\{R_2 \leq 2 - x \mid R_1 = x\}dx + \int_2^\infty xe^{-x}(0)\, dx$$

Given $R_1 = x$, R_2 is uniformly distributed between 0 and x; thus

$$P\{R_2 \leq 2 - x \mid R_1 = x\} = \frac{2 - x}{x}, \qquad 1 \leq x \leq 2$$

(see Figure 4.2.1). Therefore

$$P\{R_1 + R_2 \leq 2\} = \int_0^1 xe^{-x}\, dx + \int_1^2 xe^{-x}\left(\frac{2 - x}{x}\right) dx = 1 - 2e^{-1} + e^{-2} \blacktriangleleft$$

▶ **Example 3.**   Let R_1 be a discrete random variable, taking on the values $x_1, x_2, \dots$ with probabilities $p(x_1), p(x_2), \dots$. If $R_1 = x_i$, a random variable R_2 is observed, where R_2 has density f_i. What is $P\{(R_1, R_2) \in C\}$?

This is not quite the situation we considered in Section 4.1, since R_1 is discrete. However, the theorem of total probability should still be in force. R_1 takes the value x_i with probability $p(x_i)$; given that $R_1 = x_i$, the probability that $R_2 \in B$ is $P_{x_i}(B) = \int_B f_i(y)\, dy$. Thus we should have

$$P\{R_1 \in A, R_2 \in B\} = \sum_{x_i \in A} p(x_i) \int_B f_i(y)\, dy \qquad (4.2.2)$$

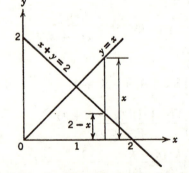

FIGURE 4.2.1   Conditional Probability Calculation.

and, more generally,

$$P\{(R_1, R_2) \in C\} = \sum_{x_i} p(x_i) \int_{C_{x_i}} f_i(y) \, dy \qquad (4.2.3)$$

In fact, if we take $\Omega = E^2$, $\mathscr{F} = $ Borel sets, $R_1(x, y) = x$, $R_2(x, y) = y$, it turns out that there is a unique probability measure on $\mathscr{F}$ satisfying (4.2.2) for all Borel subsets A, B of E^1; P is given explicitly by (4.2.3). ◄

PROBLEMS

1. Derive formula (4.2.1). HINT: in $\Gamma(r) = \int_0^\infty t^{r-1} e^{-t} \, dt$, let $t = x^2$. Then write $\Gamma(r)\Gamma(s)$ as a double integral and switch to polar coordinates.

2. In Example 2, what are the sets C and C_x in (4.1.2)? What is $P_x(C_x)$?

3. In Example 3, suppose that R_1 takes on positive integer values $1, 2, \ldots$ with probabilities $p_1, p_2, \ldots$ ($p_i \geq 0$, $\sum_{i=1}^\infty p_i = 1$). If $R_1 = n$, R_2 is selected according to the density $f_n(x) = ne^{-nx}$, $x \geq 0$; $f_n(x) = 0$, $x < 0$. Find the probability that $4 \leq R_1 + R_2 \leq 6$.

4. In Example 3 we specified $P_{x_i}(B)$ to be interpreted intuitively as the probability that $R_2 \in B$, given that $R_1 = x_i$. This, plus the specification of $p(x_i)$, $i = 1, 2, \ldots$, determines the probability measure P. Use (4.2.2) to show that if $p(x_i) > 0$ then $P\{R_2 \in B \mid R_1 = x_i\} = P_{x_i}(B)$, thus justifying the intuition. In order words, the conditional probability as computed from the probability measure P coincides with the original specification.

5. A number R_1 is chosen with density $f_1(x) = 1/x^2$, $x \geq 1$; $f_1(x) = 0$, $x < 1$. If $R_1 = x$, let R_2 be uniformly distributed between 0 and x. Find the distribution and density functions of R_2.

4.3 CONDITIONAL DENSITY FUNCTIONS

We have seen that specification of the distribution or density function of a random variable R_1, together with $P_x(B)$ (for all real x and Borel subsets B of E^1), interpreted intuitively as the conditional probability that $R_2 \in B$, given $R_1 = x$, determines the probability of all events of the form $\{(R_1, R_2) \in C\}$. However, this has not resolved the difficulty of defining conditional probabilities given events of probability 0. If we are *given* random variables R_1 and R_2 with a particular joint distribution function, we can ask whether it is possible to define in a meaningful way the conditional probability $P\{R_2 \in B \mid R_1 = x\}$, even though the event $\{R_1 = x\}$ may have probability 0 for some, in fact perhaps for all, x. We now consider this question in the case in which R_1 and R_2 have a joint density f.

A reasonable approach to the conditional probability $P\{R_2 \in B \mid R_1 = x_0\}$ is to look at $P\{R_2 \in B \mid x_0 - h < R_1 < x_0 + h\}$ and let $h \to 0$. Now

$$P\{x_0 - h < R_1 < x_0 + h, R_2 \in B\} = \int_{x_0-h}^{x_0+h} \int_B f(x, y) \, dy \, dx$$

which for small h should look like $2h \int_B f(x_0, y) \, dy$. But $P\{x_0 - h < R_1 < x_0 + h\}$ looks like $2h f_1(x_0)$ for small h, where $f_1(x) = \int_{-\infty}^{\infty} f(x, y) \, dy$ is the density of R_1. Thus, as $h \to 0$, it appears that under appropriate conditions $P\{R_2 \in B \mid x - h < R_1 < x + h\}$ should approach $\int_B [f(x, y)/f_1(x)] \, dy$, so that we find conditional probabilities involving R_2, given $R_1 = x$, by integrating $f(x, y)/f_1(x)$ with respect to y.

We are led to define the *conditional density* of R_2 given $R_1 = x$ (or, for short, the conditional density of R_2 given R_1) as

$$h(y \mid x) = \frac{f(x, y)}{f_1(x)} \qquad (4.3.1)$$

Since $\int_{-\infty}^{\infty} f(x, y) \, dy = f_1(x)$ (see Section 2.7), we have $\int_{-\infty}^{\infty} h(y \mid x) \, dy = 1$, so that $h(y \mid x)$, regarded as a function of y, is a legitimate density.

Notice that the conditional density is defined only when $f_1(x) > 0$. However, we may essentially ignore those (x, y) at which the conditional density is not defined. For let $S = \{(x, y): f_1(x) = 0\}$. We can show that $P\{(R_1, R_2) \in S\} = 0$.

$$P\{(R_1, R_2) \in S\} = \iint_S f(x, y) \, dx \, dy = \int_{\{x:f_1(x)=0\}} \int_{-\infty}^{\infty} f(x, y) \, dy \, dx$$

$$= \int_{\{x:f_1(x)=0\}} f_1(x) \, dx = 0$$

We define the conditional probability that R_2 belongs to the Borel set B, given that $R_1 = x$, as

$$P_x(B) = P\{R_2 \in B \mid R_1 = x\} = \int_B h(y \mid x) \, dy \qquad (4.3.2)$$

We can ask whether this is a sensible definition of conditional probability. We have set up our own ground rules to answer this question: "sensible" means that the theorem of total probability holds. Let us check that in fact (4.1.1) [and hence (4.1.2)] holds. We have

$$P\{R_1 \in A, R_2 \in B\} = \int_{x \in A} \int_{y \in B} f(x, y) \, dx \, dy$$

$$= \int_{x \in A} f_1(x) \left[\int_{y \in B} h(y \mid x) \, dy \right] dx = \int_A P_x(B) f_1(x) \, dx$$

which is (4.1.1).

We have seen that if (R_1, R_2) has density $f(x, y)$ and R_1 has density $f_1(x)$ we have a conditional density $h(y \mid x) = f(x, y)/f_1(x)$ for R_2, given $R_1 = x$. Let us reverse this process. Suppose that we observe a random variable R_1 with density $f_1(x)$; if $R_1 = x$, we observe a random variable R_2 with density $h(y \mid x)$. If we accept the continuous version of the theorem of total probability, we may calculate the joint distribution function of R_1 and R_2 using (4.1.1).

$$F(x_0, y_0) = P\{R_1 \le x_0, R_2 \le y_0\} = \int_{-\infty}^{x_0} P\{R_2 \le y_0 \mid R_1 = x\} f_1(x) \, dx$$

$$= \int_{-\infty}^{x_0} \left[\int_{-\infty}^{y_0} h(y \mid x) \, dy \right] f_1(x) \, dx = \int_{-\infty}^{x_0} \int_{-\infty}^{y_0} f_1(x) h(y \mid x) \, dy \, dx$$

Thus (R_1, R_2) has a density given by $f(x, y) = f_1(x) h(y \mid x)$, in agreement with (4.3.1).

To summarize: We may look at the formula $f(x, y) = f_1(x) h(y \mid x)$ in two ways.

1. If (R_1, R_2) has density $f(x, y)$, we have a natural notion of conditional probability.

$$P_x(B) = P\{R_2 \in B \mid R_1 = x\} = \int_B h(y \mid x) \, dy$$

2. If R_1 has density $f_1(x)$, and whenever $R_1 = x$ we select R_2 with density $h(y \mid x)$, then in the natural formulation of this problem (R_1, R_2) has density $f(x, y) = f_1(x) h(y \mid x)$.

In both cases "natural" indicates that (4.1.1), the continuous version of the theorem of total probability, is required to hold.

We may extend these results to higher dimensions. For example, if (R_1, R_2, R_3, R_4) has density $f(x_1, x_2, x_3, x_4)$, we define (say) the conditional density of (R_3, R_4) given (R_1, R_2), as

$$h(x_3, x_4 \mid x_1, x_2) = \frac{f(x_1, x_2, x_3, x_4)}{f_{12}(x_1, x_2)}$$

where

$$f_{12}(x_1, x_2) = \int_{-\infty}^{\infty} \int_{-\infty}^{\infty} f(x_1, x_2, x_3, x_4) \, dx_3 \, dx_4$$

The conditional probability that (R_3, R_4) belongs to the two-dimensional Borel set B, given that $R_1 = x_1, R_2 = x_2$, is defined by

$$P_{x_1 x_2}(B) = P\{(R_3, R_4) \in B \mid R_1 = x_1, R_2 = x_2\}$$

$$= \iint_B h(x_3, x_4 \mid x_1, x_2) \, dx_3 \, dx_4$$

The appropriate version of the theorem of total probability is

$$P\{(R_1, R_2) \in A, (R_3, R_4) \in B\} = \iint_A P_{x_1 x_2}(B) f_{12}(x_1, x_2)\, dx_1\, dx_2$$

If (R_1, R_2) has density $f_{12}(x_1, x_2)$, and having observed $R_1 = x_1$, $R_2 = x_2$, we select (R_3, R_4) with density $h(x_3, x_4 \mid x_1, x_2)$, then (R_1, R_2, R_3, R_4) must have density $f(x_1, x_2, x_3, x_4) = f_{12}(x_1, x_2) h(x_3, x_4 \mid x_1, x_2)$.

Let us do some examples.

▶ **Example 1.** We arrive at a bus stop at time $t = 0$. Two buses A and B are in operation. The arrival time R_1 of bus A is uniformly distributed between 0 and t_A minutes, and the arrival time R_2 of bus B is uniformly distributed between 0 and t_B minutes, with $t_A \leq t_B$. The arrival times are independent. Find the probability that bus A will arrive first.

We are looking for the probability that $R_1 < R_2$. Since R_1 and R_2 are independent (and have a joint density), the conditional density of R_2 given R_1 is

$$\frac{f(x, y)}{f_1(x)} = f_2(y) = \frac{1}{t_B}, \qquad 0 \leq y \leq t_B$$

If bus A arrives at x, $0 \leq x \leq t_A$, it will be first provided that bus B arrives between x and t_B. This happens with probability $(t_B - x)/t_B$. Thus

$$P\{R_1 < R_2 \mid R_1 = x\} = 1 - \frac{x}{t_B}, \qquad 0 \leq x \leq t_A$$

By (4.1.2),

$$P\{R_1 < R_2\} = \int_{-\infty}^{\infty} P\{R_1 < R_2 \mid R_1 = x\} f_1(x)\, dx$$

$$= \int_0^{t_A} \left(1 - \frac{x}{t_B}\right) \frac{1}{t_A}\, dx = 1 - \frac{t_A}{2t_B}$$

[Formally, taking the sample space as E^2, we have $C = \{R_1 < R_2\} = \{(x, y): x < y\}$, $C_x = \{y: x < y\}$, $P_x(C_x) = P\{R_1 < R_2 \mid R_1 = x\} = 1 - x/t_B$, $0 \leq x \leq t_A$.]

Alternatively, we may simply use the joint density:

$$P\{R_1 < R_2\} = \iint_{x < y} f(x, y)\, dx\, dy$$

$$= \text{the shaded area in Figure 4.3.1, divided by the total area } t_A t_B$$

$$= 1 - \frac{t_A^2/2}{t_A t_B} = 1 - \frac{t_A}{2t_B}$$

as before. ◀

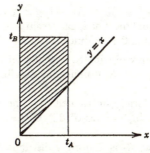

FIGURE 4.3.1 Bus Problem.

▶ **Example 2.** Let R_0 be a nonnegative random variable with density $f_0(\lambda) = e^{-\lambda}$, $\lambda \geq 0$. If $R_0 = \lambda$, we take n independent observations R_1, $R_2, \ldots, R_n$, each R_i having the exponential density $f_\lambda(y) = \lambda e^{-\lambda y}$, $y \geq 0$ ($= 0$ for $y < 0$). Find the conditional density of R_0 given $(R_1, R_2, \ldots, R_n)$.

Here we have specified $f_0(\lambda)$, the density of R_0, and the conditional density of $(R_1, R_2, \ldots, R_n)$ given R_0, namely,

$$h(x_1, x_2, \ldots, x_n \mid \lambda) = f_\lambda(x_1) f_\lambda(x_2) \cdots f_\lambda(x_n) \qquad \text{by the independence assumption}$$

$$= \lambda^n e^{-\lambda x}, \qquad x = \sum_{i=1}^{n} x_i$$

The joint density of $R_0, R_1, \ldots, R_n$ is therefore

$$f(\lambda, x_1, \ldots, x_n) = f_0(\lambda) h(x_1, \ldots, x_n \mid \lambda) = \lambda^n e^{-\lambda(1+x)}$$

The joint density of $R_1, \ldots, R_n$ is given by

$$g(x_1, \ldots, x_n) = \int_{-\infty}^{\infty} f(\lambda, x_1, \ldots, x_n) \, d\lambda = \int_0^{\infty} \lambda^n e^{-\lambda(1+x)} \, d\lambda$$

$$= \text{(with } y = \lambda(1 + x)) \int_0^{\infty} \frac{y^n e^{-y}}{(1 + x)^{n+1}} \, dy = \frac{n!}{(1 + x)^{n+1}}$$

Thus the conditional density of R_0 given $(R_1, \ldots, R_n)$ is

$$h(\lambda \mid x_1, \ldots, x_n) = \frac{f(\lambda, x_1, \ldots, x_n)}{g(x_1, \ldots, x_n)} = \frac{1}{n!} \lambda^n e^{-\lambda(1+x)} (1 + x)^{n+1},$$

$$\lambda, x_1, \ldots, x_n \geq 0, \, x = x_1 + \cdots + x_n \quad ◀$$

PROBLEMS

1. Let (R_1, R_2) have density $f(x, y) = e^{-y}$, $0 \leq x \leq y$, $f(x, y) = 0$ elsewhere. Find the conditional density of R_2 given R_1, and $P\{R_2 \leq y \mid R_1 = x\}$, the *conditional distribution function* of R_2 given $R_1 = x$.

2. Let (R_1, R_2) have density $f(x, y) = k |x|$, $-1 \leq x \leq 1$, $-1 \leq y \leq x$; $f(x, y) = 0$ elsewhere. Find k; also find the individual densities of R_1 and R_2, the conditional density of R_2 given R_1, and the conditional density of R_1 given R_2.

3. (a) If (R_1, R_2) is uniformly distributed over the set $C = \{(x, y): x^2 + y^2 \leq 1\}$, show that, given $R_1 = x$, R_2 is uniformly distributed between $-(1 - x^2)^{1/2}$ and $+(1 - x^2)^{1/2}$.

   (b) Let (R_1, R_2) be uniformly distributed over the arbitrary two-dimensional Borel set C [i.e., $P(B) = $ (area of $B \cap C$)/area of C ($=$ area B/area C if $B \subset C$)].

   Show that given $R_1 = x$, R_2 is uniformly distributed on $C_x = \{y: (x, y) \in C\}$. In other words, $h(y \mid x)$ is constant for $y \in C_x$, and 0 for $y \notin C_x$.

4. In Problem 1, let $R_3 = R_2 - R_1$. Find the conditional density of R_3 given $R_1 = x$. Also find $P\{1 \leq R_3 \leq 2 \mid R_1 = x\}$.

5. Suppose that (R_1, R_2) has density f and $R_3 = g(R_1, R_2)$. You are asked to compute the conditional distribution function of R_3, given $R_1 = x$; that is, $P\{R_3 \leq z \mid R_1 = x\}$. How would you go about it?

4.4 CONDITIONAL EXPECTATION

In the preceding sections we considered situations in which two successive observations are made, the second observation depending on the result of the first. The essential ingredient in such problems is the quantity $P_x(B)$, defined for real x and Borel sets B, to be interpreted as the conditional probability that the second observation will fall into B, given that the first observation takes the value x: for short, $P\{R_2 \in B \mid R_1 = x\}$. In particular, we may define the *conditional distribution function* of R_2 given $R_1 = x$, as $F_2(y \mid x) = P\{R_2 \leq y \mid R_1 = x\}$.

If R_1 and R_2 have a joint density, this can be computed from the conditional density of R_2 given R_1: $F_2(y_0 \mid x) = \int_{-\infty}^{y_0} h(y \mid x)\, dy$.

In any case, for each real x we have a probability measure P_x defined on the Borel subsets of E^1. Now if $R_1 = x$ and we observe R_2, there should be an average value associated with R_2, that is, a conditional expectation of R_2 given that $R_1 = x$. How should this be computed? Let us try to set up an appropriate model. We are observing a single random variable R_2, so let $\Omega = E^1$, $\mathscr{F} = $ Borel sets, $R_2(y) = y$. We are not concerned with the probability that $R_2 \in B$, but instead with the probability that $R_2 \in B$, *given that $R_1 = x$*. In other words, the appropriate probability measure is P_x. The expectation of R_2, computed with respect to P_x, is called the *conditional expectation of R_2 given that $R_1 = x$* (or, for short, the conditional expectation of R_2 given R_1), written $E(R_2 \mid R_1 = x)$.

Note that if g is a (piecewise continuous) function from E^1 to E^1, then $g(R_2)$ is also a random variable (see Section 2.7), so that we may also talk about the conditional expectation of $g(R_2)$ given $R_1 = x$, written $E[g(R_2) \mid R_1 = x]$. In particular, if there is a conditional density of R_2 given $R_1 = x$, then, by Theorem 2 of Section 3.1,

$$E[g(R_2) \mid R_1 = x] = \int_{-\infty}^{\infty} g(y)h(y \mid x)\, dy \qquad (4.4.1)$$

if $g \geq 0$ or if the integral is absolutely convergent.

There is an immediate extension to n dimensions. For example, if there is a conditional density of (R_4, R_5) given (R_1, R_2, R_3), then

$$E[g(R_4, R_5) \mid R_1 = x_1, R_2 = x_2, R_3 = x_3]$$
$$= \int_{-\infty}^{\infty} \int_{-\infty}^{\infty} g(x_4, x_5)h(x_4, x_5 \mid x_1, x_2, x_3)\, dx_4\, dx_5$$

Note also that conditional probability can be obtained from conditional expectation. If in (4.4.1) we take $g(y) = I_B(y) = 1$ if $y \in B$, and $= 0$ if $y \notin B$, then

$$E[g(R_2) \mid R_1 = x] = E[I_B(R_2) \mid R_1 = x] = \int_{-\infty}^{\infty} I_B(y)h(y \mid x)\, dy$$
$$= \int_B h(y \mid x)\, dy = P\{R_2 \in B \mid R_1 = x\}$$

We have seen previously that $P\{R_2 \in B\} = E[I_{\{R_2 \in B\}}]$. We now have a similar result under the condition that $R_1 = x$. [Notice that $I_B(R_2) = I_{\{R_2 \in B\}}$; for $I_B(R_2(\omega)) = 1$ iff $R_2(\omega) \in B$, that is, iff $I_{\{R_2 \in B\}}(\omega) = 1$.]

Let us consider again the examples of Section 4.2.

▶ **Example 1.** R_1 is uniformly distributed between 0 and 1; if $R_1 = x$, R_2 is the number of heads in n tosses of a coin with probability x of heads.

Given that $R_1 = x$, R_2 has a binomial distribution with parameters n and x: $P\{R_2 = k \mid R_1 = x\} = \binom{n}{k}x^k(1-x)^{n-k}$. It follows that $E(R_2 \mid R_1 = x)$ is the average number of successes in n Bernoulli trials, with probability x of success on a particular trial, namely, nx. ◀

▶ **Example 2.** R_1 has density $f_1(x) = xe^{-x}$, $x \geq 0$, $f_1(x) = 0$, $x < 0$. The conditional density of R_2 given $R_1 = x$ is uniform over $[0, x]$. It follows that, for $x > 0$,

$$E(R_2 \mid R_1 = x) = \int_{-\infty}^{\infty} yh(y \mid x)\, dy = \int_0^x y\frac{1}{x}\, dy = \tfrac{1}{2}x$$

Similarly,

$$E[e^{R_2} \mid R_1 = x] = \int_{-\infty}^{\infty} e^y h(y \mid x)\, dy = \int_0^x e^y \frac{1}{x}\, dy = \frac{e^x - 1}{x} \quad ◀$$

▶ **Example 3.** R_1 is discrete, with $p(x_i) = P\{R_1 = x_i\}$, $i = 1, 2, \ldots$. Given $R_1 = x_i$, R_2 has density f_i; that is,

$$P\{R_2 \in B \mid R_1 = x_i\} = \int_B f_i(y) \, dy$$

Thus

$$E[g(R_2) \mid R_1 = x_i] = \int_{-\infty}^{\infty} g(y) f_i(y) \, dy \quad ◀$$

Now let us consider a slightly different case.

▶ **Example 4.** Let R_1 and R_2 be discrete random variables. If $R_1 = x$, then R_2 will take the value y with probability

$$p(y \mid x) = P\{R_2 = y \mid R_1 = x\} = \frac{p_{12}(x, y)}{p_1(x)}$$

where

$$p_{12}(x, y) = P\{R_1 = x, R_2 = y\}, \quad p_1(x) = P\{R_1 = x\}$$

$p(y \mid x)$, which is defined provided that $p_1(x) > 0$, will be called the *conditional probability function* of R_2 given $R_1 = x$ (or the conditional probability function of R_2 given R_1, for short). We may find the probability that $R_2 \in B$ given $R_1 = x$ by summing the conditional probability function.

$$P_x(B) = P\{R_2 \in B \mid R_1 = x\} = \frac{P\{R_1 = x, R_2 \in B\}}{P\{R_1 = x\}} = \frac{\sum_{y \in B} p_{12}(x, y)}{p_1(x)}$$

$$= \sum_{y \in B} p(y \mid x)$$

Thus, given that $R_1 = x$, the probabilities of events involving R_2 are found from the probability function $p(y \mid x)$, y real. Therefore the conditional expectation of $g(R_2)$ given $R_1 = x$ is

$$E[g(R_2) \mid R_1 = x] = \sum_y g(y) p(y \mid x) \qquad (4.4.2)$$

In particular,

$$E(R_2 \mid R_1 = x) = \sum_y y p(y \mid x) \quad ◀$$

There is a feature common to all these examples. In each case the expectation of R_2 (or of a function of R_2) can be expressed as a weighted average of conditional expectations. Let us look at Example 4 first. With probability $p_1(x)$, R_1 takes the value x; if $R_1 = x$, the average value of R_2 is $E(R_2 \mid R_1 = x)$. By analogy with the theorem of total probability, it is reasonable to expect that

$$E(R_2) = \sum_x p_1(x) E(R_2 \mid R_1 = x)$$

To justify this, write

$$E(R_2) = \sum_y y p_2(y) = \sum_y y P\{R_2 = y\} = \sum_y y \sum_x P\{R_1 = x, R_2 = y\}$$

by (2.7.2)

$$= \sum_{x,y} y P\{R_1 = x\} P\{R_2 = y \mid R_1 = x\} = \sum_x p_1(x)[\sum_y y p(y \mid x)]$$

This is the desired result.

In Example 1 the probability that R_1 will lie in an interval about x is $f_1(x)\, dx = dx$; given that $R_1 = x$, the average value of R_2 is $E(R_2 \mid R_1 = x) = nx$. We expect that

$$E(R_2) = \int_{-\infty}^{\infty} f_1(x) E(R_2 \mid R_1 = x)\, dx$$

To verify this, notice that we calculated in Section 4.2 that

$$P\{R_2 = k\} = \frac{1}{n+1}, \qquad k = 0, 1, \ldots, n$$

Thus

$$E(R_2) = \sum_{k=0}^{n} k P\{R_2 = k\} = \frac{1}{n+1}(1 + 2 + \cdots + n) = \frac{1}{n+1}\frac{(n+1)n}{2} = \frac{n}{2}$$

But

$$\int_{-\infty}^{\infty} f_1(x) E(R_2 \mid R_1 = x)\, dx = \int_0^1 nx\, dx = \frac{n}{2}$$

In Example 2, the joint density of R_1 and R_2 is

$$f(x, y) = f_1(x) h(y \mid x) = \frac{xe^{-x}}{x} = e^{-x}, \qquad x \geq 0, 0 \leq y \leq x$$

Now

$$E(R_2) = \int_{-\infty}^{\infty} \int_{-\infty}^{\infty} y f(x, y)\, dx\, dy$$

[*Notice that we need not compute* $f_2(y)$ *explicitly;* instead we simply regard R_2 as a function of R_1 and R_2; that is, we set $g(R_1, R_2) = R_2$ and compute

$$E[g(R_1, R_2)] = \int_{-\infty}^{\infty} \int_{-\infty}^{\infty} g(x, y) f(x, y)\, dx\, dy]$$

Thus

$$E(R_2) = \int_0^{\infty} e^{-x}\left[\int_0^x y\, dy\right] dx = \int_0^{\infty} \tfrac{1}{2}x^2 e^{-x}\, dx = \tfrac{1}{2}\Gamma(3) = 1$$

But

$$\int_{-\infty}^{\infty} f_1(x) E(R_2 \mid R_1 = x)\, dx = \int_0^{\infty} xe^{-x}(\tfrac{1}{2}x)\, dx = 1$$

In Example 3 we have [see (4.2.2)]

$$P\{R_2 \in B\} = \sum_i p(x_i) \int_B f_i(y)\, dy = \int_B \left[\sum_i p(x_i) f_i(y) \right] dy$$

so that R_2 has density

$$f_2(y) = \sum_i p(x_i) f_i(y) \tag{4.4.3}$$

Thus

$$E(R_2) = \int_{-\infty}^{\infty} y f_2(y)\, dy = \sum_i p(x_i) \int_{-\infty}^{\infty} y f_i(y)\, dy$$

and consequently

$$E(R_2) = \sum_i p(x_i) E(R_2 \mid R_1 = x_i)$$

as expected.

Results of the form

$$E(R_2) = \sum_i p(x_i) E(R_2 \mid R_1 = x_i) \tag{4.4.4}$$

or

$$E(R_2) = \int_{-\infty}^{\infty} f_1(x) E(R_2 \mid R_1 = x)\, dx \tag{4.4.5}$$

are called versions of the *theorem of total expectation*.

In the situations we are considering, conditional expectations are derived ultimately from a given set of probabilities $P_x(B) = P\{R_2 \in B \mid R_1 = x\}$. In such cases it turns out that if $E(R_2)$ exists, (4.4.4) will hold if R_1 is discrete, and (4.4.5) will hold if R_1 is absolutely continuous.

Notice that $E(R_2 \mid R_1 = x)$ will in general depend on x and hence may be written as $g(x)$; $\int_{-\infty}^{\infty} g(x) f_1(x)\, dx$ in (4.4.5) [or $\sum_x g(x) p(x)$ in (4.4.4)] is then the expectation of $g(R_1)$. Thus (4.4.4) and (4.4.5) may be rephrased as follows.

The expectation of the conditional expectation of R_2 given R_1 is the (over-all) expectation of R_2.

▶ **Example 5.** Let R be a random variable with the distribution function shown in Figure 4.4.1. Find $E(R^3)$.

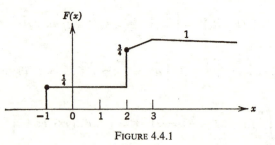

FIGURE 4.4.1

If R were discrete we would compute

$$E(R^3) = \sum_x x^3 p_R(x)$$

and if R were absolutely continuous we would compute

$$E(R^3) = \int_{-\infty}^{\infty} x^3 f_R(x)\, dx$$

In this case, however, R falls into neither category. We are going to show how to use the theorem of total expectation to compute $E(R^3)$.

We have $P\{R = -1\} = 1/4$, $P\{R = 2\} = 3/4 - 1/4 = 1/2$, $P\{R = x\} = 0$ for other values of x. Let F_1 be a step function that is 0 for $x < -1$ and has a jump of 1/4 at $x = -1$ and a jump of 1/2 at $x = 2$. Subtract F_1 from F to obtain a continuous function F_2 that can be represented as an integral of a nonnegative function f_2. F_1 is called the "discrete part" of F, and F_2 the "absolutely continuous part" (see Figure 4.4.2). F_1 and F_2 are monotone, right-continuous functions, and they approach zero as $x \to -\infty$. However, they approach limits that are less than 1 as $x \to \infty$, so that they cannot be regarded as distribution functions of random variables. However, $(4/3)F_1$ and $4F_2$ are legitimate distribution functions.

We shall show that

$$E(R^3) = \sum_x x^3 p_R(x) + \int_{-\infty}^{\infty} x^3 f_2(x)\, dx$$

Consider the following random experiment. With probability 3/4 ($= F_1(\infty) = \sum_x p_R(x)$, where $p_R(x) = P\{R = x\}$), pick a number in accordance

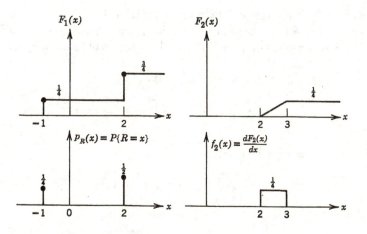

FIGURE 4.4.2 Discrete and Absolutely Continuous Parts of a Distribution Function.

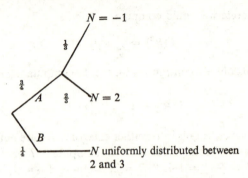

FIGURE 4.4.3   Tree Diagram for Example 5.

with $(4/3)F_1$; that is, pick -1 with probability $1/3$ and 2 with probability $2/3$. With probability $1/4$ $[= F_2(\infty)]$, pick a number in accordance with F_2, that is, one uniformly distributed between 2 and 3 (see Figure 4.4.3).

If N is the resulting number, then, by the theorem of total probability,

$$P\{N \le x\} = P(A)P\{N \le x \mid A\} + P(B)P\{N \le x \mid B\}$$

where A and B correspond to the two possible results at the first stage of the experiment. Thus

$$F_N(x) = \tfrac{3}{4}(\tfrac{4}{3}F_1(x)) + \tfrac{1}{4}(4F_2(x)) = F_1(x) + F_2(x) = F(x)$$

Therefore F_N is the original distribution function F.

Since N and R have the same distribution function, we expect that $E(N^3) = E(R^3)$. Now we may compute $E(N^3)$ by the theorem of total expectation.

$$E(N^3) = P(A)E(N^3 \mid A) + P(B)E(N^3 \mid B)$$

$$= \tfrac{3}{4}[(-1)^3\tfrac{1}{3} + 2^3 \tfrac{2}{3}] + \tfrac{1}{4}\int_2^3 x^3 \, dx = \tfrac{15}{4} + \tfrac{65}{16} = \tfrac{125}{16}$$

Notice that this may be expressed as

$$(-1)^3 \tfrac{1}{4} + 2^3 \tfrac{1}{2} + \int_2^3 x^3 \tfrac{1}{4} \, dx$$

that is,

$$E(R^3) = \sum_x x^3 p_R(x) + \int_{-\infty}^{\infty} x^3 f_2(x) \, dx$$

More generally, the expectation of a function of R may be computed by

$$E[g(R)] = \sum_x g(x)p_R(x) + \int_{-\infty}^{\infty} g(x)f_2(x)\,dx \qquad (4.4.6)\dagger$$

if $g \geq 0$ or if both the series and the integral are absolutely convergent. ◀

▶ **Example 6.** Let R be a random variable on a given probability space, and A an event with $P(A) > 0$. Formulate the proper definition of the conditional expectation of R, given that A has occurred.

This actually is not a new concept. If I_A is the indicator of A, we are looking for the expectation of R, given that $I_A = 1$. Let the experiment be performed independently n times, n very large, and let R_i be the value of R obtained on trial i, $i = 1, 2, \ldots, n$. Renumber the trials so that A occurs on the first k trials, and A^c on the last $n - k$ [k will be approximately $nP(A)$]. The average value of R, considering only those trials on which A occurs, is

$$\frac{R_1 + \cdots + R_k}{k} = \left(\frac{1}{n}\sum_{j=1}^{n} R_j I_j\right)\frac{n}{k}$$

where $I_j = 1$ if A occurs on trial j; $I_j = 0$ if A does not occur on trial j. In other words, I_j is simply the jth observation of I_A. It appears that $1/n \sum_{j=1}^{n} R_j I_j$ approximates the expectation of RI_A; since k/n approximates $P(A)$, we are led to define the *conditional expectation of R given A* as

$$E(R \mid A) = \frac{E(RI_A)}{P(A)} \qquad \text{if } P(A) > 0 \qquad (4.4.7)$$

Let us check that (4.4.7) agrees with previous results when R is discrete. By (4.4.2),

$$E(R \mid I_A = 1) = \sum_y yP\{R = y \mid I_A = 1\} = \sum_{y \neq 0} yP\{R = y \mid I_A = 1\}$$

But if $y \neq 0$,

$$P\{R = y \mid I_A = 1\} = \frac{P\{R = y, I_A = 1\}}{P\{I_A = 1\}} = \frac{P\{RI_A = y\}}{P(A)}$$

Thus

$$E(R \mid I_A = 1) = \frac{1}{P(A)} \sum_{y \neq 0} yP\{RI_A = y\} = \frac{E(RI_A)}{P(A)}$$

† The reader may recognize this as the Riemann-Stieltjes integral $\int_{-\infty}^{\infty} g(x)\,dF(x)$. Alternatively, if one differentiates F formally to obtain $f = f_2$ plus "impulses" or "delta functions" at -1 and 2 of strength $1/4$ and $1/2$, respectively, and then evaluates $\int_{-\infty}^{\infty} g(x)f(x)\,dx$, (4.4.6) is obtained.

Let us look at another special case. For any random variable R and event A with $P(A) > 0$, we may define the *conditional distribution function of R given A* in a natural way, namely,

$$F_R(x \mid A) = P\{R \leq x \mid A\} = \frac{P(A \cap \{R \leq x\})}{P(A)} \qquad (4.4.8)$$

Now assume that R has density f and A is of the form $\{R \in B\}$ for some Borel set B. Then

$$P(A \cap \{R \leq x_0\}) = P\{R \in B, R \leq x_0\} = \int_{\substack{x \in B \\ x \leq x_0}} f(x)\, dx$$

$$= \int_{x \leq x_0} f(x) I_B(x)\, dx$$

Thus (4.4.8) becomes

$$F_R(x_0 \mid A) = \int_{-\infty}^{x_0} \frac{f(x)}{P(A)} I_B(x)\, dx$$

In other words, there is a *conditional density of R given A*, namely,

$$f_R(x \mid A) = \frac{f(x)}{P(A)} I_B(x) = \frac{f(x)}{P(A)} \qquad \text{if } x \in B$$

$$= 0 \qquad \text{if } x \notin B \qquad (4.4.9)$$

We may then compute the conditional expectation of R given A.

$$E(R \mid A) = \int_{-\infty}^{\infty} x f_R(x \mid A)\, dx = \int_{-\infty}^{\infty} \frac{x I_B(x)}{P(A)} f(x)\, dx$$

$$= \frac{E(R I_B(R))}{P(A)} \qquad \text{by Theorem 2 of Section 3.1}$$

But
$$I_B(R) = I_{\{R \in B\}} \qquad \text{by the discussion preceding Example 1}$$
$$= I_A$$

Thus
$$E(R \mid A) = \frac{E(R I_A)}{P(A)}$$

in agreement with (4.4.7).

REMARK. (4.4.8) and (4.4.9) extend to n dimensions. The conditional distribution function of $(R_1, \ldots, R_n)$ given A is $F_{12\cdots n}(x_1, \ldots, x_n \mid A) = P\{R_1 \leq x_1, \ldots, R_n \leq x_n \mid A\}$. If $(R_1, \ldots, R_n)$ has density f and $A = \{(R_1, \ldots, R_n) \in B\}$, there is a conditional density of

$(R_1, \ldots, R_n)$ given A.

$$f_R(x_1, \ldots, x_n \mid A) = \frac{f(x_1, \ldots, x_n)}{P(A)} I_B(x_1, \ldots, x_n)$$

The argument is essentially the same as above. ◄

PROBLEMS

1. Let (R_1, R_2) have density $f(x, y) = 8xy, 0 \le y \le x \le 1; f(x, y) = 0$ elsewhere.
   (a) Find the conditional expectation of R_2 given $R_1 = x$, and the conditional expectation of R_1 given $R_2 = y$.
   (b) Find the conditional expectation of R_2^4 given $R_1 = x$.
   (c) Find the conditional expectation of R_2 given $A = \{R_1 \le 1/2\}$.

2. In Example 2 of Section 4.3, find the conditional expectation of R_0^{-n}, given $R_1 = x_1, \ldots, R_n = x_n$.

3. Let (R_1, R_2) be uniformly distributed over the parallelogram with vertices $(0, 0), (2, 0), (3, 1), (1, 1)$. Find $E(R_2 \mid R_1 = x)$.

4. If a single die is tossed independently n times, find the average number of 2's, given that the number of 1's is k.

5. Let R_1 and R_2 be independent random variables, each uniformly distributed between 0 and 2.
   (a) Find the conditional probability that $R_1 \ge 1$, given that $R_1 + R_2 \le 3$.
   (b) Find the conditional expectation of R_1, given that $R_1 + R_2 \le 3$.

6. Let $B_1, B_2, \ldots$ be mutually exclusive, exhaustive events, with $P(B_n) > 0$, $n = 1, 2, \ldots$, and let R be a random variable. Establish the following version of the theorem of total expectation:

$$E(R) = \sum_{n=1}^{\infty} P(B_n)E(R \mid B_n)$$

[if $E(R)$ exists].

7. Of the 100 people in a certain village, 50 always tell the truth, 30 always lie, and 20 always refuse to answer. A single unbiased die is tossed. If the result is $1, 2, 3,$ or 4, a sample of size 30 is taken *with replacement*. If the result is 5 or 6, a sample of size 30 is taken *without replacement*. A random variable R is defined as follows:
   $R = 1$ if the resulting sample contains 10 people of each category.
   $R = 2$ if the sample is taken with replacement and contains 12 liars.
   $R = 3$ otherwise.
   Find $E(R)$.

8. Let R_1 and R_2 be independent random variables, each uniformly distributed between 0 and 1. Define

$$R_3 = g(R_1, R_2) = R_1 \quad \text{if } R_1^2 + R_2^2 \leq 1$$
$$R_3 = 2 \quad \text{if } R_1^2 + R_2^2 > 1$$

(a) Find $F_3(z)$ and compute $E(R_3)$ from this.

(b) Compute $E(R_3)$ from $\int_{-\infty}^{\infty} \int_{-\infty}^{\infty} g(x, y) f_{12}(x, y) \, dx \, dy$.

(c) Compute $E(R_3 \mid R_1^2 + R_2^2 \leq 1)$ and $E(R_3 \mid R_1^2 + R_2^2 > 1)$; then find $E(R_3)$ by using the theorem of total expectation.

9. The density for the time T required for the failure of a light bulb is $f(x) = \lambda e^{-\lambda x}$, $x \geq 0$. Find the conditional density function of $T - t_0$, given that $T > t_0$, and interpret the result intuitively.

10. Let R_1 and R_2 be independent random variables, each uniformly distributed between 0 and 1. Find the conditional expectation of $(R_1 + R_2)^2$ given $R_1 - R_2$.

11. Let R_1 and R_2 be independent random variables, each with density $f(x) = (1/2)e^{-x}$, $x \geq 0$; $f(x) = 1/2$, $-1 \leq x \leq 0$; $f(x) = 0$, $x < -1$. Let $R_3 = R_1^2 + R_2^2$. Find $E(R_3 \mid R_1 = x)$.

12. Let R_1 be a discrete random variable; if $R_1 = x$, let R_2 have a conditional density $h(y \mid x)$. Define the conditional probability that $R_1 = x$ given that $R_2 = y$ as

$$P\{R_1 = x \mid R_2 = y\} = \frac{P\{R_1 = x\}h(y \mid x)}{\sum_{x'} P\{R_1 = x'\}h(y \mid x')}$$

(cf. Bayes' Theorem).

(a) Interpret this definition intuitively by considering $P\{R_1 = x \mid y < R_2 < y + dy\}$.

(b) Show that the definition is natural in the sense that the appropriate version of the theorem of total probability is satisfied:

$$P\{R_1 \in A, R_2 \in B\} = \int_B f_2(y) P\{R_1 \in A \mid R_2 = y\} \, dy$$

where

$$P\{R_1 \in A \mid R_2 = y\} = \sum_{x \in A} P\{R_1 = x \mid R_2 = y\}$$
$$f_2(y) = \sum_x P\{R_1 = x\}h(y \mid x)$$

[see (4.4.3)].

13. If R_1 is absolutely continuous and R_2 discrete, and $p(y \mid x) = P\{R_2 = y \mid R_1 = x\}$ is specified, show that there is a conditional density of R_1 given R_2, namely,

$$h(x \mid y) = \frac{f_1(x)p(y \mid x)}{p_2(y)}$$

where

$$p_2(y) = P\{R_2 = y\} = \int_{-\infty}^{\infty} f_1(x)p(y \mid x) \, dx$$

14. Let R be uniformly distributed between 0 and 1. If $R = \lambda$, a coin with probability of heads λ is tossed independently n times. If $R_1, \ldots, R_n$ are the results of the tosses ($R_i = 1$ for a head, $R_i = 0$ for a tail), find the conditional density of R given $(R_1, \ldots, R_n)$, and the conditional expectation of R given $(R_1, \ldots, R_n)$.

15. (Hypothesis testing) Consider the following experiment. Throw a coin with probability p of heads. If the coin comes up heads, observe a random variable R with density $f_0(x)$; if the coin comes up tails, let R have density $f_1(x)$. Suppose that we are not told the result of the coin toss, but only the value of R, and we have to guess whether or not the coin came up heads. We do this by means of a *decision scheme*, which is simply a Borel set S of real numbers with the interpretation that if $R = x$ and $x \in S$, we decide for tails, that is, f_1, and if $x \notin S$ we decide for heads, that is, f_0.

    (a) Find the over-all probability of error in terms of p, f_0, f_1, and S. [There are two types of errors: if the actual density is f_0 and we decide for f_1 (type 1 error), and if the actual density is f_1 and we decide for f_0 (type 2 error).]

    (b) For a given p, f_0, f_1, find the S that makes the over-all probability of error a minimum. Apply the results to the case in which f_i is the normal density with mean m_i and variance σ^2, $i = 0, 1$.

REMARK. A physical model for part (b) is the following. The input R to a radar receiver is of the form $\theta + N$, where θ (the signal) and N (the noise) are independent random variables, with $P\{\theta = m_0\} = p$, $P\{\theta = m_1\} = 1 - p$, and N normally distributed with mean 0 and variance σ^2. If $\theta = m_i$ ($i = 0$ corresponds to a head in the above discussion, and $i = 1$ to a tail), then R is normal with mean m_i and variance σ^2; thus f_i is the conditional density of R given $\theta = m_i$. We are trying to determine the actual value of the signal with as low a probability of error as possible.

16. Let R be the number of successes in n Bernoulli trials, with probability p of success on a given trial. Find the conditional expectation of R, given that $R \geq 2$.

17. Let R_1 be uniformly distributed between 0 and 10, and define R_2 by

$$R_2 = R_1{}^2 \quad \text{if } 0 \leq R_1 \leq 6$$
$$= 3 \quad \text{if } 6 < R_1 \leq 10$$

Find the conditional expectation of R_2 given that $2 \leq R_2 \leq 4$.

18. Consider the following two-stage random experiment.

    (i) A circle of radius R and center at $(0, 0)$ is selected, where R has density $f_R(z) = e^{-z}, z \geq 0; f_R(z) = 0, z < 0$.

    (ii) A point (R_1, R_2) is chosen, where (R_1, R_2) is uniformly distributed inside the circle selected in step (i).

    (a) If $D = (R_1{}^2 + R_2{}^2)^{1/2}$ is the distance of the resulting point from the origin, find $E(D)$.

    (b) Find the conditional density of R given $R_1 = x$, $R_2 = y$. (Leave the answer in the form of an integral.)

19. (An estimation problem) The input R to a radar receiver is of the form $\theta + N$, where θ (the signal) and N (the noise) are independent random variables with finite mean and variance. The value of R is observed, and then an estimate of θ is made, say, $\theta^* = d(R)$, where d is a function from the reals to the reals. We wish to choose the estimate so that $E[(\theta^* - \theta)^2]$ is as small as possible.

(a) Show that $d(x)$ is the conditional expectation $E(\theta \mid R = x)$. (Assume that R is either absolutely continuous or discrete.)

(b) Let $\theta = \pm 1$ with equal probability, and let N be uniformly distributed between -2 and $+2$. Find $d(x)$ and the minimum value of $E[(\theta^* - \theta)^2]$.

20. A number θ is chosen at random with density $f_\theta(x) = e^{-x}$, $x \geq 0$; $f_\theta(x) = 0$, $x < 0$. If θ takes the value λ, a random variable R is observed, where R has the Poisson distribution with parameter λ. For example, R might be the number of radioactive particles (or particles with some other distinguishing characteristic) passing through a counting device in a given time interval, where the average number of such particles is selected randomly. The value of R is observed and an estimate of θ is made, say $\theta^* = d(R)$. The argument of Problem 19, which applies in any situation when one makes an estimate $\theta^* = d(R)$ of a parameter θ, and when the distribution function of R depends on θ, shows that the estimate that minimizes $E[(\theta^* - \theta)^2]$ is $d(x) = E(\theta \mid R = x)$. Find $d(x)$ in this case.

REMARK.   Problems 15, 19, and 20 illustrate some techniques of statistics. This subject will be taken up systematically in Chapter 8.

4.5   APPENDIX: THE GENERAL CONCEPT OF CONDITIONAL EXPECTATION

By shifting our viewpoint slightly, we may regard a conditional expectation as a random variable defined on the given probability space. For example, suppose that $E(R_2 \mid R_1 = x) = x^2$. We may then say that, having observed R_1, the average value of R_2 is R_1^2. We adopt the notation $E(R_2 \mid R_1) = R_1^2$. In general, if $E(R_2 \mid R_1 = x) = g(x)$, we define $E(R_2 \mid R_1) = g(R_1)$. Then $E(R_2 \mid R_1)$ is a function defined on Ω; its value at the point ω is $g(R_1(\omega))$.

Let us see what happens to the theorem of total expectation in this notation. If, for example,

$$E(R_2) = \int_{-\infty}^{\infty} f_1(x) E(R_2 \mid R_1 = x)\, dx = \int_{-\infty}^{\infty} f_1(x) g(x)\, dx$$

then $E(R_2) = E[g(R_1)]$; in other words,

$$E(R_2) = E[E(R_2 \mid R_1)] \tag{4.5.1}$$

The expectation of the conditional expectation of R_2 given R_1 is the expectation of R_2.

Let us develop this a bit further. Let A be a Borel subset of E^1. Then, assuming that (4.5.1) holds for the random variable $R_2 I_{\{R_1 \in A\}}$, we have

$$E(R_2 I_{\{R_1 \in A\}}) = E[E(R_2 I_{\{R_1 \in A\}} \mid R_1)]$$

But having observed R_1, $R_2 I_{\{R_1 \in A\}}$ will be R_2 if $R_1 \in A$, and 0 otherwise; thus we expect intuitively that

$$E(R_2 I_{\{R_1 \in A\}} \mid R_1) = I_{\{R_1 \in A\}} E(R_2 \mid R_1)$$

It appears reasonable to expect, then, that

$$E(R_2 I_{\{R_1 \in A\}}) = E[I_{\{R_1 \in A\}} E(R_2 \mid R_1)] \qquad \text{for all Borel subsets } A \text{ of } E^1 \qquad (4.5.2)$$

It turns out that if R_1 is an arbitrary random variable and R_2 a random variable whose expectation exists, there is a random variable R, of the form $g(R_1)$ for some Borel measurable function g, such that

$$E(R_2 I_{\{R_1 \in A\}}) = E[I_{\{R_1 \in A\}} R] \qquad \text{for all Borel subsets } A \text{ of } E^1$$

We set $R = E(R_2 \mid R_1)$. Furthermore, R is essentially unique: if $R' = g'(R_1)$ for some Borel measurable function g', and R' also satisfies (4.5.2), then $R = R'$ except perhaps on a set of probability 0.

In the cases considered in this chapter, the conditional expectations all satisfy (4.5.2) (which is just a restatement of the theorem of total expectation), and thus the examples of the chapter are consistent with the general notion of conditional expectation.

5

Characteristic Functions

5.1 INTRODUCTION

In Chapter 2 we examined the problem of finding probabilities of the form $P\{(R_1, \ldots, R_n) \in B\}$, where $R_1, \ldots, R_n$ were random variables on a given probability space. If $(R_1, \ldots, R_n)$ has density f, then

$$P\{(R_1, \ldots, R_n) \in B\} = \int \cdots \int_B f(x_1, \ldots, x_n)\, dx_1 \cdots dx_n$$

In general, the evaluation of integrals of this type is quite difficult, if it is possible at all. In this chapter we describe an approach to a particular class of problems, those involving sums of independent random variables, which avoids integration in n dimensions. The approach is similar in spirit to the application of Fourier or Laplace transforms to a differential equation.

Let R be a random variable on a given probability space. We introduce the *characteristic function* of R, defined by

$$M_R(u) = E(e^{-iuR}), \qquad u \text{ real} \tag{5.1.1}$$

Here we meet complex-valued random variables for the first time. A complex-valued random variable on $(\Omega, \mathscr{F}, P)$ is a function T from Ω to the complex numbers C, such that the real part T_1 and the imaginary part T_2 of T are (real-valued) random variables. Thus $T(\omega) = T_1(\omega) + iT_2(\omega)$,

154

$\omega \in \Omega$. We define the *expectation* of T as the complex number $E(T) = E(T_1) + iE(T_2)$; $E(T)$ is defined only if $E(T_1)$ and $E(T_2)$ are both finite. In the present case we have $M_R(u) = E(\cos uR) - iE(\sin uR)$; since the cosine and the sine are ≤ 1 in absolute value, all expectations are finite. Thus M_R is a function from the reals to the complex numbers. If R has density f_R we obtain

$$M_R(u) = \int_{-\infty}^{\infty} e^{-iux} f_R(x)\, dx \qquad (5.1.2)$$

which is the *Fourier transform* of f_R.

It will be convenient in many computations to use a Laplace rather than a Fourier transform. The *generalized characteristic function* of R is defined by

$$N_R(s) = E(e^{-sR}) \qquad s \text{ complex}\dagger \qquad (5.1.3)$$

$N_R(s)$ is defined only for those s such that $E(e^{-sR})$ is finite. If s is imaginary, that is, if $s = iu$, u real, then $N_R(s) = M_R(u)$, so that $N_R(s)$ is defined at least for s on the imaginary axis. There will be situations in which $N_R(s)$ is not defined for any s off the imaginary axis, and other situations in which $N_R(s)$ is defined for all s.

If R has density f_R, we obtain

$$N_R(s) = \int_{-\infty}^{\infty} e^{-sx} f_R(x)\, dx \qquad (5.1.4)$$

This is the (two-sided) *Laplace transform* of f_R.

The basic fact about characteristic functions is the following.

Theorem 1. Let $R_1, \ldots, R_n$ be independent random variables on a given probability space, and let $R_0 = R_1 + \cdots + R_n$. If $N_{R_i}(s)$ is finite for all $i = 1, 2, \ldots, n$, then $N_{R_0}(s)$ is finite, and

$$N_{R_0}(s) = N_{R_1}(s) N_{R_2}(s) \cdots N_{R_n}(s)$$

In particular, if we set $s = iu$, we obtain

$$M_{R_0}(u) = M_{R_1}(u) M_{R_2}(u) \cdots M_{R_n}(u)$$

Thus the characteristic function of a sum of independent random variables is the product of the characteristic functions.

$\dagger$ In doing most of the examples in this chapter, the student will not come to grief if he regards s as a real variable and replaces statements such as "$a < \operatorname{Re} s < b$" by "$a < s < b$." Also, a comment about notation. We have taken $E(e^{-iuR})$ as the definition of the characteristic function rather than the more usual $E(e^{iuR})$ in order to preserve a notational symmetry between Fourier and Laplace transforms ($\int e^{-sx}f(x)\, dx$, not $\int e^{sx}f(x)\, dx$, is the standard notation for Laplace transform). Since u ranges over all real numbers, this change is of no essential significance.

PROOF.

$$E(e^{-sR_0}) = E(e^{-s(R_1+\cdots+R_n)}) = E\left[\prod_{k=1}^{n} e^{-sR_k}\right] = \prod_{k=1}^{n} E(e^{-sR_k})$$

by independence.

We have glossed over one point in this argument. If we take $n = 2$ for simplicity, we have complex-valued random variables $V = V_1 + iV_2$ and $W = W_1 + iW_2$ ($V = e^{-sR_1}$, $W = e^{-sR_2}$), where, by Theorem 2 of Section 2.7, V_j and W_k are independent ($j, k = 1, 2$), and all expectations are finite. We must show that $E(VW) = E(V)E(W)$, which we have proved only in the case when V and W are real-valued and independent. However, there is no difficulty.

$$\begin{aligned}
E(VW) &= E[V_1W_1 - V_2W_2 + i(V_1W_2 + V_2W_1)] \\
&= E(V_1)E(W_1) - E(V_2)E(W_2) + i(E(V_1)E(W_2) + E(V_2)E(W_1)) \\
&= [E(V_1) + iE(W_1)][E(V_2) + iE(W_2)] = E(V)E(W)
\end{aligned}$$

The proof for arbitrary n is more cumbersome, but the idea is exactly the same.

Thus we may find the characteristic function of a sum of independent random variables without any n-dimensional integration. However, this technique will not be of value unless it is possible to recover the distribution function from the characteristic function. In fact we have the following result, which we shall not prove.

Theorem 2 (Correspondence Theorem). *If $M_{R_1}(u) = M_{R_2}(u)$ for all u, then*

$$F_{R_1}(x) = F_{R_2}(x) \qquad \text{for all } x$$

For computational purposes we need some facts about the Laplace transform. Let f be a piecewise continuous function from E^1 to E^1 (not necessarily a density) and L_f its Laplace transform:

$$L_f(s) = \int_{-\infty}^{\infty} f(x)e^{-sx}\, dx$$

Laplace Transform Properties

1. If there are real numbers K_1 and K_2 and nonnegative real numbers A_1 and A_2 such that $|f(x)| \le A_1 e^{K_1 x}$ for $x \ge 0$, and $|f(x)| \le A_2 e^{K_2 x}$ for $x \le 0$, then $L_f(s)$ is finite for $K_1 < \text{Re } s < K_2$. This follows, since

$$\int_0^{\infty} |f(x)e^{-sx}|\, dx \le \int_0^{\infty} A_1 e^{(K_1-a)x}\, dx$$

and

$$\int_{-\infty}^0 |f(x)e^{-sx}|\, dx \le \int_{-\infty}^0 A_2 e^{(K_2-a)x}\, dx$$

where $a = \text{Re } s$. The integrals are finite if $K_1 < a < K_2$. Thus the class of functions whose Laplace transform can be taken is quite large.

2. If $g(x) = f(x - a)$ and L_f is finite at s, then L_g is also finite at s and $L_g(s) = e^{-as} L_f(s)$. This follows, since

$$\int_{-\infty}^{\infty} f(x - a)e^{-sx}\, dx = e^{-as} \int_{-\infty}^{\infty} f(x - a)e^{-s(x-a)}\, d(x - a)$$

3. If $h(x) = f(-x)$ and L_f is finite at s, then L_h is finite at $-s$ and $L_h(-s) = L_f(s)$ [or $L_h(s) = L_f(-s)$ if L_f is finite at $-s$]. To verify this, write

$$\int_{-\infty}^{\infty} h(x)e^{sx}\, dx = (\text{with } y = -x) \int_{-\infty}^{\infty} f(y)e^{-sy}\, dy$$

4. If $g(x) = e^{-ax}f(x)$ and L_f is finite at s, then L_g is finite at $s - a$ and $L_g(s - a) = L_f(s)$ [or $L_g(s) = L_f(s + a)$ if L_f is finite at $s + a$]. For

$$\int_{-\infty}^{\infty} e^{-ax}e^{-(s-a)x}f(x)\, dx = \int_{-\infty}^{\infty} e^{-sx}f(x)\, dx$$

We now construct a very brief table of Laplace transforms for use in the examples. In Table 5.1.1, $u(x)$ is the *unit step function*, defined by $u(x) = 1$,

Table 5.1.1   Laplace Transforms

$f(x)$		$L_f(s)$	Region of Convergence
$u(x)$		$1/s$	$\text{Re } s > 0$
$e^{-ax}u(x)$		$1/(s + a)$	$\text{Re } s > -a$
$x^n e^{-ax}u(x),$	$n = 0, 1, \ldots$	$n!/(s + a)^{n+1}$	$\text{Re } s > -a$
$x^\alpha e^{-ax}u(x),$	$\alpha > -1$	$\Gamma(\alpha + 1)/(s + a)^{\alpha+1}$	$\text{Re } s > -a$

$x \geq 0; u(x) = 0, x < 0$. If we verify the last entry in the table the others will follow. Now

$$\int_0^{\infty} x^\alpha e^{-ax}e^{-sx}\, dx = [\text{with } y = (s + a)x] \int_0^{\infty} \frac{y^\alpha e^{-y}}{(s + a)^{\alpha+1}}\, dy$$

$$= \frac{\Gamma(\alpha + 1)}{(s + a)^{\alpha+1}}\dagger$$

† Strictly speaking, these manipulations are only valid for s *real* and $> -a$. However, one can show that under the hypothesis of property 1 L_f is analytic for $K_1 < \text{Re } s < K_2$. In the present case $K_1 = -a$ and K_2 can be taken arbitrarily large, so that L_f is analytic for $\text{Re } s > -a$. Now $L_f(s) = \Gamma(\alpha + 1)/(s + a)^{\alpha+1}$ for s real and $> -a$, and therefore, by the identity theorem for analytic functions, the formula holds for all s with $\text{Re } s > -a$. This technique, which allows one to treat certain complex integrals as if the integrands were real-valued, will be used several times without further comment.

REMARK. $u(x)$ and $-u(-x)$ have the same Laplace transform $1/s$, but the regions of convergence are disjoint:

$$\int_{-\infty}^{\infty} u(x)e^{-sx}\, dx = \int_{0}^{\infty} e^{-sx}\, dx = \frac{1}{s}, \qquad \text{Re } s > 0$$

and

$$\int_{-\infty}^{\infty} -u(-x)e^{-sx}\, dx = \int_{-\infty}^{0} -e^{-sx}\, dx = \frac{1}{s}, \qquad \text{Re } s < 0$$

This indicates that any statement about Laplace transforms should be accompanied by some information about the region of convergence.

We need the following result in doing examples; the proof is measure-theoretic and will be omitted.

5. Let R be an absolutely continuous random variable. If h is a nonnegative (piecewise continuous) function and $L_h(s)$ is finite and coincides with the generalized characteristic function $N_R(s)$ for all s on the line Re $s = a$, then h is the density of R.

5.2 EXAMPLES

We are going to examine some typical problems involving sums of independent random variables. We shall use the result, to be justified in Example 6, that if $R_1, R_2, \ldots, R_n$ are independent, each absolutely continuous, then $R_1 + \cdots + R_n$ is also absolutely continuous.

In all examples $N_i(s)$ will denote the generalized characteristic function of the random variable R_i.

▶ **Example 1.** Let R_1 and R_2 be independent random variables, with R_1 uniformly distributed between -1 and $+1$, and R_2 having the exponential density $e^{-y}u(y)$. Find the density of $R_0 = R_1 + R_2$.

We have

$$N_1(s) = \int_{-1}^{1} \tfrac{1}{2}e^{-sx}\, dx = \frac{1}{2s}(e^s - e^{-s}), \qquad \text{all } s$$

$$N_2(s) = \int_{0}^{\infty} e^{-sy}e^{-y}\, dy = \frac{1}{s+1} \qquad \text{Re } s > -1$$

Thus, by Theorem 1 of Section 5.1,

$$N_0(s) = N_1(s)N_2(s) = \frac{1}{2s(s+1)}(e^s - e^{-s})$$

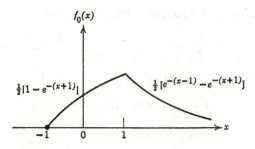

FIGURE 5.2.1

at least for Re $s > -1$. To find a function with this Laplace transform, we use *partial fraction expansion* of the rational function part of $N_0(s)$:

$$\frac{1}{2s(s+1)} = \frac{1}{2s} - \frac{1}{2(s+1)}$$

Now from Table 5.1.1, $u(x)$ has transform $1/s$ (Re $s > 0$) and $e^{-x}u(x)$ has transform $1/(s+1)$ (Re $s > -1$). Thus $(1/2)(1-e^{-x})\,u(x)$ has transform $1/2s(s+1)$ (Re $s > 0$). By property 2 of Laplace transforms (Section 5.1), $(1/2)(1-e^{-(x+1)})u(x+1)$ has transform $e^s/2s(s+1)$ and $(1/2)(1-e^{-(x-1)})u(x-1)$ has transform $e^{-s}/2s(s+1)$ (Re $s > 0$). Thus a function h whose transform is $N_0(s)$ for Re $s > 0$ is

$$h(x) = \tfrac{1}{2}(1 - e^{-(x+1)})u(x+1) - \tfrac{1}{2}(1 - e^{-(x-1)})u(x-1)$$

By property 5 of Laplace transforms, h is the density of R_0; for a sketch, see Figure 5.2.1. ◄

▶ **Example 2.** Let $R_0 = R_1 + R_2 + R_3$, where R_1, R_2, and R_3 are independent with densities $f_1(x) = f_2(x) = e^x u(-x), f_3(x) = e^{-(x-1)}u(x-1)$. Find the density of R_0.

We have

$$N_1(s) = N_2(s) = \int_{-\infty}^{0} e^x e^{-sx}\,dx = \frac{1}{1-s}, \qquad \text{Re } s < 1$$

and

$$N_3(s) = \int_{1}^{\infty} e^{-(x-1)}e^{-sx}\,dx = \frac{e^{-s}}{s+1}, \qquad \text{Re } s > -1$$

Thus

$$N_0(s) = N_1(s)N_2(s)N_3(s) = \frac{e^{-s}}{(s-1)^2(s+1)}, \qquad -1 < \text{Re } s < 1$$

We expand the rational function in partial fractions.

$$\frac{1}{(s-1)^2(s+1)} = \frac{A}{(s-1)^2} + \frac{B}{s-1} + \frac{C}{s+1}$$

The coefficients may be found as follows.

$$A = [(s-1)^2 G(s)]_{s=1} = \tfrac{1}{2}$$

$$B = \left[\frac{d}{ds}((s-1)^2 G(s))\right]_{s=1} = -\tfrac{1}{4}$$

$$C = [(s+1)G(s)]_{s=-1} = \tfrac{1}{4}$$

From Table 5.1.1, the transform of $xe^{-x}u(x)$ is $1/(s+1)^2$, Re $s > -1$. By Laplace transform property 3, the transform of $-xe^{x}u(-x)$ is $1/(1-s)^2$, Re $s < 1$. The transform of $e^{-x}u(x)$ is $1/(s+1)$, Re $s > -1$, so that, again by property 3, the transform of $e^{x}u(-x)$ is $1/(1-s)$, Re $s < 1$.

Thus the transform of

$$-\tfrac{1}{2}xe^{x}u(-x) + \tfrac{1}{4}e^{x}u(-x) + \tfrac{1}{4}e^{-x}u(x)$$

is

$$G(s) = \frac{1/2}{(s-1)^2} - \frac{1/4}{s-1} + \frac{1/4}{s+1}, \qquad -1 < \text{Re } s < 1$$

By property 2, the transform of

$$h(x) = [\tfrac{1}{4} - \tfrac{1}{2}(x-1)]e^{x-1}u(-(x-1)) + \tfrac{1}{4}e^{-(x-1)}u(x-1)$$

is

$$e^{-s}G(s) = N_0(s), \qquad -1 < \text{Re } s < 1$$

By property 5, h is the density of R_0 (see Figure 5.2.2). ◄

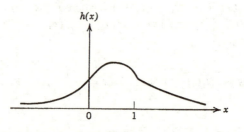

$$h(x)$$

FIGURE 5.2.2

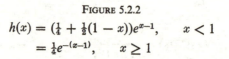

$$h(x) = (\tfrac{1}{4} + \tfrac{1}{2}(1-x))e^{x-1}, \qquad x < 1$$
$$= \tfrac{1}{4}e^{-(x-1)}, \qquad x \geq 1$$

▶ **Example 3.** Let R have the *Cauchy density;* that is,

$$f_R(x) = \frac{1}{\pi(1 + x^2)}, \qquad -\infty < x < \infty$$

The characteristic function of R is

$$M_R(u) = \int_{-\infty}^{\infty} e^{-iux} f_R(x) \, dx$$

[In this case $N_R(s)$ is finite only for s on the imaginary axis.] $M_R(u)$ turns out to be $e^{-|u|}$. This may be verified by complex variable methods (see Problem 9), but instead we give a rough sketch of another attack. If the characteristic function of a random variable R is integrable, that is,

$$\int_{-\infty}^{\infty} |M_R(u)| \, du < \infty$$

it turns out that R has a density and in fact f_R is given by the *inverse Fourier transform.*

$$f_R(x) = \frac{1}{2\pi} \int_{-\infty}^{\infty} M_R(u) e^{iux} \, du \qquad (5.2.1)$$

In the present case

$$\int_{-\infty}^{\infty} e^{-|u|} \, du = \int_{-\infty}^{0} e^{u} \, du + \int_{0}^{\infty} e^{-u} \, du = 2 < \infty$$

and thus the density corresponding to $e^{-|u|}$ is

$$\frac{1}{2\pi} \int_{-\infty}^{\infty} e^{-|u|} e^{iux} \, du = \frac{1}{2\pi} \int_{-\infty}^{0} e^{u(1+ix)} \, du + \frac{1}{2\pi} \int_{0}^{\infty} e^{-u(1-ix)} \, du$$

$$= \frac{1}{2\pi} \left[\frac{1}{1 + ix} + \frac{1}{1 - ix} \right] = \frac{1}{\pi(1 + x^2)}$$

Thus the Cauchy density in fact corresponds to the characteristic function $e^{-|u|}$.

This argument has a serious gap. We started with the assumption that $e^{-|u|}$ was the characteristic function of some random variable, and deduced from this that the random variable must have density $1/\pi(1 + x^2)$. We must establish that $e^{-|u|}$ is in fact a characteristic function (see Problem 8).

Now let $R_0 = R_1 + \cdots + R_n$, where the R_i are independent, each with the Cauchy density. Let us find the density of R_0. We have

$$M_0(u) = M_1(u) M_2(u) \cdots M_n(u) = (e^{-|u|})^n = e^{-n|u|}$$

If instead we consider R_0/n, we obtain

$$M_{R_0/n}(u) = E[e^{-iuR_0/n}] = M_0\left(\frac{u}{n}\right) = e^{-|u|}$$

Thus R_0/n has the Cauchy density. Now if $R_2 = nR_1$, then $f_2(y) = (1/n)f_1(y/n)$ (see Section 2.4), and so the density of R_0 is

$$f_0(y) = \frac{1}{n\pi(1 + y^2/n^2)} = \frac{n}{\pi(y^2 + n^2)}$$

REMARKS.

1. The arithmetic average R_0/n of a sequence of independent Cauchy distributed random variables has the same density as each of the components. There is no convergence of the arithmetic average to a constant, as we might expect physically. The trouble is that $E(R)$ does not exist.

2. If R has the Cauchy density and $R_1 = c_1R$, $R_2 = c_2R$, c_1, c_2 constant and > 0, then

$$M_1(u) = E(e^{-iuR_1}) = E(e^{-iuc_1R}) = M_R(c_1u) = e^{-c_1|u|}$$

and similarly

$$M_2(u) = e^{-c_2|u|}$$

Thus, if $R_0 = R_1 + R_2 = (c_1 + c_2)R$,

$$M_0(u) = e^{-(c_1+c_2)|u|}$$

which happens to be $M_1(u)M_2(u)$. This shows that if the characteristic function of the sum of two random variables is the product of the characteristic functions, the random variables need not be independent.

3. If R has the Cauchy density and $R_1 = \theta R$, $\theta > 0$, then by the calculation performed before Remark 1, R_1 has density $f_1(y) = \theta/\pi(y^2 + \theta^2)$ and (as in Remark 2) characteristic function $M_1(u) = e^{-\theta|u|}$. A random variable with this density is said to be of the *Cauchy type with parameter* θ or to have the *Cauchy density with parameter* θ. The formula for $M_1(u)$ shows immediately that if $R_1, \ldots, R_n$ are independent and R_i is of the Cauchy type with parameter θ_i, $i = 1, \ldots, n$, then $R_1 + \cdots + R_n$ is of the Cauchy type with parameter $\theta_1 + \cdots + \theta_n$. ◄

▶ **Example 4.** If $R_1, R_2, \ldots, R_n$ are independent and normally distributed, then $R_0 = R_1 + \cdots + R_n$ is also normally distributed.

We first show that if R is normally distributed with mean m and variance σ^2, then

$$N_R(s) = e^{-sm} e^{s^2\sigma^2/2} \qquad \text{(all } s) \qquad (5.2.2)$$

Now

$$N_R(s) = \int_{-\infty}^{\infty} e^{-sx} f_R(x)\, dx = \int_{-\infty}^{\infty} e^{-sx} \frac{1}{\sqrt{2\pi}\,\sigma} e^{-(x-m)^2/2\sigma^2}\, dx$$

Let $y = (x - m)/\sqrt{2}\,\sigma$ and complete the square to obtain

$$N_R(s) = \frac{1}{\sqrt{\pi}} e^{-sm} \int_{-\infty}^{\infty} \exp\left[-\left(y^2 + s\sqrt{2}\,\sigma y + \frac{s^2\sigma^2}{2}\right)\right] e^{s^2\sigma^2/2}\, dy$$

$$= \frac{1}{\sqrt{\pi}} e^{-sm} e^{s^2\sigma^2/2} \int_{-\infty}^{\infty} e^{-t^2}\, dt = e^{-sm} e^{s^2\sigma^2/2} \qquad \text{by (2.8.2)}$$

(See the footnote on page 157.) Now if $E(R_i) = m_i$, $\text{Var } R_i = \sigma_i^2$, then

$$N_0(s) = N_1(s)N_2(s) \cdots N_n(s) = e^{-s(m_1+\ldots+m_n)} e^{s^2(\sigma_1^2+\ldots+\sigma_n^2)/2}$$

But this is the characteristic function of a normally distributed random variable, and the result follows. Note that $m_0 = m_1 + \cdots + m_n$, $\sigma_0^2 = \sigma_1^2 + \cdots + \sigma_n^2$, as we should expect from the results of Section 3.3. ◄

▶ **Example 5.** Let R have the Poisson distribution.

$$p_R(k) = \frac{e^{-\lambda}\lambda^k}{k!}, \qquad k = 0, 1, \ldots$$

We first show that the generalized characteristic function of R is

$$N_R(s) = \exp\left[\lambda(e^{-s} - 1)\right] \qquad \text{(all } s) \qquad (5.2.3)$$

We have

$$N_R(s) = E(e^{-sR}) = \sum_{k=0}^{\infty} e^{-sk} p_R(k) = \sum_{k=0}^{\infty} \frac{e^{-\lambda}\lambda^k}{k!} e^{-sk}$$

$$= e^{-\lambda} \sum_{k=0}^{\infty} \frac{(\lambda e^{-s})^k}{k!} = e^{-\lambda} \exp(\lambda e^{-s})$$

as asserted.

We now show that if $R_1, \ldots, R_n$ are independent random variables, each with the Poisson distribution, then $R_0 = R_1 + \cdots + R_n$ also has the poisson distribution.

If R_i has the Poisson distribution with parameter λ_i, then

$$N_0(s) = N_1(s)N_2(s) \cdots N_n(s) = \exp\left[(\lambda_1 + \cdots + \lambda_n)(e^{-s} - 1)\right]$$

This is the characteristic function of a Poisson random variable, and the result follows. Note that if R has the Poisson distribution with parameter

λ, then $E(R) = \text{Var } R = \lambda$ (see Problem 8, Section 3.2). Thus the result that the parameter of R_0 is $\lambda_1 + \cdots + \lambda_n$ is consistent with the fact that $E(R_0) = E(R_1) + \cdots + E(R_n)$ and $\text{Var } R_0 = \text{Var } R_1 + \cdots + \text{Var } R_n$. ◄

▶ **Example 6.**   In certain situations (especially when the Laplace transforms cannot be expressed in closed form) it may be convenient to use a convolution procedure rather than the transform technique to find the density of a sum of independent random variables. The method is based on the following result.

Convolution Theorem.   *Let R_1 and R_2 be independent random variables, having densities f_1 and f_2, respectively. Let $R_0 = R_1 + R_2$. Then R_0 has a density given by*

$$f_0(z) = \int_{-\infty}^{\infty} f_2(z - x) f_1(x) \, dx = \int_{-\infty}^{\infty} f_1(z - y) f_2(y) \, dy \qquad (5.2.4)$$

(Intuitively, the probability that R_1 lies in $(x, x + dx]$ is $f_1(x) \, dx$; given that $R_1 = x$, the probability that R_0 lies in $(z, z + dz]$ is the probability that R_2 lies in $(z - x, z - x + dz]$, namely, $f_2(z - x) \, dz$. Integrating with respect to x, we obtain the result that the probability that R_0 lies in $(z, z + dz]$ is

$$dz \int_{-\infty}^{\infty} f_2(z - x) f_1(x) \, dx$$

Since the probability is $f_0(z) \, dz$, (5.2.4) follows.)

PROOF.   To prove the convolution theorem, observe that

$$F_0(z) = P\{R_1 + R_2 \le z\} = \iint_{x+y \le z} f_1(x) f_2(y) \, dx \, dy$$

$$= \int_{-\infty}^{\infty} \left[\int_{-\infty}^{z-x} f_2(y) \, dy \right] f_1(x) \, dx$$

Let $y = u - x$ to obtain

$$\int_{-\infty}^{\infty} \left[\int_{-\infty}^{z} f_2(u - x) \, du \right] f_1(x) \, dx = \int_{-\infty}^{z} \left[\int_{-\infty}^{\infty} f_1(x) f_2(u - x) \, dx \right] du$$

This proves the first relation of (5.2.4); the other follows by a symmetrical argument.

We consider a numerical example. Let $f_1(x) = 1/x^2$, $x \ge 1$; $f_1(x) = 0$, $x < 1$. Let $f_2(y) = 1$, $0 \le y \le 1$; $f_2(y) = 0$ elsewhere. If $z < 1$, $f_0(z) = 0$; if $1 \le z \le 2$,

$$f_0(z) = \int_{-\infty}^{\infty} f_1(x) f_2(z - x) \, dx = \int_{1}^{z} \frac{1}{x^2} \, dx = 1 - \frac{1}{z}$$

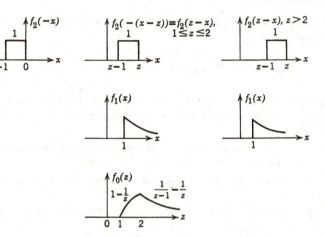

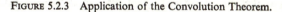

FIGURE 5.2.3   Application of the Convolution Theorem.

If $z > 2$,

$$f_0(z) = \int_{z-1}^{z} \frac{1}{x^2}\, dx = \frac{1}{z-1} - \frac{1}{z}$$

(see Figure 5.2.3).

REMARK.   The successive application of the convolution theorem shows that if $R_1, \ldots, R_n$ are independent, each absolutely continuous, then $R_1 + \cdots + R_n$ is absolutely continuous. ◄

PROBLEMS

1. Let R_1, R_2, and R_3 be independent random variables, each uniformly distributed between -1 and $+1$. Find and sketch the density function of the random variable $R_0 = R_1 + R_2 + R_3$.

2. Two independent random variables R_1 and R_2 each have the density function $f(x) = 1/3$, $-1 \le x < 0$; $f(x) = 2/3$, $0 \le x \le 1$; $f(x) = 0$ elsewhere. Find and sketch the density function of $R_1 + R_2$.

3. Let $R = R_1^2 + \cdots + R_n^2$, where $R_1, \ldots, R_n$ are independent, and each R_i is normal with mean 0 and variance 1. Show that the density of R is

$$f(x) = \frac{1}{2^{n/2}\Gamma(n/2)} x^{(n/2)-1} e^{-x/2}, \qquad x \ge 0$$

(R is said to have the "chi-square" distribution with n "degrees of freedom.")

4. A random variable R is said to have the "gamma distribution" if its density is, for some $\alpha, \beta > 0$,

$$f(x) = \frac{x^{\alpha-1}e^{-x/\beta}}{\Gamma(\alpha)\beta^{\alpha}}, x \geq 0; f(x) = 0, x < 0$$

Show that if R_1 and R_2 are independent random variables, each having the gamma distribution with the same β, then $R_1 + R_2$ also has the gamma distribution.

5. If $R_1, \ldots, R_n$ are independent nonnegative random variables, each with density $\lambda e^{-\lambda x}u(x)$, find the density of $R_0 = R_1 + \cdots + R_n$.

6. Let θ be uniformly distributed between $-\pi/2$ and $\pi/2$. Show that $\tan \theta$ has the Cauchy density.

7. Let R have density $f(x) = 1 - |x|, |x| \leq 1; f(x) = 0, |x| > 1$. Show that $M_R(u) = 2(1 - \cos u)/u^2$.

*8. (a) Suppose that f is the density of a random variable and the associated characteristic function M is real-valued, nonnegative, and integrable. Show that $kf(u), -\infty < u < \infty$, is the characteristic function of a random variable with density $kM(x)/2\pi$, where k is chosen so that $kf(0) = 1$, that is,

$$\int_{-\infty}^{\infty} [kM(x)/2\pi] \, dx = 1$$

(b) Use part (a) to show that the following are characteristic functions of random variables: (i) $e^{-|u|}$, (ii) $M(u) = 1 - |u|, |u| \leq 1; M(u) = 0, |u| > 1$.

*9. Use the calculus of residues to evaluate the characteristic function of the Cauchy density.

10. Calculate the characteristic function of the normal $(0, 1)$ random variable as follows. Differentiate

$$M(u) = \int_{-\infty}^{\infty} (\cos ux) \frac{e^{-x^2/2}}{\sqrt{2\pi}} \, dx$$

under the integral sign; then integrate by parts to obtain $M'(u) = -uM(u)$. Solve the resulting differential equation to obtain $M(u) = e^{-u^2/2}$. From this, find the characteristic function of a random variable that is normal with mean m and variance σ^2.

5.3 PROPERTIES OF CHARACTERISTIC FUNCTIONS

Let R be a random variable with characteristic function M and generalized characteristic function N. We shall establish several properties of M and N.

1. $M(0) = N(0) = 1$.

This follows, since $M(0) = N(0) = E(e^0)$.

2. $|M(u)| \leq 1$ for all u.

If R has a density f, we have

$$|M(u)| = \left| \int_{-\infty}^{\infty} e^{-iux} f(x) \, dx \right| \leq \int_{-\infty}^{\infty} |e^{-iux} f(x)| \, dx = \int_{-\infty}^{\infty} f(x) \, dx = 1$$

The general case can be handled by replacing $f(x) \, dx$ by $dF(x)$, where F is the distribution function of R. This involves Riemann-Stieltjes integration, which we shall not enter into here.

3. If R has a density f, and f is even, that is, $f(-x) = f(x)$ for all x, then $M(u)$ is real-valued for all u. For

$$M(u) = \int_{-\infty}^{\infty} f(x) \cos ux \, dx - i \int_{-\infty}^{\infty} f(x) \sin ux \, dx$$

Since $f(x)$ is an even function of x and $\sin ux$ is an odd function of x, $f(x) \sin ux$ is odd; hence the second integral is 0.

It turns out that the assertion that $M(u)$ is real for all u is equivalent to the statement that R has a symmetric distribution, that is, $P\{R \in B\} = P\{R \in -B\}$ for every Borel set B. ($-B = \{-x : x \in B\}$.)

4. If R is a discrete random variable taking on only integer values, then $M(u + 2\pi) = M(u)$ for all u.

To see this, write

$$M(u) = E(e^{-iuR}) = \sum_{n=-\infty}^{\infty} p_n e^{-iun} \tag{5.3.1}$$

where $p_n = P\{R = n\}$. Since $e^{-iun} = e^{-i(u+2\pi)n}$, the result follows.

Note that the p_n are the coefficients of the *Fourier series* of M on the interval $[0, 2\pi]$. If we multiply (5.3.1) by e^{iuk} and integrate, we obtain

$$p_k = \frac{1}{2\pi} \int_0^{2\pi} M(u) e^{iuk} \, du \tag{5.3.2}$$

We come now to the important *moment-generating property*. Suppose that $N(s)$ can be expanded in a power series about $s = 0$.

$$N(s) = \sum_{k=0}^{\infty} a_k s^k$$

where the series converges in some neighborhood of the origin. This is just the Taylor expansion of N; hence the coefficients must be given by

$$a_k = \frac{1}{k!} \frac{d^k N(s)}{ds^k} \bigg]_{s=0}$$

But if R has density f and we can differentiate $N(s) = \int_{-\infty}^{\infty} e^{-sx} f(x)\, dx$ under the integral sign, we obtain $N'(s) = \int_{-\infty}^{\infty} -xe^{-sx} f(x)\, dx$; if we can differentiate k times, we find that

$$N^{(k)}(s) = \int_{-\infty}^{\infty} (-1)^k x^k e^{-sx} f(x)\, dx$$

Thus

$$N^{(k)}(0) = (-1)^k E(R^k) \qquad (5.3.3)$$

and hence

$$a_k = \frac{(-1)^k}{k!} E(R^k)$$

The precise statement is as follows.

5. If $N_R(s)$ is analytic at $s = 0$ (i.e., expandable in a power series in a neighborhood of $s = 0$), then all moments of R are finite, and

$$N_R(s) = \sum_{k=0}^{\infty} \frac{(-1)^k}{k!} E(R^k) s^k \qquad (5.3.4)$$

within the radius of convergence of the series. In particular, (5.3.3) holds for all k.

We shall not give a proof of (5.3.4). The above remarks make it at least plausible; further evidence is presented by the following argument. If R has density f, then

$$N(s) = \int_{-\infty}^{\infty} e^{-sx} f(x)\, dx$$

$$= \int_{-\infty}^{\infty} \left(1 - sx + \frac{s^2 x^2}{2!} - \frac{s^3 x^3}{3!} + \cdots + \frac{(-1)^k s^k x^k}{k!} + \cdots \right) f(x)\, dx$$

If we are allowed to integrate term by term, we obtain (5.3.4).

Let us verify (5.3.4) for a numerical example. Let $f(x) = e^{-x} u(x)$, so that $N(s) = 1/(s + 1)$, Re $s > -1$. We have a power series expansion for $N(s)$ about $s = 0$.

$$\frac{1}{1 + s} = 1 - s + s^2 - s^3 + \cdots + (-1)^k s^k + \cdots \qquad |s| < 1$$

Equation 5.3.4 indicates that we should have $(-1)^k E(R^k)/k! = (-1)^k$, or $E(R^k) = k!$ To check this, notice that

$$E(R^k) = \int_0^{\infty} x^k e^{-x}\, dx = \Gamma(k + 1) = k!$$

REMARK. Let R be a discrete random variable taking on only nonnegative integer values. In the generalized characteristic function

$$N(s) = \sum_{k=0}^{\infty} p_k e^{-sk}, \qquad p_k = P\{R = k\}$$

make the substitution $z = e^{-s}$. We obtain

$$A(z) = N(s)]_{z=e^{-s}} = E(z^R) = \sum_{k=0}^{\infty} p_k z^k$$

A is called the *generating function* of R; it is finite at least for $|z| \leq 1$, since $\sum_{k=0}^{\infty} p_k = 1$.

We consider generating functions in detail in connection with the random walk problem in Chapter 6.

PROBLEMS

1. Could $[2/(s + 1)] - (1/s)(1 - e^{-s})$(Re $s > 0$) be the generalized characteristic function of an (absolutely continuous) random variable? Explain.

2. If the density of a random variable R is zero for $x \notin$ the finite interval $[a, b]$, show that $N_R(s)$ is finite for all s.

3. We have stated that if $M_R(u)$ is integrable, R has a density [see (5.2.1)]. Is the converse true?

4. Let R have a *lattice distribution*; that is, R is discrete and takes on the values $a + nd$, where a and d are fixed real numbers and n ranges over the integers. What can be said about the characteristic function of R?

5. If R has the Poisson distribution with parameter λ, calculate the mean and variance of R by differentiating $N_R(s)$.

5.4 THE CENTRAL LIMIT THEOREM

The weak law of large numbers states that if, for each n, $R_1, R_2, \ldots, R_n$ are independent random variables with finite expectations and uniformly bounded variances, then, for every $\varepsilon > 0$,

$$P\left\{\left|\frac{1}{n}\sum_{i=1}^{n}(R_i - ER_i)\right| \geq \varepsilon\right\} \to 0 \qquad \text{as } n \to \infty$$

In particular, if the R_i are independent observations of a random variable R (with finite mean m and finite variance σ^2), then

$$P\left\{\left|\frac{1}{n}\sum_{i=1}^{n}R_i - m\right| \geq \varepsilon\right\} \to 0 \qquad \text{as } n \to \infty$$

The central limit theorem gives further information; it says roughly that for large n, the sum $R_1 + \cdots + R_n$ of n independent random variables is approximately normally distributed, under wide conditions on the individual R_i.

To make the idea of "approximately normal" more precise, we need the notion of *convergence in distribution*. Let $R_1, R_2, \ldots$ be random variables with distribution functions $F_1, F_2, \ldots$, and let R be a random variable with distribution function F. We say that the sequence $R_1, R_2, \ldots$ converges in distribution to R (notation: $R_n \overset{d}{\longrightarrow} R$) iff $F_n(x) \to F(x)$ at all points x at which F is continuous.

To see the reason for the restriction to continuity points of F, consider the following example.

▶ **Example 1.** Let R_n be uniformly distributed between 0 and $1/n$ (see Figure 5.4.1). Intuitively, as $n \to \infty$, R_n approximates more and more closely a random variable R that is identically 0. But $F_n(x) \to F(x)$ when $x \neq 0$, but not at $x = 0$, since $F_n(0) = 0$ for all n, and $F(0) = 1$. Since $x = 0$ is not a continuity point of F, we have $R_n \overset{d}{\longrightarrow} R$. ◀

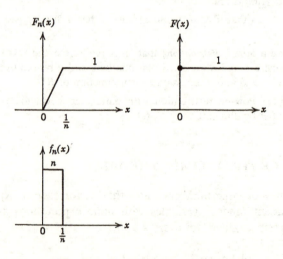

FIGURE 5.4.1   Convergence in Distribution.

REMARK. The type of convergence involved in the weak law of large numbers is called *convergence in probability*. The sequence R_1, $R_2, \ldots$ is said to converge in probability to R (notation: $R_n \xrightarrow{P} R$) iff for every $\varepsilon > 0$, $P\{|R_n - R| \geq \varepsilon\} \to 0$ as $n \to \infty$. Intuitively, for large n, R_n is very likely to be very close to R. Thus the weak law of large numbers states that $(1/n) \sum_{i=1}^{n} (R_i - E(R_i)) \xrightarrow{P} 0$; in the case in which $E(R_i) = m$ for all i, we have

$$\frac{1}{n} \sum_{i=1}^{n} R_i \xrightarrow{P} m$$

The relation between convergence in probability and convergence in distribution is outlined in Problem 1.

The basic result about convergence in distribution is the following.

Theorem 1. *The sequence R_1, $R_2, \ldots$ converges in distribution to R if and only if $M_n(u) \to M(u)$ for all u, where M_n is the characteristic function of R_n, and M is the characteristic function of R.*

The proof is measure-theoretic, and will be omitted.

Thus, in order to show that a sequence converges in distribution to a normal random variable, it suffices to show that the corresponding sequence of characteristic functions converges to a normal characteristic function. This is the technique that will be used to prove the main theorem, which we now state.

Theorem 2. (Central Limit Theorem). *For each n, let R_1, $R_2, \ldots, R_n$ be independent random variables on a given probability space. Assume that the R_i all have the same density function f (and characteristic function M) with finite mean m and finite variance $\sigma^2 > 0$, and finite third moment as well. Let*

$$T_n = \frac{\sum\limits_{j=1}^{n} R_j - nm}{\sqrt{n}\,\sigma}$$

$(= [S_n - E(S_n)]/\sigma(S_n)$, where $S_n = R_1 + \cdots + R_n$ and $\sigma(S_n)$ is the standard deviation of S_n) so that T_n has mean 0 and variance 1. Then T_1, $T_2, \ldots$ converge in distribution to a random variable that is normal with mean 0 and variance 1.

*Before giving the proof, we need some preliminaries.

Theorem 3. *Let f be a complex-valued function on E^1 with n continuous derivatives on the interval $V = (-b, b)$. Then, on V,*

$$f(u) = \sum_{k=0}^{n-1} \frac{f^{(k)}(0)u^k}{k!} + u^n \int_0^1 \frac{(1-t)^{n-1}}{(n-1)!} f^{(n)}(tu)\, dt$$

Thus, if $|f^{(n)}| \leq M$ on V,

$$f(u) = \sum_{k=0}^{n-1} \frac{f^{(k)}(0)u^k}{k!} + \frac{\theta\, |u|^n}{n!} \qquad \textit{where } |\theta| \leq M \ (\theta \textit{ depends on } u)$$

PROOF.   Using integration by parts, we obtain

$$\int_0^u f^{(n)}(t) \frac{(u-t)^{n-1}}{(n-1)!}\, dt = \left[\frac{(u-t)^{n-1}}{(n-1)!} f^{(n-1)}(t) \right]_0^u + \int_0^u f^{(n-1)}(t) \frac{(u-t)^{n-2}}{(n-2)!}\, dt$$

$$= -\frac{f^{(n-1)}(0)u^{n-1}}{(n-1)!} + \int_0^u f^{(n-1)}(t) \frac{(u-t)^{n-2}}{(n-2)!}\, dt$$

$$= -\frac{f^{(n-1)}(0)u^{n-1}}{(n-1)!} - \frac{f^{(n-2)}(0)u^{n-2}}{(n-2)!}$$

$$+ \int_0^u f^{(n-2)}(t) \frac{(u-t)^{n-3}}{(n-3)!}\, dt$$

$$= -\sum_{k=1}^{n-1} \frac{f^{(k)}(0)u^k}{k!} + \int_0^u f'(t)\, dt \qquad \text{by iteration}$$

Thus

$$f(u) = \sum_{k=0}^{n-1} \frac{f^{(k)}(0)u^k}{k!} + \int_0^u f^{(n)}(t) \frac{(u-t)^{n-1}}{(n-1)!}\, dt$$

The change of variables $t = ut'$ in the above integral yields the desired expression for $f(u)$.

Now if

$$I = u^n \int_0^1 \frac{(1-t)^{n-1}}{(n-1)!} f^{(n)}(tu)\, dt$$

then

$$|I| \leq M\, |u|^n \int_0^1 \frac{(1-t)^{n-1}}{(n-1)!}\, dt = \frac{M\, |u|^n}{n!}$$

Let $\theta = In!/|u|^n$; then $|\theta| \leq M$ and the result follows.

Theorem 4.

(a) $e^{iy} = 1 + iy + \dfrac{\theta y^2}{2}$, $\quad |\theta| \leq 1$

$e^{iy} = 1 + iy - \dfrac{y^2}{2} + \dfrac{\theta_1 |y|^3}{6}$, $\quad |\theta_1| \leq 1$

where y is an arbitrary real number, θ, θ_1 depending on y.

PROOF. This is immediate from Theorem 3.

(b) *If z is a complex number and $|z| \leq 1/2$, then $\ln(1+z) = z + \theta |z|^2$, where $|\theta| \leq 1$, θ depending on z. (Take $\ln 1 = 0$ to determine the branch of the logarithm.)*

PROOF.

$$\ln(1+z) = z - \frac{z^2}{2} + \frac{z^3}{3} - \frac{z^4}{4} + \cdots$$

$$= z + z^2 \left(-\frac{1}{2} + \frac{z}{3} - \frac{z^2}{4} + \cdots \right)$$

Now

$$\left| -\tfrac{1}{2} + \tfrac{1}{3}z - \tfrac{1}{4}z^2 + \cdots \right| \leq \tfrac{1}{2} + \tfrac{1}{3}(\tfrac{1}{2}) + \tfrac{1}{4}(\tfrac{1}{2})^2 + \cdots \leq \sum_{k=1}^{\infty} (\tfrac{1}{2})^k = 1$$

Since $z^2 = |z|^2 e^{i \arg(z^2)}$ and $|e^{i \arg(z^2)}| = 1$, the result follows.

PROOF OF THEOREM 2. By Theorem 1, we must show that $M_{T_n}(u) \to e^{-u^2/2}$, the characteristic function of a normal $(0, 1)$ random variable. Now we may assume without loss of generality that $m = 0$. For if we have proved the theorem under this assumption, then write

$$T_n = \frac{\sum_{j=1}^{n} (R_j - m)}{\sqrt{n}\, \sigma}$$

The random variables $R_j - m$ have mean 0 and variance σ^2, and the result will follow.

The characteristic function of $\sum_{j=1}^{n} R_j$ is $(M(u))^n$; hence the characteristic function of T_n is

$$E(e^{-iuT_n}) = E\left[\exp\left(-i \frac{u}{\sqrt{n}\,\sigma} \sum_{j=1}^{n} R_j \right) \right] = \left[M\left(\frac{u}{\sqrt{n}\,\sigma} \right) \right]^n$$

Now

$$M\left(\frac{u}{\sqrt{n}\,\sigma}\right) = \int_{-\infty}^{\infty} e^{-iux/\sqrt{n}\sigma} f(x)\,dx$$

$$= \int_{-\infty}^{\infty}\left(1 - \frac{iux}{\sqrt{n}\,\sigma} - \frac{u^2 x^2}{2n\sigma^2} + \frac{\theta_1\,|u|^3\,|x|^3}{6n^{3/2}\sigma^3}\right) f(x)\,dx \quad \text{by Theorem 4a}$$

But

$$\int_{-\infty}^{\infty} xf(x)\,dx = m = 0, \qquad \int_{-\infty}^{\infty} x^2 f(x)\,dx = \sigma^2$$

Thus

$$\left[M\left(\frac{u}{\sqrt{n}\,\sigma}\right)\right]^n = \left(1 - \frac{u^2}{2n} + \frac{c}{n^{3/2}}\right)^n$$

where c depends on u. Take logarithms to obtain, by Theorem 4b,

$$n\ln\left(1 - \frac{u^2}{2n} + \frac{c}{n^{3/2}}\right) = n\left(-\frac{u^2}{2n} + \frac{c}{n^{3/2}} + \theta\left|-\frac{u^2}{2n} + \frac{c}{n^{3/2}}\right|^2\right) \to -\frac{u^2}{2}$$

$$\text{as } n \to \infty$$

Thus

$$\left[M\left(\frac{u}{\sqrt{n}\,\sigma}\right)\right]^n \to e^{-u^2/2}$$

which is the desired result.*

REMARK. Convergence in distribution of $T_n = [S_n - E(S_n)]/\sigma(S_n)$, $S_n = \sum_{j=1}^{n} R_j$, $\sigma(S_n) =$ standard deviation of S_n, can be established under conditions much more general than those given in Theorem 2. For example, the finiteness of the third moment is not necessary in Theorem 2; neither is the assumption that the R_i have a density. We give two other sufficient conditions for normal convergence.

1. The R_i are uniformly bounded; that is, there is a constant k such that $|R_i| \le k$ for all i, and also Var $(S_n) \to \infty$.
   The requirement that Var $S_n \to \infty$ is necessary, for otherwise we could take R_1 to have an arbitrary distribution function, and $R_n = 0$ for $n \ge 2$. Then $S_n = R_1$,

$$T_n = \frac{S_n - ES_n}{\sigma(S_n)} = \frac{R_1 - .ER_1}{\sigma(R_1)}$$

Thus the functions F_{T_n} are the same for all n and hence in general cannot approach

$$F^*(x) = \int_{-\infty}^{x} \frac{1}{\sqrt{2\pi}} e^{-t^2/2}\,dt$$

which is the normal distribution function with mean 0 and variance 1.

2. (The Liapounov condition)

$$\frac{\sum_{k=1}^{n} E\,|R_k - ER_k|^{2+\delta}}{|\sigma(S_n)|^{2+\delta}} \to 0 \qquad \text{for some } \delta > 0$$

REMARK.   For each n, let $R_1, R_2, \ldots, R_n$ be independent random variables, where all R_i have the same distribution function, with finite mean m, finite variance σ^2, and also finite third moment. It can be shown that there is a positive number k such that $|F_{T_n}(x) - F^*(x)| \le k/\sqrt{n}$ for all x and all n. It follows that, for large n, S_n is approximately normal with mean nm and variance $n\sigma^2$, in the sense that for all x,

$$\left| F_{S_n}(x) - \int_{-\infty}^{x} \frac{1}{\sqrt{2\pi n}\ \sigma} e^{-(t-nm)^2/2n\sigma^2}\, dt \right| \le \frac{k}{\sqrt{n}}$$

For

$$F_{S_n}(x) = P\{S_n \le x\} = P\left\{\frac{S_n - nm}{\sqrt{n}\ \sigma} \le \frac{x - nm}{\sqrt{n}\ \sigma}\right\}$$

and this differs from $F^*((x - nm)/\sqrt{n}\ \sigma)$ by $\le k/\sqrt{n}$. But

$$F^*\left(\frac{x - nm}{\sqrt{n}\ \sigma}\right) = \int_{-\infty}^{(x-nm)/\sqrt{n}\ \sigma} \frac{1}{\sqrt{2\pi}} e^{-t^2/2}\, dt$$

$$= \left(\text{set } t = \frac{y - nm}{\sqrt{n}\ \sigma}\right)\frac{1}{\sqrt{2\pi n}\ \sigma} \int_{-\infty}^{x} e^{-(y-nm)^2/2n\sigma^2}\, dy$$

and the result follows.

In particular, let R be the number of successes in n Bernoulli trials; then (Example 1, Section 3.5) $R = R_1 + \cdots + R_n$, where the R_i are independent, and $P\{R_i = 1\} = p$, $P\{R_i = 0\} = 1 - p$. Thus, for large n, R is approximately normal with mean np and variance $np(1 - p)$, in the sense described above.

PROBLEMS

1. Show that $R_n \xrightarrow{P} R$ implies $R_n \xrightarrow{d} R$, as follows. Let F_n be the distribution function of R_n, and F the distribution function of R.
   (a) If $\epsilon > 0$, show that

$$P\{R_n \le x\} \le P\{|R_n - R| \ge \epsilon\} + P\{R \le x + \epsilon\}$$

and

$$P\{R \le x - \epsilon\} \le P\{|R_n - R| \ge \epsilon\} + P\{R_n \le x\}$$

Conclude that

$$F(x - \epsilon) - P\{|R_n - R| \geq \epsilon\} \leq F_n(x) \leq P\{|R_n - R| \geq \epsilon\} + F(x + \epsilon)$$

(b) If $R_n \xrightarrow{P} R$ and F is continuous at x, show that $F_n(x) \to F(x)$.

2. Give an example of a sequence $R_1, R_2, \ldots$ that converges in distribution to a random variable R, but does not converge in probability to R.

3. If R_n converges in distribution to a *constant* c, that is, $\lim_{n\to\infty} F_n(x) = 1$ for $x > c$, and $= 0$ for $x < c$, show that $R_n \xrightarrow{P} c$.

4. Let $R_1, R_2, \ldots$ be random variables such that $P\{R_n = e^n\} = 1/n$, $P\{R_n = 0\} = 1 - 1/n$. Show that $R_n \xrightarrow{P} 0$, but $E(R_n{}^k) \to \infty$ as $n \to \infty$, for any fixed $k > 0$.

5. Two candidates, A and B, are running for President and it is desired to predict the outcome of the election. Assume that n people are selected independently and at random and asked their preference. Suppose that the probability of selecting a voter who favors A in any particular observation is p. (p is fixed but unknown.) Let Q_n be the relative frequency of "A" voters in the sample; that is,

$$Q_n = \frac{\text{number of ``}A\text{'' voters in sample}}{\text{size of sample}}$$

(a) We wish to choose n large enough so that $P\{|Q_n - p| \leq .001\} \geq .99$ for all possible values of p. In other words, we wish to predict A's percentage of the vote to within .1%, with 99% confidence. Estimate the minimum value of n.

(b) Estimate the minimum value of n if we wish to predict A's percentage to within 1%, with 95% confidence. (Use the central limit theorem.)

6. (a) Show that the normal density function (with mean 0, variance 1) satisfies the inequality

$$\int_x^\infty \frac{1}{\sqrt{2\pi}} e^{-t^2/2}\, dt \leq \frac{1}{\sqrt{2\pi}\, x} e^{-x^2/2}, \qquad x > 0$$

HINT: show that

$$\frac{1}{\sqrt{2\pi}\, x} e^{-x^2/2} = \frac{1}{\sqrt{2\pi}} \int_x^\infty e^{-t^2/2}\left(1 + \frac{1}{t^2}\right) dt$$

by differentiating both sides.

(b) Show that

$$\int_x^\infty \frac{1}{\sqrt{2\pi}} e^{-t^2/2}\, dt \sim \frac{1}{\sqrt{2\pi}\, x} e^{-x^2/2}$$

in the sense that the ratio of the two sides approaches 1 as $x \to \infty$.

7. Consider a container holding $n = 10^6$ molecules. In the steady state it is reasonable that there be roughly as many molecules on the left side as on the right. Assume that the molecules are dropped independently and at random into the

container and that each molecule may fall with equal probability on the left or right side.

If R is the number of molecules on the right side of the container, we may invoke the central limit theorem to justify the physical assumption that for the purpose of calculating $P\{a \leq R \leq b\}$ we may regard R as normally distributed with mean $np = n/2$ and variance $np(1 - p) = n/4$.

Use Problem 6 to bound $P\{|R - n/2| > .005n\}$, the probability of a fluctuation about the mean of more than $\pm.5\%$ of the total number of molecules.

8. Let R be the number of successes in 10,000 Bernoulli trials, with probability of success .8 on a given trial. Use the central limit theorem to estimate $P\{7940 \leq R \leq 8080\}$.

6

Infinite Sequences of
Random Variables

6.1 INTRODUCTION

We have not yet encountered any situation in which it is necessary to consider an infinite collection of random variables, all defined on the same probability space. In the central limit theorem, for example, the basic underlying hypothesis is "For each n, let $R_1, \ldots, R_n$ be independent random variables." As n changes, the underlying probability space may change, but this is of no consequence, since a convergence in distribution statement is a statement about convergence of a sequence of real-valued functions on E^1. If $R_1, \ldots, R_n$ are independent, with distribution functions $F_1, \ldots, F_n$, and $T_n = (S_n - E(S_n))/\sigma(S_n)$, $S_n = R_1 + \cdots + R_n$, the distribution function of T_n is completely determined by the F_i, and the validity of a statement about convergence in distribution of T_n is also determined by the F_i, regardless of the construction of the underlying space.

However, there are occasions when it is necessary to consider an infinite number of random variables defined on the same probability space. For example, consider the following random experiment. We start at the origin on the real line, and flip a coin independently over and over again. If the result of the first toss is heads, we take one step to the right (i.e., from $x = 0$

to $x = 1$), and if the result is tails, we move one step to the left (to $x = -1$). We continue the process; if we are at $x = k$ after n trials, then at trial $n + 1$ we move to $x = k + 1$ if the $(n + 1)$th toss results in heads, or to $x = k - 1$ if it results in tails. We ask, for example, for the probability of eventually returning to the origin.

Now the position S_n after n steps is the sum $R_1 + \cdots + R_n$ of n independent random variables, where $P\{R_k = 1\} = p = $ probability of heads, $P\{R_k = -1\} = 1 - p$. We are looking for $P\{S_n = 0$ for some $n > 0\}$.

We must decide what probability space we are considering. If we are interested only in the first n trials, there is no problem. We simply have a sequence of n Bernoulli trials, and we have considered the assignment of probabilities in detail. However, the difficulty is that the event $\{S_n = 0$ for some $n > 0\}$ involves infinitely many trials. We must take $\Omega = E^\infty = $ all *infinite* sequences $(x_1, x_2, \ldots)$ of real numbers. (In this case we may restrict the x_i to be ± 1, but it is convenient to allow arbitrary x_i so that the discussion will apply to the general problem of assigning probabilities to events involving infinitely many random variables.)

We have the problem of specifying the sigma field $\mathscr{F}$ and the probability measure P. The physical description of the problem has determined all probabilities involving finitely many R_i; that is, we know $P\{(R_1, \ldots, R_n) \in B\}$ for each positive integer n and n-dimensional Borel set B. What we would like to conclude is that a reasonable specification of probabilities involving finitely many R_i determines the probability of events involving all the R_i. For example, consider $\{$all $R_i = 1\}$. This event may be expressed as

$$\bigcap_{n=1}^{\infty} \{R_1 = 1, \ldots, R_n = 1\}$$

The sets $\{R_1 = 1, \ldots, R_n = 1\}$ form a contracting sequence; hence

$$P\{\text{all } R_i = 1\} = \lim_{n \to \infty} P\{R_1 = 1, \ldots, R_n = 1\} = \lim_{n \to \infty} p^n = 0 \quad \text{if } p < 1$$

As another example,

$\{R_n = 1$ for infinitely many $n\}$

$$= \{\text{for every } n, \text{ there exists } k \geq n \text{ such that } R_k = 1\}$$

$$= \bigcap_{n=1}^{\infty} \bigcup_{k=n}^{\infty} \{R_k = 1\}$$

Thus

$$P\{R_n = 1 \text{ for infinitely many } n\} = \lim_{n \to \infty} P\left[\bigcup_{k=n}^{\infty} \{R_k = 1\}\right]$$

$$= \lim_{n \to \infty} \lim_{m \to \infty} P\left[\bigcup_{k=n}^{m} \{R_k = 1\}\right]$$

Thus again the probability is determined once we know the probabilities of all events involving finitely many R_i.

We now sketch the general situation. Let $\Omega = E^{\infty}$. A set of the form $\{(x_1, x_2, \ldots): (x_1, \ldots, x_n) \in B_n\}$, where $B_n \subset E^n$, is called a *cylinder* with base B_n, a *measurable cylinder* if B_n is a Borel subset of E^n.

Suppose that for each n we specify a probability measure P_n on the Borel subsets of E^n; $P_n(B_n)$ is to be interpreted as $P\{(R_1, \ldots, R_n) \in B_n\}$, where $R_i(x_1, x_2, \ldots) = x_i$.

Suppose, for example, that we have specified P_5. Then P_k, $k < 5$, is determined. In particular, in the discrete case we have

$$P\{(R_1, R_2, R_3) \in B_3\}$$
$$= \sum_{\substack{(x_1, x_2, x_3) \in B_3, \\ -\infty < x_4 < \infty, \\ -\infty < x_5 < \infty}} P\{R_1 = x_1, R_2 = x_2, R_3 = x_3, R_4 = x_4, R_5 = x_5\}$$

and in the absolutely continuous case we have

$$P\{(R_1, R_2, R_3) \in B_3\} = \int \cdots \int_{\substack{(x_1, x_2, x_3) \in B_3 \\ -\infty < x_4 < \infty, \\ -\infty < x_5 < \infty}} f(x_1, x_2, x_3, x_4, x_5) \, dx_1 \, dx_2 \, dx_3 \, dx_4 \, dx_5$$

In general, once P_n is given, P_k, $k < n$, is determined. But we have specified P_k, $k < n$, at the beginning; if our assignment of probabilities is to make sense, the original P_k must agree with that derived from P_n, $n > k$.

If, for all $n = 1, 2, \ldots$ and all $k < n$, the probability measure P_k originally specified agrees with that derived from P_n, $n > k$, we say that the probability measures are *consistent*. Under the consistency hypothesis, the *Kolmogorov extension theorem* states that there is a unique probability measure P on $\mathscr{F} = $ the smallest sigma field of subsets of Ω containing the measurable cylinders, such that

$$P \text{ (the measurable cylinder with base } B_n) = P_n(B_n)$$

for all $n = 1, 2, \ldots$ and all Borel subsets B_n of E^n.

In other words, a consistent specification of finite dimensional probabilities determines the probabilities of events involving all the R_i.

We now consider the case in which the R_i are discrete. Here we determine probabilities involving $(R_1, \ldots, R_n)$ by prescribing the joint probability function

$$p_{12\cdots n}(x_1, \ldots, x_n) = P\{R_1 = x_1, \ldots, R_n = x_n\}$$

We may then derive the joint probability function of $R_1, \ldots, R_k$:

$$P\{R_1 = x_1, \ldots, R_k = x_k\} = \sum_{x_{k+1}, \ldots, x_n} P\{R_1 = x_1, \ldots, R_n = x_n\} \quad (6.1.1)$$

If this coincides with the given $p_{12\cdots k}$ (for all n and all $k < n$) we say that the system of joint probability functions is consistent. If we sum (6.1.1) over $(x_1, \ldots, x_k) \in B_k$, we find that consistency of the joint probability functions is equivalent to consistency of the associated probability measures P_n. Thus in the discrete case the essential point is the consistency of the joint probability functions. In particular, suppose that we require that for each n, $R_1, \ldots, R_n$ be independent, with R_i having a specified probability function p_i. Then (6.1.1) becomes

$$P\{R_1 = x_1, \ldots, R_k = x_k\} = \sum_{x_{k+1}, \ldots, x_n} p_1(x_1) \cdots p_n(x_n) = p_1(x_1) \cdots p_k(x_k)$$

and thus the joint probability functions are consistent. The point we are making here is that there is a unique probability measure on $\mathscr{F}$ such that the random variables $R_1, R_2, \ldots$ are independent, each with a specified probability function. In other words, the statement "Let $R_1, R_2, \ldots$ be independent random variables, where R_i is discrete and has probability function p_i," is unambiguous.

In the absolutely continuous case, probabilities involving $R_1, \ldots, R_n$ are determined by the joint density function $f_{12\cdots n}$. The joint density of $R_1, \ldots, R_k$ is then given by

$$g(x_1, \ldots, x_k) = \int_{-\infty}^{\infty} \cdots \int_{-\infty}^{\infty} f_{12\cdots n}(x_1, \ldots, x_n) \, dx_{k+1} \cdots dx_n \quad (6.1.2)$$

If this coincides with the given $f_{12\cdots k}$ ($n, k = 1, 2, \ldots, k < n$), we say that the system of joint densities is consistent. By integrating (6.1.2) over Borel sets $B_k \subset E^k$, we find that consistency of joint density functions is equivalent to consistency of the associated probability measures P_n. In particular, if we require that for each n, $R_1, \ldots, R_n$ be independent, with R_i having a specified density function f_i, then (6.1.2) becomes

$$g(x_1, \ldots, x_k) = \int_{-\infty}^{\infty} \cdots \int_{-\infty}^{\infty} f_1(x_1) \cdots f_n(x_n) \, dx_{k+1} \cdots dx_n$$
$$= f_1(x_1) \cdots f_k(x_k)$$

Therefore the joint density functions are consistent, and the statement "Let $R_1, R_2, \ldots$ be independent random variables, where R_i is absolutely continuous with density function f_i," is unambiguous.

PROBLEMS

1. By working directly with the probability measures P_n, give an argument shorter than the one above to show that the statement "Let $R_1, R_2, \ldots$ be independent random variables, where R_i is absolutely continuous with density f_i," is unambiguous.

2. If $R_1, R_2, \ldots$ are independent, with $P\{R_i = 1\} = p$, $P\{R_i = -1\} = 1 - p$, as at the beginning of the section, find $P\{R_n = 1$ for infinitely many $n\}$; also, find $P\{\lim_{n \to \infty} R_n = 1\}$. (Assume $0 < p < 1$.)

6.2 THE GAMBLER'S RUIN PROBLEM

Suppose that a gambler starts with a capital of x dollars and plays a sequence of games against an opponent with $b - x$ dollars. At each trial he wins a dollar with probability p, and loses a dollar with probability $q = 1 - p$. (The trials are assumed independent, with $0 < p < 1$, $0 < x < b$.) The process continues until the gambler's capital reaches 0 (ruin) or b (victory). We wish to find $h(x)$, the probability of eventual ruin when the initial capital is x.

Let $A = \{$eventual ruin$\}$, $B_1 = \{$win on trial 1$\}$, $B_2 = \{$lose on trial 1$\}$. By the theorem of total probability,

$$P(A) = P(B_1)P(A \mid B_1) + P(B_2)P(A \mid B_2)$$

We are given that $P(B_1) = p$, $P(B_2) = q$; $P(A)$ is the unknown probability $h(x)$. Now if the gambler wins at the first trial, his capital is then $x + 1$; thus $P(A \mid B_1)$ is the probability of eventual ruin, starting at $x + 1$, that is, $h(x + 1)$. Similarly, $P(A \mid B_2) = h(x - 1)$. Thus

$$h(x) = ph(x + 1) + qh(x - 1), x = 1, 2, \ldots, b - 1 \qquad (6.2.1)$$

[The intuition behind the argument leading to (6.2.1) is compelling; however, a formal proof involves concepts not treated in this book, and will be omitted.]

We have not yet found $h(x)$, but we know that it satisfies (6.2.1), a linear homogeneous *difference equation* with constant coefficients. The boundary conditions are $h(0) = 1$, $h(b) = 0$. To see this, note that if $x = 1$, then with probability p the gambler wins on trial 1; his probability of eventual ruin is then $h(2)$. With probability q he loses on trial 1, and then he is already ruined. In other words, if (6.2.1) is to be satisfied at $x = 1$, we must have $h(0) = 1$. Similarly, examination of (6.2.1) at $x = b - 1$ shows that $h(b) = 0$.

The difference equation may be put into the standard form

$$ph(x + 2) - h(x + 1) + qh(x) = 0,$$

$$x = 0, 1, \ldots, b - 2, h(0) = 1, h(b) = 0$$

It is solved in the same way as the analogous differential equation

$$p\frac{d^2y}{dx^2} - \frac{dy}{dx} + qy = 0$$

We assume an exponential solution; for convenience, we take $h(x) = \lambda^x$ ($= e^{x \ln \lambda}$). Then $p\lambda^{x+2} - \lambda^{x+1} + q\lambda^x = \lambda^x(p\lambda^2 - \lambda + q) = 0$. Since λ^x is never 0, the only allowable λ's are the roots of the *characteristic equation* $p\lambda^2 - \lambda + q = 0$, namely,

$$\lambda = \frac{1}{2p}(1 \pm \sqrt{1 - 4pq})$$

Now

$$(p - q)^2 = p^2 - 2pq + q^2 = p^2 + 2pq + q^2 - 4pq = (p + q)^2 - 4pq$$
$$= 1 - 4pq \tag{6.2.2}$$

Hence

$$\lambda = \frac{1}{2p}(1 \pm |p - q|)$$

The two roots are

$$\lambda_1 = \frac{1 + p - q}{2p} = 1, \qquad \lambda_2 = \frac{1 + q - p}{2p} = \frac{q}{p}$$

CASE 1. $p \neq q$. Then λ_1 and λ_2 are distinct; hence

$$h(x) = A\lambda_1^x + C\lambda_2^x = A + C\left(\frac{q}{p}\right)^x$$
$$h(0) = A + C = 1$$
$$h(b) = A + C\left(\frac{q}{p}\right)^b = 0$$

Solving, we obtain

$$A = \frac{-(q/p)^b}{1 - (q/p)^b} \qquad C = \frac{1}{1 - (q/p)^b}$$

Therefore

$$h(x) = \frac{(q/p)^x - (q/p)^b}{1 - (q/p)^b} \tag{6.2.3}$$

CASE 2. $p = q = 1/2$. Then $\lambda_1 = \lambda_2 = \lambda = 1$, a repeated root. In such a case (just as in the analogous differential equation) we may construct two linearly independent solutions by taking λ^x and $x\lambda^x$; that is,

$$h(x) = A\lambda^x + Cx\lambda^x = A + Cx$$
$$h(0) = A = 1$$
$$h(b) = A + Cb = 0 \qquad \text{so } C = -\frac{1}{b}$$

Thus

$$h(x) = 1 - \frac{x}{b} = \frac{b - x}{b} \tag{6.2.4}$$

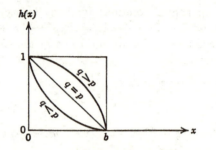

FIGURE 6.2.1   Probability of Eventual Ruin.

so that the probability of eventual ruin is the ratio of the adversary's capital to the total capital. A sketch of $h(x)$ in the various cases is shown in Figure 6.2.1.

Similarly, let $g(x)$ be the probability of eventual victory, starting with a capital of x dollars. We cannot conclude immediately that $g(x) = 1 - h(x)$, since there is the possibility that the game will never end; that is, the gambler's fortune might oscillate forever within the limits $x = 1$ and $x = b - 1$. However, we can show that this event has probability 0, as follows. By the same reasoning as that leading to (6.2.1), we obtain

$$g(x) = pg(x + 1) + qg(x - 1) \qquad (6.2.5)$$

The boundary conditions are now $g(0) = 0$, $g(b) = 1$. But we may verify that $g(x) = 1 - h(x)$ satisfies (6.2.5) with the given boundary conditions; since the solution is unique (see Problem 1), we must have $g(x) = 1 - h(x)$; that is, *the game ends with probability 1.*

We should mention, at least in passing, the probability space we are working in. We take $\Omega = E^\infty$, $\mathscr{F}$ = the smallest sigma field containing the measurable cylinders, $R_i(x_1, x_2, \ldots) = x_i$, $i = 1, 2, \ldots$, P the probability measure determined by the requirement that $R_1, R_2, \ldots$ be independent, with $P\{R_i = 1\} = p$, $P\{R_i = -1\} = q$. Thus R_i is the gambler's net gain on trial i, and $x + \sum_{i=1}^{n} R_i$ is his capital after n trials. We are looking for $h(x) = P\{$for some n, $x + \sum_{i=1}^{n} R_i = 0$, $0 < x + \sum_{i=1}^{k} R_i < b$, $k = 1, 2, \ldots, n - 1\}$.

A sequence of random variables of the form $x + \sum_{i=1}^{n} R_i, n = 1, 2, \ldots$, where the R_i are independent and have the same distribution function (or, more generally, $R_0 + \sum_{i=1}^{n} R_i$, $n = 1, 2, \ldots$, where $R_0, R_1, R_2, \ldots$ are independent and $R_1, R_2, \ldots$ have the same distribution function), is called a *random walk*, a *simple* random walk if $R_i (i \geq 1)$ takes on only the values ± 1. The present case may be regarded as a simple random walk with *absorbing barriers* at 0 and b, since when the gambler's fortune reaches either of these figures, the game ends, and we may as well regard his capital as forever frozen.

We wish to investigate the effect of removing one or both of the barriers. Let $h_b(x)$ be the probability of eventual ruin starting from x, when the total capital is b. It is reasonable to expect that $\lim_{b \to \infty} h_b(x)$ should be the probability of eventual ruin when the gambler has the misfortune of playing against an adversary with infinite capital. Let us verify this.

Consider the simple random walk with only the barrier at $x = 0$ present; that is, the adversary has infinite capital. If the gambler starts at $x > 0$, his probability $h^*(x)$ of eventual ruin is

$$P(A) = P\{\text{for some positive integer } b, 0 \text{ is reached before } b\}$$

Let $A_b = \{0 \text{ is reached before } b\}$. The sets A_b, $b = 1, 2, \ldots$ form an expanding sequence whose union is A; hence

$$P(A) = \lim_{b \to \infty} P(A_b)$$

But

$$P(A_b) = h_b(x)$$

Consequently

$$h^*(x) = \lim_{b \to \infty} h_b(x)$$

$$= 1 \quad \text{if } q \geq p$$

$$= \left(\frac{q}{p}\right)^x \quad \text{if } q < p, \text{ by } (6.2.3) \text{ and } (6.2.4) \ (x = 1, 2, \ldots) \quad (6.2.6)$$

Thus, in fact, $\lim_{b \to \infty} h_b(x)$ is the probability $h^*(x)$ of eventual ruin when the adversary has infinite capital; $1 - h^*(x)$ is the probability that the origin will never be reached, that is, that the game will never end. If $q < p$, then $h^*(x) < 1$, and so there is a positive probability that the game will go on forever.

Finally, consider a simple random walk starting at 0, with no barriers. Let r be the probability of eventually returning to 0. Now if $R_1 = 1$ (a win on trial 1), there will be a return to 0 with probability $h^*(1)$. If $R_1 = -1$ (a loss on trial 1), the probability of eventually reaching 0 is found by evaluating $h^*(1)$ with q and p interchanged, that is, 1 for $q \leq p$, and p/q if $p < q$.

Thus, if $q \leq p$,

$$r = p\left(\frac{q}{p}\right) + q(1) = 2q$$

If $p < q$,

$$r = p(1) + q\left(\frac{p}{q}\right) = 2p$$

One expression covers both of these cases, namely,

$$r = 1 - |p - q|$$
$$= 1 \quad \text{if } p = q = \tfrac{1}{2}$$
$$< 1 \quad \text{if } p \neq q \tag{6.2.7}$$

PROBLEMS

1. Show that the difference equation arising from the gambler's ruin problem has a unique solution subject to given boundary conditions at $x = 0$ and $x = b$.

2. In the gambler's ruin problem, let $D(x)$ be the average duration of the game when the initial capital is x. Show that $D(x) = p(1 + D(x + 1)) + q(1 + D(x - 1))$, $x = 1, 2, \ldots, b - 1$ [the boundary conditions are $D(0) = D(b) = 0$].

3. Show that the solution to the difference equation of Problem 2 is

$$D(x) = \frac{x}{q - p} - \frac{(b/(q - p))(1 - (q/p)^x)}{1 - (q/p)^b} \quad \text{if } p \neq q$$
$$= x(b - x) \quad \text{if } p = q \doteq 1/2$$

[$D(x)$ can be shown to be finite, so that the usual method of solution applies; see Problem 4, Section 7.4.]

REMARK. If $D_b(x)$ is the average duration of the game when the total capital is b, then $\lim_{b \to \infty} D_b(x)$ ($= \infty$ if $p \geq q$, $= x/q - p$ if $p < q$) can be interpreted as the average length of time required to reach 0 when the adversary has infinite capital.

4. In a simple random walk starting at 0 (with no barriers), show that the average length of time required to return to the origin is infinite. (Corollary: A couple decides to have children until the number of boys equals the number of girls. The average number of children is infinite.)

5. Consider the simple random walk starting at 0. If $b > 0$, find the probability that $x = b$ will eventually be reached.

6.3 COMBINATORIAL APPROACH TO THE RANDOM WALK; THE REFLECTION PRINCIPLE

In this section we obtain, by combinatorial methods, some explicit results connected with the simple random walk. We assume that the walk starts at 0, with no barriers; thus the position at time n is $S_n = \sum_{k=1}^{n} R_k$, where $R_1, R_2, \ldots$ are independent random variables with $P\{R_k = 1\} = p$,

$P\{R_k = -1\} = q = 1 - p$. We may regard the R_k as the results of an infinite sequence of Bernoulli trials; we call an occurrence of $R_k = 1$ a "success," and that of $R_k = -1$ a "failure."

Suppose that among the first n trials there are exactly a successes and b failures $(a + b = n)$; say $a > b$. We ask for the (conditional) probability that the process will always be positive at times $1, 2, \ldots, n$, that is,

$$P\{S_1 > 0, S_2 > 0, \ldots, S_n > 0 \mid S_n = a - b\} \qquad (6.3.1)$$

(Notice that the only way that S_n can equal $a - b$ is for there to be a successes and b failures in the first n trials; for if x is the number of successes and y the number of failures, then $x + y = n = a + b$, $x - y = a - b$; hence $x = a$, $y = b$. Thus $\{S_n = a - b\} = \{a$ successes, b failures in the first n trials$\}$.)

Now (6.3.1) may be written as

$$\frac{P\{S_1 > 0, \ldots, S_n > 0, S_n = a - b\}}{P\{S_n = a - b\}} \qquad (6.3.2)$$

A favorable outcome in the numerator corresponds to a path from $(0, 0)$ to $(n, a - b)$ that always lies above the axis,† and a favorable outcome in the denominator to an arbitrary path from $(0, 0)$ to $(n, a - b)$ (see Figure 6.3.1). Thus (6.3.2) becomes

$$\frac{p^a q^b [\text{the number of paths from } (0, 0) \text{ to } (n, a - b) \text{ that are always above } 0]}{p^a q^b [\text{the total number of paths from } (0, 0) \text{ to } (n, a - b)]}$$

A path from $(0, 0)$ to $(n, a - b)$ is determined by selecting a positions out of n for the successes to occur; the total number of paths is $\binom{n}{a} = \binom{a+b}{a}$. To count the number of paths lying above 0, we reason as follows (see Figure 6.3.2).

Let A and B be points above the axis. Given any path from A to B that touches or crosses the axis, reflect the segment between A and the first zero point T, as shown. We get a path from A' to B, where A' is the reflection of A. Conversely, given *any* path from A' to B, the path must reach the axis at some point T. Reflecting the segment from A' to T, we obtain a path from A to B that touches or crosses the axis. The correspondence thus established is one-to-one; hence

the number of paths from A to B that touch or cross the axis

$$= \textit{the total number of paths from A' to B}$$

---

† Terminology: For the purpose of determining whether or not a path lies above the axis (or touches it, crosses it, etc.), the end points are not included in the path.

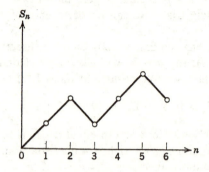

FIGURE 6.3.1   A Path in the Random Walk.

$$n = 6$$
$$a = 4, \qquad b = 2$$
$$P\{R_1 = R_2 = 1, R_3 = -1, R_4 = R_5 = 1, R_6 = -1\} = p^4 q^2;$$
this is one contribution to
$$P\{S_1 > 0, \ldots, S_5 > 0, S_6 = 2\}$$

This is called the *reflection principle*. Now
the number of paths from $(0, 0)$ to $(n, a - b)$ lying entirely above the axis

= the number from $(1, 1)$ to $(n, a - b)$ that *neither touch nor cross* the axis (since R_1 must be $+1$ in this case)

= the total number from $(1, 1)$ to $(n, a - b)$ — the number from $(1, 1)$ to $(n, a - b)$ that either touch or cross the axis

= the total number from $(1, 1)$ to $(n, a - b)$ — the total number from $(1, -1)$ to $(n, a - b)$ (by the reflection principle)

$$= \binom{n - 1}{a - 1} - \binom{n - 1}{a}$$

[Notice that in a path from $(1, 1)$ to $(n, a - b)$ there are x successes and y failures, where $x + y = n - 1 = a + b - 1$, $x - y = a - b - 1$, so

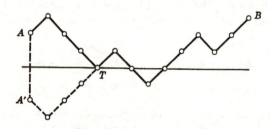

FIGURE 6.3.2   Reflection Principle.

$x = a - 1$, $y = b$. Similarly, a path from $(1, -1)$ to $(n, a - b)$ must have a successes and $b - 1$ failures.]

$$= \frac{(n-1)!}{(a-1)!\, b!} - \frac{(n-1)!}{a!\,(b-1)!} = \frac{n!}{a!\, b!}\left(\frac{a}{n} - \frac{b}{n}\right)$$

Thus

if $a + b = n$, the number of paths from $(0, 0)$ to $(n, a - b)$

$$\text{lying entirely above the axis} = \left(\frac{a-b}{n}\right)\binom{n}{a} \qquad (6.3.3)$$

Therefore

$$P\{S_1 > 0, \ldots, S_n > 0 \mid S_n = a - b\} = \frac{a-b}{n} = \frac{a-b}{a+b} \qquad (6.3.4)$$

REMARK.   This problem is equivalent to the *ballot problem:* In an election with two candidates, candidate 1 receives a votes and candidate 2 receives b votes, with $a > b$, $a + b = n$. The ballots are shuffled and counted one by one. The probability that candidate 1 will lead throughout the balloting is $(a - b)/(a + b)$. [Each possible sequence of ballots corresponds to a path from $(0, 0)$ to $(n, a - b)$; a sequence in which candidate 1 is always ahead corresponds to a path from $(0, 0)$ to $(n, a - b)$ that is always above the axis.]

We now compute

$$h_j = P\{\text{the first return to 0 occurs at time } j\}$$

Since h_j must be 0 for j odd, we may as well set $j = 2n$. Now

$$h_{2n} = P\{S_1 \neq 0, \ldots, S_{2n-1} \neq 0, S_{2n} = 0\}$$

and thus h_{2n} is the number of paths from $(0, 0)$ to $(2n, 0)$ lying above the axis, times 2 (to take into account paths lying below the axis), times $p^n q^n$, the probability of each path (see Figure 6.3.3).

The number of paths from $(0, 0)$ to $(2n, 0)$ lying above the axis is the number from $(0, 0)$ to $(2n - 1, 1)$ lying above the axis, which, by (6.3.3), is $\binom{2n-1}{a}(a-b)/(2n-1)$ (where $a + b = 2n - 1$, $a - b = 1$, hence $a = n$,

FIGURE 6.3.3   A First Return to 0 at Time $2n$.

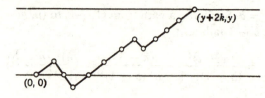

FIGURE 6.3.4   Computation of First Passage Times.

$b = n - 1$), that is,

$$\binom{2n-1}{n}\frac{1}{2n-1} = \frac{(2n-2)!}{n!\,(n-1)!} = \frac{1}{n}\binom{2n-2}{n-1}$$

Thus

$$h_{2n} = \frac{2}{n}\binom{2n-2}{n-1}(pq)^n \tag{6.3.5}$$

We now compute probabilities of *first passage times*, that is, $P\{$the first passage through $y > 0$ takes place at time $r\}$. The only possible values of r are of the form $y + 2k$, $k = 0, 1, \ldots$; hence we are looking for

$$h_{y+2k}^{(y)} = P\{\text{the first passage through } y > 0 \text{ occurs at time } y + 2k\}$$

To do the computation in an effortless manner, see Figure 6.3.4. If we look at the path of Figure 6.3.4 backward, it always lies below y and travels a vertical distance y in time $y + 2k$. Thus the number of favorable paths is the number of paths from $(0, 0)$ to $(y + 2k, y)$ that lie above the axis; that is, by (6.3.3),

$$\binom{y+2k}{a}\frac{a-b}{y+2k} \qquad \text{where } a + b = y + 2k,\ a - b = y$$

Thus $a = y + k$, $b = k$. Consequently

$$h_{y+2k}^{(y)} = \frac{y}{y+2k}\binom{y+2k}{k}p^{y+k}q^{k} \tag{6.3.6}$$

PROBLEMS

The first five problems refer to the simple random walk starting at 0, with no barriers.

1. Show that $P\{S_1 \geq 0, S_2 \geq 0, \ldots, S_{2n-1} \geq 0, S_{2n} = 0\} = u_{2n}/(n+1)$, where $u_{2n} = P\{S_{2n} = 0\}$ is the probability of n successes (and n failures) in $2n$ Bernoulli trials, that is, $\binom{2n}{n}(pq)^n$.

2. Let $p = q = 1/2$.
   (a) Show that $h_{2n} = u_{2n-2}/2n$, where $u_{2n} = \binom{2n}{n}(1/2)^{2n}$.
   (b) Show that $u_{2n}/u_{2n-2} = 1 - 1/2n$; hence $h_{2n} = u_{2n-2} - u_{2n}$.

3. If $p = q = 1/2$, show that $P\{S_1 \neq 0, \ldots, S_{2n} \neq 0\}$, the probability of no return to the origin in the first $2n$ steps, is $u_{2n} = \binom{2n}{n}2^{-2n}$. Show also that $P\{S_1 \neq 0, \ldots, S_{2n-1} \neq 0\} = h_{2n} + u_{2n}$.

4. If $p = q = 1/2$, show that $P\{S_1 \geq 0, \ldots, S_{2n} \geq 0\} = \binom{2n}{n}2^{-2n}$.

5. If $p = q = 1/2$, show that the average length of time required to return to the origin is infinite, by using Stirling's formula to find the asymptotic expression for h_{2n}, and then showing that $\sum_{n=1}^{\infty} nh_{2n} = \infty$.

6. Two players each toss an unbiased coin independently n times. Show that the probability that each player will have the same number of heads after n tosses is $(1/2)^{2n} \sum_{k=0}^{n} \binom{n}{k}^2$.

7. By looking at Problem 6 in a slightly different way, show that $\sum_{k=0}^{n} \binom{n}{k}^2 = \binom{2n}{n}$.

8. A spider and a fly are situated at the corners of an n by n grid, as shown in Figure P.6.3.8. The spider walks only north or east, the fly only south or west;

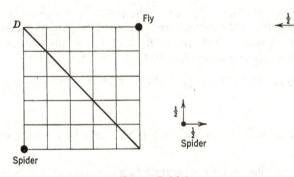

FIGURE P.6.3.8

they take their steps simultaneously, to an adjacent vertex of the grid.
   (a) Show that if they meet, the point of contact must be on the diagonal D.
   (b) Show that if the successive steps are independent, and equally likely to go in each of the two possible directions, the probability that they will meet is $\binom{2n}{n}(1/2)^{2n}$.

6.4 GENERATING FUNCTIONS

Let $\{a_n, n \geq 0\}$ be a bounded sequence of real numbers. The *generating function* of the sequence is defined by

$$A(z) = \sum_{n=0}^{\infty} a_n z^n, \quad z \text{ complex}$$

The series converges at least for $|z| < 1$. If R is a discrete random variable taking on only nonnegative integer values, and $P\{R = n\} = a_n$, $n = 0$, $1, \ldots$, then $A(z)$ is called the *generating function of R*. Note that $A(z) = \sum_{n=0}^{\infty} z^n P\{R = n\} = E(z^R)$, the characteristic function of R with z replacing e^{-iu}.

We have seen that the characteristic function of a sum of independent random variables is the product of the characteristic functions. An analogous result holds for generating functions.

Theorem 1. *Let $\{a_n\}$ and $\{b_n\}$ be bounded sequences of real numbers. Let $\{c_n\}$ be the convolution of $\{a_n\}$ and $\{b_n\}$, defined by*

$$c_n = \sum_{k=0}^{n} a_k b_{n-k} \left(= \sum_{j=0}^{n} b_j a_{n-j} \right)$$

Then $C(z) = \sum_{n=0}^{\infty} c_n z^n$ is convergent at least for $|z| < 1$, and

$$C(z) = A(z)B(z)$$

PROOF. Suppose first that $a_n = P\{R_1 = n\}$, $b_n = P\{R_2 = n\}$, where R_1 and R_2 are independent nonnegative integer-valued random variables. Then $c_n = P\{R_1 + R_2 = n\}$, since $\{R_1 + R_2 = n\}$ is the disjoint union of the events $\{R_1 = k, R_2 = n - k\}$, $k = 0, 1, \ldots, n$. Thus

$$C(z) = E(z^{R_1+R_2}) = E(z^{R_1}z^{R_2}) = E(z^{R_1})E(z^{R_2}) = A(z)B(z)$$

In general,

$$\sum_{n=0}^{\infty} c_n z^n = \sum_{n=0}^{\infty}\sum_{k=0}^{n} a_k b_{n-k} z^n = \sum_{k=0}^{\infty}\left(\sum_{n=k}^{\infty} b_{n-k}z^{n-k}\right)a_k z^k = A(z)B(z)$$

We have seen that under appropriate conditions the moments of a random variable can be obtained from its characteristic function. Similar results hold for generating functions. Let $A(z)$ be the generating function of the random variable R; restrict z to be real and between 0 and 1. We show that

$$E(R) = A'(1) \tag{6.4.1}$$

where

$$A'(1) = \lim_{z \to 1} A'(z)$$

If $E(R)$ is finite, then the variance of R is given by

$$\text{Var } R = A''(1) + A'(1) - [A'(1)]^2 \tag{6.4.2}$$

To establish (6.4.1) and (6.4.2), notice that

$$A(z) = \sum_{n=0}^{\infty} a_n z^n, \qquad a_n = P\{R = n\}$$

Thus

Thus

$$A'(z) = \sum_{n=1}^{\infty} na_n z^{n-1}$$

Let $z \to 1$ to obtain

$$A'(1) = \sum_{n=1}^{\infty} na_n = E(R)$$

proving (6.4.1). Similarly,

$$A''(z) = \sum_{n=1}^{\infty} n(n-1)a_n z^{n-2}, \quad \text{so } A''(1) = E(R^2) - E(R)$$

Therefore

$$\text{Var } R = E(R^2) - [E(R)]^2 = A''(1) + A'(1) - [A'(1)]^2$$

which is (6.4.2).

Now consider the simple random walk starting at 0, with no barriers. Let $u_n = P\{S_n = 0\}$, h_n = the probability that the first return to 0 will occur at time $n = P\{S_1 \neq 0, \dots, S_{n-1} \neq 0, S_n = 0\}$. Let

$$U(z) = \sum_{n=0}^{\infty} u_n z^n, \quad H(z) = \sum_{n=0}^{\infty} h_n z^n$$

(For the remainder of this section, z is restricted to real values.)

If we are at the origin at time n, the first return to 0 must occur at some time $k, k = 1, 2, \dots, n$. If the first return to 0 occurs at time k, we must be at the origin after $n - k$ additional steps. Since the events {first return to 0 at time k}, $k = 1, 2, \dots, n$, are disjoint, we have

$$u_n = \sum_{k=1}^{n} h_k u_{n-k}, \quad n = 1, 2, \dots$$

Let us write this as

$$u_n = \sum_{k=0}^{n} h_k u_{n-k}, \quad n = 1, 2, \dots$$

This will be valid provided that we define $h_0 = 0$. Now $u_0 = 1$, since the walk starts at the origin, but $h_0 u_0 = 0$. Thus we may write

$$v_n = \sum_{k=0}^{n} h_k u_{n-k}, \quad n = 0, 1, \dots \tag{6.4.3}$$

where $v_n = u_n, n \geq 1$; $v_0 = 0 = u_0 - 1$.

Since $\{v_n\}$ is the convolution of $\{h_n\}$ and $\{u_n\}$, Theorem 1 yields

$$V(z) = H(z)U(z)$$

But

$$V(z) = \sum_{n=0}^{\infty} v_n z^n = \sum_{n=1}^{\infty} u_n z^n = U(z) - 1$$

Thus

$$U(z)(1 - H(z)) = 1 \qquad (6.4.4)$$

We may use (6.4.4) to find the h_n explicitly. For

$$U(z) = \sum_{n=0}^{\infty} u_n z^n = \sum_{n=0}^{\infty} u_{2n} z^{2n} = \sum_{n=0}^{\infty} \binom{2n}{n} (pq)^n z^{2n}$$

This can be put into a closed form, as follows. We claim that

$$\binom{2n}{n} = \binom{-1/2}{n}(-4)^n \qquad (6.4.5)$$

where, for any real number α, $\binom{\alpha}{n}$ denotes

$$\frac{\alpha(\alpha - 1) \cdots (\alpha - n + 1)}{n!}$$

To see this, write

$$\binom{-1/2}{n} = \frac{(-1/2)(-3/2) \cdots [-(2n - 1)/2]}{n!}$$

$$= (-1)^n \frac{1 \cdot 3 \cdot 5 \cdots (2n - 1)}{n! \, 2^n} \frac{n!}{n!}$$

$$= \frac{1 \cdot 3 \cdot 5 \cdots (2n - 1)}{n!} \frac{(2/2)(4/2)(6/2) \cdots (2n/2)}{n! \, 2^n} (-1)^n$$

$$= \frac{(2n)!}{n! \, n!} \frac{(-1)^n}{4^n}$$

proving (6.4.5). Thus

$$U(z) = \sum_{n=0}^{\infty} \binom{-1/2}{n}(-4pqz^2)^n = (1 - 4pqz^2)^{-1/2}$$

by the binomial theorem. By (6.4.4) we have

$$H(z) = 1 - \frac{1}{U(z)} = 1 - (1 - 4pqz^2)^{1/2} \qquad (6.4.6)$$

This may be expanded by the binomial theorem to obtain the h_n (see Problem 1); of course the results will agree with (6.3.5), obtained by the combinatorial approach of the preceding section. Notice that we must have the positive square root in (6.4.6), since $H(0) = h_0 = 0$.

Some useful information may be gathered without expanding $H(z)$. Observe that $H(1) = \sum_{n=0}^{\infty} h_n$ is the probability of eventual return to 0.

By (6.4.6),

$$H(1) = 1 - (1 - 4pq)^{1/2}$$
$$= 1 - |p - q| \qquad \text{by (6.2.2)}$$

This agrees with the result (6.2.7) obtained previously.

Now assume $p = q = 1/2$, so that there is a return to 0 with probability 1. We show that the average length of time required to return to the origin is infinite. For if T is the time of first return to 0, then

$$E(T) = \sum_{n=1}^{\infty} nP\{T = n\} = \sum_{n=1}^{\infty} nh_n = H'(1)$$

as in (6.4.1). By (6.4.6),

$$H'(z) = -\frac{d}{dz}(1 - z^2)^{1/2} = z(1 - z^2)^{-1/2} \to \infty \qquad \text{as } z \to 1$$

Thus $E(T) = \infty$, as asserted (see Problem 5, Section 6.3, for another approach).

PROBLEMS

1. Expand (6.4.6) by the binomial theorem to obtain the h_n.

2. Solve the difference equation $a_{n+1} - 3a_n = 4$ by taking the generating function of both sides to obtain

$$A(z) = \frac{4z}{(1 - z)(1 - 3z)} + \frac{a_0}{1 - 3z}$$

Expand in partial fractions and use a geometric series expansion to find a_n.

3. Let $A(z)$ be the generating function of the sequence $\{a_n\}$; assume that $\sum_{n=0}^{\infty} |a_n - a_{n-1}| < \infty$. Show that if $\lim_{n \to \infty} a_n$ exists, the limit is

$$\lim_{z \to 1} (1 - z)A(z)$$

4. If R is a random variable with generating function $A(z)$, find the generating function of $R + k$ and kR, where k is a nonnegative integer. If $F(n) = P\{R \le n\}$, find the generating function of $\{F(n)\}$.

5. Let $R_1, R_2, \ldots$ be independent random variables, with $P\{R_i = 1\} = p$, $P\{R_i = 0\} = q = 1 - p$, $i = 1, 2, \ldots$. Thus we have an infinite sequence of Bernoulli trials; $R_i = 1$ corresponds to a success on trial i, and $R_i = 0$ is a failure. (Assume $0 < p < 1$.) Let R be the number of trials required to obtain the first success.
   (a) Show that $P\{R = k\} = q^{k-1}p$, $k = 1, 2, \ldots$.
   (b) Use generalized characteristic functions to show that $E(R) = 1/p$, Var $R = (1 - p)/p^2$; check the result by calculating the generating function of R and using (6.4.1) and (6.4.2). R is said to have the *geometric* distribution.

6. With the R_i as in Problem 5, let N_r be the number of trials required to obtain the rth success ($r = 1, 2, \ldots$).

(a) Show that $P\{N_r = k\} = \binom{k-1}{r-1}p^r q^{k-r}$, $k = r, r+1, \ldots$

$$= \binom{-r}{k-r}p^r(-q)^{k-r}, \quad k = r, r+1, \ldots$$

where $\binom{-r}{j}$ is defined as $(-r)(-r-1)\cdots(-r-j+1)/j!$, $j = 1, 2, \ldots$, $\binom{-r}{0} = 1$.

(b) Let $T_1 =$ the number of trials required to obtain the first success, $T_2 =$ the number of trials following the first success up to and including the second success, $\ldots$, $T_r =$ the number of trials following the $(r-1)$st success up to and including the rth success (thus $N_r = T_1 + \cdots + T_r$). Show that the T_i are independent, each with the geometric distribution.

(c) Show that $E(N_r) = r/p$, $\mathrm{Var}\, N_r = r(1-p)/p^2$. Find the characteristic function and the generating function of N_r. N_r is said to have the *negative binomial* distribution.

7. With the R_i as in Problem 5, let R be the length of the run (of either successes or failures) started by the first trial. Find $P\{R = k\}$, $k = 1, 2, \ldots$, and $E(R)$.

8. In Problem 6, find the joint probability functions of N_1 and N_2; also find (in a relatively effortless manner) the correlation coefficient between N_1 and N_2.

6.5   THE POISSON RANDOM PROCESS

We now consider a mathematical model that fits a wide variety of physical phenomena. Let $T_1, T_2, \ldots$ by a sequence of independent random variables, where each T_i is absolutely continuous with density $f(x) = \lambda e^{-\lambda x}$, $x \geq 0$; $f(x) = 0$, $x < 0$ (λ is a fixed positive constant). Let $A_n = T_1 + \cdots + T_n$, $n = 1, 2, \ldots$. We may think of A_n as the arrival time of the nth customer at a serving counter, so that T_n is the waiting time between the arrival of the $(n-1)$st customer and the arrival of the nth customer. Equally well, A_n may be regarded as the time at which the nth call is made at a telephone exchange, the time at which the nth component fails on an assembly line, or the time at which the nth electron arrives at the anode of a vacuum tube.

If $t \geq 0$, let R_t be the number of customers that have arrived up to and including time t; that is, $R_t = n$ if $A_n \leq t < A_{n+1}$ ($n = 0, 1, \ldots$; define $A_0 = 0$). A sketch of $(R_t, t \geq 0)$ is given in Figure 6.5.1.

Thus we have a *family* of random variables R_t, $t \geq 0$ (not just a sequence, but instead a random variable defined for each nonnegative real number). A family of random variables R_t, where t ranges over an arbitrary set I, is called a *random process* or *stochastic process*. Note that if I is the set of positive integers, the random process becomes a sequence of random variables; if I is a finite set, we obtain a finite collection of random variables; and if I consists of only one element, we obtain a single random variable. Thus the concept of a random process includes all situations studied previously.

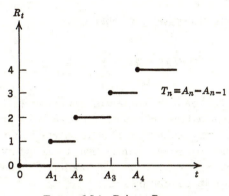

FIGURE 6.5.1   Poisson Process.

If the outcome of the experiment is ω, we may regard $(R_t(\omega),\ t \in I)$ as a real-valued function defined on I. In Figure 6.5.1 what is actually sketched is $R_t(\omega)$ versus t, $t \in I =$ the nonnegative reals, for a particular ω. Thus, roughly speaking, we have a "random function," that is, a function that depends on the outcome of a random experiment.

The particular process introduced above is called the *Poisson process* since, for each $t > 0$, R_t has the Poisson distribution with parameter λt. Let us verify this.

If k is a nonnegative integer,

$$P\{R_t \leq k\} = P\{\text{at most } k \text{ customers have arrived by time } t\}$$
$$= P\{(k + 1)\text{st customer arrives after time } t\}$$
$$= P\{T_1 + \cdots + T_{k+1} > t\}$$

But $A_{k+1} = T_1 + \cdots + T_{k+1}$ is the sum of $k + 1$ independent random variables, each with generalized characteristic function $\int_0^\infty \lambda e^{-\lambda x} e^{-sx}\, dx = \lambda/(s + \lambda)$, $\operatorname{Re} s > -\lambda$; hence A_{k+1} has the generalized characteristic function $[\lambda/(s + \lambda)]^{k+1}$, $\operatorname{Re} s > -\lambda$. Thus (see Table 5.1.1) the density of A_{k+1} is

$$f_{A_{k+1}}(x) = \frac{1}{k!}\,\lambda^{k+1} x^k e^{-\lambda x} u(x) \qquad (6.5.1)$$

where u is the unit step function. Thus

$$P\{R_t \leq k\} = P\{T_1 + \cdots + T_{k+1} > t\}$$
$$= \int_t^\infty \frac{1}{k!}\,\lambda^{k+1} x^k e^{-\lambda x}\, dx$$
$$= \int_t^\infty \frac{1}{k!}\,\lambda^{k+1} x^k\, d\left(\frac{-e^{-\lambda x}}{\lambda}\right)$$
$$= \frac{1}{k!}\,(\lambda t)^k e^{-\lambda t} + \int_t^\infty \frac{1}{(k-1)!}\,\lambda^k x^{k-1} e^{-\lambda x}\, dx \quad \text{(integrate by parts)}$$

Integrating by parts successively, we obtain

$$P\{R_t \leq k\} = \sum_{i=0}^{k} e^{-\lambda t}\frac{(\lambda t)^i}{i!}$$

Hence R_t has the Poisson distribution with parameter λt.

Now the mean of a Poisson random variable is its parameter, so that $E(R_t) = \lambda t$. Thus λ may be interpreted as the average number of customers arriving per second. We should expect that λ^{-1} is the average number of seconds per customer, that is, the average waiting time between customers. This may be verified by computing $E(T_i)$.

$$E(T_i) = \int_0^\infty \lambda x e^{-\lambda x}\, dx = \frac{1}{\lambda}$$

We now establish an important feature of the Poisson process. Intuitively, if we arrive at the serving counter at time t and the last customer to arrive came at time $t - h$, the distribution function of the length of time we must wait for the arrival of the next customer does not depend on h, and in fact coincides with the distribution function of T_1. Thus we are essentially starting from scratch at time t; the process does not remember that h seconds have elapsed between the arrival of the last customer and the present time.

If W_t is the waiting time from t to the arrival of the next customer, we wish to show that $P\{W_t \leq z\} = P\{T_1 \leq z\}$, $z \geq 0$. We have

$P\{W_t \leq z\} = P\{$for some $n = 1, 2, \ldots$, the nth customer arrives in

$(t, t + z]$ and the $(n + 1)$st customer arrives after time $t + z\}$

$$= P\left[\bigcup_{n=1}^{\infty} \{t < A_n \leq t + z < A_{n+1}\}\right] \qquad (6.5.2)$$

(see Figure 6.5.2). To justify (6.5.2), notice that if $t < A_n \leq t + z < A_{n+1}$ for some n, then $W_t \leq z$. Conversely, if $W_t \leq z$, then some customer arrives in $(t, t + z]$ and hence there will be a last customer to arrive in the interval. (If not, then $\sum_{n=1}^\infty T_n < \infty$; but this event has probability 0; see Problem 1.) Now $P\{t < A_n \leq t + z < A_{n+1}\} = P\{t < A_n \leq t + z, A_n + T_{n+1} > t + z\}$.

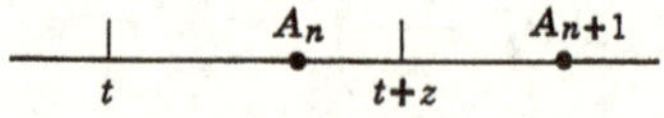

FIGURE 6.5.2

Since $A_n(= T_1 + \cdots + T_n)$ and T_{n+1} are independent, we obtain, by (6.5.1),

$$P\{t < A_n \leq t + z < A_{n+1}\} = \iint\limits_{\substack{t < x \leq t+z, \\ x+y > t+z}} \frac{1}{(n-1)!} \lambda^n x^{n-1} e^{-\lambda x} \lambda e^{-\lambda y} \, dx \, dy$$

$$= \frac{\lambda^n}{(n-1)!} \int_t^{t+z} x^{n-1} e^{-\lambda x} \int_{t+z-x}^\infty \lambda e^{-\lambda y} \, dy \, dx$$

$$= \frac{\lambda^n}{(n-1)!} \int_t^{t+z} x^{n-1} e^{-\lambda x} e^{-\lambda(t+z-x)} \, dx$$

$$= \frac{1}{n!} \lambda^n e^{-\lambda(t+z)} [(t+z)^n - t^n]$$

Since $\sum_{n=0}^\infty r^n/n! = e^r$, (6.5.2) yields

$$P\{W_t \leq z\} = e^{-\lambda(t+z)} [e^{\lambda(t+z)} - 1 - (e^{\lambda t} - 1)]$$

$$= 1 - e^{-\lambda z}$$

Thus W_t has the same distribution function as T_1.

Alternatively, we may write

$$P\{W_t \leq z\} = P\left[\bigcup_{n=0}^\infty \{A_n \leq t < A_{n+1} \leq t + z\} \right]$$

For if $A_n \leq t < A_{n+1} \leq t + z$ for some n, then $W_t \leq z$. Conversely, if $W_t \leq z$, then some customer arrives in $(t, t + z]$, and there must be a first customer to arrive in this interval, say customer $n + 1$. Thus $A_n \leq t < A_{n+1} \leq t + z$ for some $n \geq 0$. An argument very similar to the above shows that

$$P\{A_n \leq t < A_{n+1} \leq t + z\} = \frac{(\lambda t)^n}{n!} (e^{-\lambda t} - e^{-\lambda(t+z)})$$

and therefore $P\{W_t \leq z\} = 1 - e^{-\lambda z}$ as before. In this approach, we do not have the problem of showing that $P\left\{ \sum_{n=1}^\infty T_n < \infty \right\} = 0$.

To justify completely the statement that the process starts from scratch at time t, we may show that if $V_1, V_2, \ldots$ are the successive waiting times starting at t (so $V_1 = W_t$), then $V_1, V_2, \ldots$ are independent, and V_i and T_i have the same distribution function for all i. To see this, observe that

$$P\{V_1 \leq x_1, \ldots, V_k \leq x_k\}$$

$$= P\left[\bigcup_{n=0}^\infty \{A_n \leq t < A_{n+1} \leq t + x_1, T_{n+2} \leq x_2, \ldots, T_{n+k} \leq x_k\} \right]$$

For it is clear that the set on the right is a subset of the set on the left. Conversely, if $V_1 \leq x_1, \ldots, V_k \leq x_k$, then a customer arrives in $(t, t + x_1]$, and hence there is a first customer in this interval, say customer $n + 1$. Then $A_n \leq t < A_{n+1} \leq t + x_1$, and also $V_i = T_{n+i}$, $i = 2, \ldots, k$, as desired. Therefore

$$P\{V_1 \leq x_1, \ldots, V_k \leq x_k\} = \left[\sum_{n=0}^{\infty} P\{A_n \leq t < A_{n+1} \leq t + x_1\} \right]$$

$$\times \prod_{i=2}^{k} P\{T_{n+i} \leq x_i\}$$

$$= P\{W_t \leq x_1\} \prod_{i=2}^{k} P\{T_i \leq x_i\}$$

$$= \prod_{i=1}^{k} P\{T_i \leq x_i\}$$

Fix j and let $x_i \to \infty$, $i \neq j$, to conclude that $P\{V_j \leq x_j\} = P\{T_j \leq x_j\}$. Consequently

$$P\{V_1 \leq x_1, \ldots, V_k \leq x_k\} = \prod_{i=1}^{k} P\{V_i \leq x_i\}$$

and the result follows. In particular, the number of customers arriving in the interval $(t, t + \tau]$ has the Poisson distribution with parameter $\lambda \tau$.

The "memoryless" feature of the Poisson process is connected with a basic property of the exponential density, as the following intuitive argument shows. Suppose that we start counting at time t, and the last customer to arrive, say customer $n - 1$, came at time $t - h$. Then (see Figure 6.5.3) the probability that $W_t \leq z$ is the probability that $T_n \leq z + h$, given that we have already waited h seconds; that is, given that $T_n > h$. Thus

$$P\{W_t \leq z\} = P\{T_n \leq z + h \mid T_n > h\}$$

$$= \frac{P\{h < T_n \leq z + h\}}{P\{T_n > h\}}$$

$$= \frac{\int_h^{z+h} \lambda e^{-\lambda x} \, dx}{\int_h^{\infty} \lambda e^{-\lambda x} \, dx}$$

$$= \frac{e^{-\lambda h} - e^{-\lambda(z+h)}}{e^{-\lambda h}} = 1 - e^{-\lambda z} = P\{T_n \leq z\}$$

so that W_t and T_n have the same distribution function (for any n).

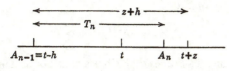

FIGURE 6.5.3

The key property of the exponential density used in this argument is

$$P\{T_n \leq z + h \mid T_n > h\} = P\{T_n \leq z\}$$

that is,

$$P\{T_n > z + h \mid T_n > h\} = P\{T_n > z\}$$

or (since $\{T_n > h, T_n > z + h\} = \{T_n > z + h\}$)

$$P\{T_n > z + h\} = P\{T_n > z\}P\{T_n > h\}$$

In fact, a positive random variable T having the property that

$$P\{T > z + h\} = P\{T > z\}P\{T > h\} \qquad \text{for all } z, h \geq 0$$

must have exponential density. For if $G(z) = P\{T > z\}$, then

$$G(z + h) = G(z)G(h)$$

Therefore

$$\frac{G(z + h) - G(z)}{h} = G(z)\left(\frac{G(h) - 1}{h}\right) = G(z)\left(\frac{G(h) - G(0)}{h}\right) \quad \text{since } G(0) = 1$$

Let $h \to 0$ to conclude that

$$G'(z) = G'(0)G(z)$$

This is a differential equation of the form

$$\frac{dy}{dx} + \lambda y = 0 \qquad (\lambda = -G'(0))$$

whose solution is $y = ce^{-\lambda x}$; that is,

$$G(z) = ce^{-\lambda z}$$

But $G(0) = c = 1$; hence the distribution function of T is

$$F_T(z) = 1 - G(z) = 1 - e^{-\lambda z}, \qquad z \geq 0$$

and thus the density of T is

$$f_T(z) = \lambda e^{-\lambda z}, \quad z \geq 0$$

as desired.

(The above argument assumes that T has a continuous density, but actually the result is true without this requirement.)

The memoryless feature of the Poisson process may be used to show that the process satisfies the *Markov property*:

If $0 \leq t_1 < t_2 < \cdots < t_{n+1}$ and $a_1, \ldots, a_{n+1}$ are nonnegative integers with $a_1 \leq \cdots \leq a_{n+1}$, then

$$P\{R_{t_{n+1}} = a_{n+1} \mid R_{t_1} = a_1, \ldots, R_{t_n} = a_n\} = P\{R_{t_{n+1}} = a_{n+1} \mid R_{t_n} = a_n\}$$

Thus the behavior of the process at the "future" time t_{n+1}, given the behavior at "past" times $t_1, \ldots, t_{n-1}$ and the "present" time t_n, depends only on the present state, or the number of customers at time t_n. For example,

$$P\{R_{t_4} = 15 \mid R_{t_1} = 0, R_{t_2} = 3, R_{t_3} = 8\} = P\{R_{t_4} = 15 \mid R_{t_3} = 8\}$$

$$= \text{the probability that exactly 7 customers will arrive between } t_3 \text{ and } t_4$$

$$= e^{-\lambda(t_4 - t_3)} \frac{[\lambda(t_4 - t_3)]^7}{7!}$$

This result is reasonable in view of the memoryless feature, but a formal proof becomes quite cumbersome and will be omitted.

We consider the Markov property in detail in the next chapter.

We close this section by describing a physical approach to the Poisson process. Suppose that we divide the interval $(t, t + \tau]$ into subintervals of length Δt, and assume that the subintervals are so small that the probability of the arrival of more than one customer in a given subinterval I is negligible. If the average number of customers arriving per second is λ, and R_I is the number of customers arriving in the subinterval I, then $E(R_I) = \lambda \Delta t$. But $E(R_I) = (0)P\{R_I = 0\} + (1)P\{R_I = 1\} = P\{R_I = 1\}$, so that with probability $\lambda \Delta t$ a customer will arrive in I, and with probability $1 - \lambda \Delta t$ no customer will arrive.

We assume that if I and J are disjoint subintervals, R_I and R_J are independent. Then we have a sequence of $n = \tau/\Delta t$ Bernoulli trials, with probability $p = \lambda \Delta t$ of success on a given trial, and with Δt very small. Thus we expect that the number $N(t, t + \tau]$ of customers arriving in $(t, t + \tau]$ should have the Poisson distribution with parameter $\lambda \tau$. Furthermore, if W_t is the waiting time from t to the arrival of the next customer, then $P\{W_t > x\} = P\{N(t, t + x] = 0\} = e^{-\lambda x}$, so that W_t has exponential density.

PROBLEMS

1. Show that $P\{\sum_{n=1}^{\infty} T_n < \infty\} = 0$.

2. Show that the probability that an even number of customers will arrive in the interval $(t, t + \tau]$ is $(1/2)(1 + e^{-2\lambda\tau})$ and the probability that an odd number of customers will arrive in this interval is $(1/2)(1 - e^{-2\lambda\tau})$.

3. (The random telegraph signal) Let $T_1, T_2, \ldots$ be independent, each with density $\lambda e^{-\lambda x} u(x)$. Define a random process R_t, $t \geq 0$, as follows.

   $R_0 = +1$ or -1 with equal probability (assume R_0 independent of the T_i)

   $$R_t = R_0, \quad 0 \leq t < T_1$$

   $$R_t = -R_0, \quad T_1 \leq t < T_1 + T_2$$

   $$R_t = R_0, \quad T_1 + T_2 \leq t < T_1 + T_2 + T_3$$

   .
   .
   .

   $$R_t = (-1)^n R_0, \quad T_1 + \cdots + T_n \leq t < T_1 + \cdots + T_{n+1}$$

   (see Figure P.6.5.3).

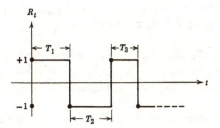

R_t

FIGURE P.6.5.3

   (a) Find the joint probability function of R_t and $R_{t+\tau}$ $(t, \tau \geq 0)$.
   (b) Find the *covariance function* of the process, defined by $K(t, \tau) = \text{Cov}(R_t, R_{t+\tau})$, $t, \tau \geq 0$.

*6.6   THE STRONG LAW OF LARGE NUMBERS

In this section we show that the arithmetic average $(R_1 + \cdots + R_n)/n$ of a sequence of independent observations of a random variable R converges with probability 1 to $E(R)$ (assumed finite). In other words, we have

$$\lim_{n \to \infty} \frac{R_1(\omega) + \cdots + R_n(\omega)}{n} = E(R)$$

for all ω, except possibly on a set of probability 0. We shall see that this is a stronger convergence statement than the weak law of large numbers.

Let $(\Omega, \mathscr{F}, P)$ be a probability space, fixed throughout the discussion. If $A_1, A_2, \ldots$ is a sequence of events, we define the *upper limit* or *limit superior* of the sequence as

$$\limsup_n A_n = \bigcap_{n=1}^{\infty} \bigcup_{k=n}^{\infty} A_k \qquad (6.6.1)$$

and the *lower limit* or *limit inferior* of the sequence as

$$\liminf_n A_n = \bigcup_{n=1}^{\infty} \bigcap_{k=n}^{\infty} A_k \qquad (6.6.2)$$

Theorem 1. $\limsup_n A_n = \{\omega \colon \omega \in A_n \text{ for infinitely many } n\}$, $\liminf_n A_n = \{\omega \colon \omega \in A_n \text{ eventually, i.e., for all but finitely many } n\}$.

Proof.   By (6.6.1), $\omega \in \limsup A_n$ iff, for every n, $\omega \in \bigcup_{k=n}^{\infty} A_k$; that is, for every n there is a $k \geq n$ such that $\omega \in A_k$, or $\omega \in A_n$ for infinitely many n. By (6.6.2), $\omega \in \liminf_n A_n$ iff, for some n, $\omega \in \bigcap_{k=n}^{\infty} A_k$; that is, for some n, $\omega \in A_k$ for all $k \geq n$, or $\omega \in A_n$ eventually.

Theorem 2.   *Let $R, R_1, R_2, \ldots$ be random variables on $(\Omega, \mathscr{F}, P)$. Denote by $\{R_n \to R\}$ the set $\{\omega \colon \lim_{n \to \infty} R_n(\omega) = R(\omega)\}$. Then $\{R_n \to R\} = \bigcap_{m=1}^{\infty} \liminf_n A_{nm}$, where $A_{nm} = \{\omega \colon |R_n(\omega) - R(\omega)| < 1/m\}$.*

Proof. $R_n(\omega) \to R(\omega)$ iff for every $m = 1, 2, \ldots$, $|R_n(\omega) - R(\omega)| < 1/m$ eventually, that is (Theorem 1), for every $m = 1, 2, \ldots$ $\omega \in \liminf_n A_{nm}$.

We say that the sequence $R_1, R_2, \ldots$ converges *almost surely* to R (notation: $R_n \xrightarrow{\text{a.s.}} R$) iff $P\{R_n \to R\} = 1$. The terminology "almost everywhere" is also used.

Theorem 3. $R_n \xrightarrow{\text{a.s.}} R$ *iff for every $\varepsilon > 0$, $P\{|R_k - R| \geq \varepsilon \text{ for at least one } k \geq n\} \to 0 \text{ as } n \to \infty$.*

Proof.   By Theorem 2, $P\{R_n \to R\} \leq P\{\liminf_n A_{nm}\}$ for every m; hence $R_n \xrightarrow{\text{a.s.}} R$ implies that $\liminf_n A_{nm}$ has probability 1 for every m. Conversely, if $P(\liminf_n A_{nm}) = 1$ for all m, then, by Theorem 2, $\{R_n \to R\}$

is a countable intersection of sets with probability 1 and hence has probability 1 (see Problem 1).

Now in (6.6.2), $\bigcap_{k=n}^{\infty} A_k$, $n = 1, 2, \ldots$ is an expanding sequence; hence $P(\liminf_n A_n) = \lim_{n \to \infty} P(\bigcap_{k=n}^{\infty} A_k)$. Thus

$$R_n \xrightarrow{\text{a.s.}} R \quad \text{iff} \quad P(\lim_n \inf A_{nm}) = 1 \quad \text{for all } m = 1, 2, \ldots$$

$$\text{iff} \quad \lim_{n \to \infty} P\left(\bigcap_{k=n}^{\infty} A_{km} \right) = 1 \quad \text{for all } m = 1, 2, \ldots$$

$$\text{iff} \quad P\left\{ |R_k - R| < \frac{1}{m} \quad \text{for all } k \geq n \right\} \to 1 \text{ as } n \to \infty$$
$$\text{for all } m = 1, 2, \ldots$$

$$\text{iff} \quad P\left\{ |R_k - R| \geq \frac{1}{m} \quad \text{for at least one } k \geq n \right\} \to 0 \text{ as } n \to \infty$$
$$\text{for all } m = 1, 2, \ldots$$

$$\text{iff} \quad P\{|R_k - R| \geq \varepsilon \} \quad \text{for at least one } k \geq n\} \to 0 \text{ as } n \to \infty$$
$$\text{for all } \varepsilon > 0$$

(see Problem 2).

COROLLARY. $R_n \xrightarrow{\text{a.s.}} R$ implies $R_n \xrightarrow{P} R$.

PROOF. $R_n \xrightarrow{P} R$ iff for every $\varepsilon > 0$, $P\{|R_n - R| \geq \varepsilon\} \to 0$ as $n \to \infty$ (see Section 5.4). Now $\{|R_k - R| \geq \varepsilon$ for at least one $k \geq n\} \supset \{|R_n - R| \geq \varepsilon\}$, so that $P\{|R_k - R| \geq \varepsilon$ for at least one $k \geq n\} \geq P\{|R_n - R| \geq \varepsilon\}$. The result now follows from Theorem 3.

For an example in which $R_n \xrightarrow{P} R$ but $R_n \xrightarrow{\text{a.s.}} R$, see Problem 3.

Theorem 4 (Borel-Cantelli Lemma). *If $A_1, A_2, \ldots$ are events in a given probability space, and $\sum_{n=1}^{\infty} P(A_n) < \infty$, then $P(\limsup_n A_n) = 0$; that is, the probability that A_n occurs for infinitely many n is 0.*

PROOF. By (6.6.1),

$$P\left(\limsup_n A_n \right) \leq P\left(\bigcup_{k=n}^{\infty} A_k \right) \quad \text{for every } n$$

$$\leq \sum_{k=n}^{\infty} P(A_k) \quad \text{by (1.3.10)}$$

$$= \sum_{k=1}^{\infty} P(A_k) - \sum_{k=1}^{n-1} P(A_k) \to 0 \quad \text{as } n \to \infty$$

since $\sum_n P(A_n) < \infty$.

Theorem 5. *If for every $\varepsilon > 0$, $\sum_{n=1}^{\infty} P\{|R_n - R| \geq \varepsilon\} < \infty$, then $R_n \xrightarrow{\text{a.s.}} R$.*

PROOF. By Theorem 4, $\limsup_n \{|R_n - R| \geq \varepsilon\}$ has probability 0 for every $\varepsilon > 0$; that is, the probability that $|R_n - R| \geq \varepsilon$ for infinitely many n is 0. Take complements to show that the probability that $|R_n - R| \geq \varepsilon$ for only finitely many n is 1; that is, with probability 1, $|R_n - R| < \varepsilon$ eventually. Since ε is arbitrary, we may set $\varepsilon = 1/m$, $m = 1, 2, \ldots$. Thus $P(\liminf_n A_{nm}) = 1$ for $m = 1, 2, \ldots$, and the result follows by Theorem 2.

Theorem 6. *If $\sum_{n=1}^{\infty} E[|R_n - R|^k] < \infty$ for some $k > 0$, then $R_n \xrightarrow{\text{a.s.}} R$.*

PROOF. $P\{|R_n - R| \geq \varepsilon\} \leq E[|R_n - R|^k]/\varepsilon^k$ by Chebyshev's inequality, and the result follows by Theorem 5.

Theorem 7 (Strong Law of Large Numbers). *Let $R_1, R_2, \ldots$ be independent random variables on a given probability space. Assume uniformly bounded fourth central moments; that is, for some positive real number M,*

$$E[|R_i - E(R_i)|^4] \leq M$$

for all i. Let $S_n = R_1 + \cdots + R_n$. Then

$$\frac{S_n - E(S_n)}{n} \xrightarrow{\text{a.s.}} 0$$

In particular, if $E(R_i) = m$ for all i, then $E(S_n) = nm$; hence

$$\frac{S_n}{n} \xrightarrow{\text{a.s.}} m$$

PROOF. Since $S_n - E(S_n) = \sum_{i=1}^{n} R_i'$, where

$$R_i' = R_i - E(R_i), \qquad E(R_i') = 0, \qquad E(|R_i'|^4) = E[|R_i - E(R_i)|^4] \leq M < \infty$$

we may assume without loss of generality that all $E(R_i)$ are 0. We show that $\sum_{n=1}^{\infty} E[(S_n/n)^4] < \infty$, and the result will follow by Theorem 6. Now

$$S_n^4 = (R_1 + \cdots + R_n)^4$$

$$= \sum_{j=1}^{n} R_j^4 + \sum_{\substack{j,k=1 \\ j<k}}^{n} \frac{4!\, R_j^2 R_k^2}{2!\,2!} + \sum_{\substack{j \neq k \\ j \neq l \\ k < l}} \frac{4!}{2!\,1!\,1!} R_j^2 R_k R_l + \sum_{j<k<l<m} 4!\, R_j R_k R_l R_m$$

$$+ \sum_{j \neq k} \frac{4!}{3!\,1!} R_j^3 R_k$$

But

$$E(R_j{}^2 R_k R_l) = E(R_j{}^2)E(R_k)E(R_l) \qquad \text{by independence}$$

$$= 0$$

Similarly,

$$E(R_j R_k R_l R_m) = E(R_j{}^3 R_k) = 0$$

Thus

$$E(S_n{}^4) = \sum_{j=1}^{n} E(R_j{}^4) + \sum_{\substack{j,k=1 \\ j<k}}^{n} 6E(R_j{}^2)E(R_k{}^2)$$

By the Schwarz inequality,

$$E(R_j{}^2) = E(R_j{}^2 \cdot 1) \le [E(R_j{}^4)E(1^2)]^{1/2} \le M^{1/2}$$

Hence

$$E(S_n{}^4) \le nM + 6\,\frac{n(n-1)}{2}\,M = (3n^2 - 2n)M < 3n^2 M$$

Consequently

$$\sum_{n=1}^{\infty} \frac{E(S_n{}^4)}{n^4} < \sum_{n=1}^{\infty} \frac{3n^2 M}{n^4} = 3M \sum_{n=1}^{\infty} \frac{1}{n^2} < \infty$$

The theorem is proved.

REMARKS. If all R_i have the same distribution function, it turns out that the hypothesis on the fourth central moments can be replaced by the assumption that the $E(R_i)$ are finite [of course, in this case $E(R_i)$ is the same for all i]. In general, the hypothesis on the fourth central moments can be replaced by the assumption that for some M and $\delta > 0$, $E[|R_i - E(R_i)|^{1+\delta}] \le M$ for all i.

Now consider an infinite sequence of Bernoulli trials, with $R_i = 1$ if there is a success on trial i, $R_i = 0$ if there is a failure on trial i. Then

$$\frac{S_n}{n} = \frac{R_1 + \cdots + R_n}{n}$$

is the relative frequency of successes in n trials and $E(R_i)$ is the probability p of success on a given trial. The strong law of large numbers says that if we regard the observation of all the R_i as one performance of the experiment, the relative frequency of successes will almost certainly converge to p. The weak law of large numbers says only that if we consider a sufficiently large but fixed n (essentially regarding observation of $R_1, \ldots, R_n$ as one performance of the experiment), the probability that the relative frequency will be within a specified distance ε of p is $> 1 - \delta$, where $\delta > 0$ is preassigned. The requirements on n will depend on ε and δ, as well as p. Recall that, by

Chebyshev's inequality,

$$P\left\{\left|\frac{S_n - E(S_n)}{n}\right| \geq \varepsilon\right\} \leq \frac{E[(S_n - ES_n)^2/n^2]}{\varepsilon^2} = \frac{\operatorname{Var} S_n}{n^2 \varepsilon^2} = \frac{\displaystyle\sum_{j=1}^{n} \operatorname{Var} R_j}{n^2 \varepsilon^2}$$

$$= \frac{p(1-p)}{n\varepsilon^2}$$

Thus

$$P\left\{\left|\frac{S_n}{n} - p\right| < \varepsilon\right\} \geq 1 - \frac{p(1-p)}{n\varepsilon^2} > 1 - \delta \qquad \text{for large enough } n$$

PROBLEMS

1. Show that a countable intersection of sets with probability 1 still has probability 1. Does this hold for an uncountable intersection?

2. If $P\{|R_k - R| \geq \varepsilon \text{ for at least one } k \geq n\} \to 0$ as $n \to \infty$ for all ε of the form $1/m$, $m = 1, 2, \ldots$, show that this holds for all $\varepsilon > 0$.

3. Let R_0 be uniformly distributed on the interval $(0, 1]$. Define the following sequence of random variables.

$$R_1 = g_1(R_0) \equiv 1$$
$$R_{21} = g_{21}(R_0) = 1 \quad \text{if } 0 < R_0 \leq \tfrac{1}{2}; \qquad R_{21} = 0 \text{ otherwise}$$
$$R_{22} = 1 \quad \text{if } \tfrac{1}{2} < R_0 \leq 1, \qquad 0 \text{ otherwise}$$
$$R_{31} = 1 \quad \text{if } 0 < R_0 \leq \tfrac{1}{3}, \qquad 0 \text{ otherwise}$$
$$R_{32} = 1 \quad \text{if } \tfrac{1}{3} < R_0 \leq \tfrac{2}{3}, \qquad 0 \text{ otherwise}$$
$$R_{33} = 1 \quad \text{if } \tfrac{2}{3} < R_0 \leq 1, \qquad 0 \text{ otherwise}$$

In general, let

$$R_{nm} = 1 \quad \text{if } \frac{m-1}{n} < R_0 \leq \frac{m}{n}, \qquad n = 1, 2, \ldots, m = 1, 2, \ldots, n,$$

$$0 \text{ otherwise}$$

(see Figure P.6.6.3 for $n = 4$).

The fact that we are using two subscripts is unimportant. We may arrange the R_{nm} as a single sequence.

$$R_1, R_{21}, R_{22}, R_{31}, R_{32}, R_{33}, R_{41}, R_{42}, R_{43}, R_{44}, \qquad \text{etc.}$$

Show that the sequence converges in probability to 0, but does not converge almost surely to 0. In fact, $P\{R_{nm} \to 0\} = 0$.

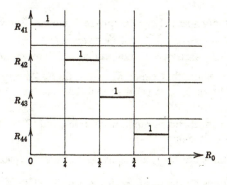

FIGURE P.6.6.3

4. Let $A_1, A_2, \ldots$ be an arbitrary sequence of events in a given probability space.
   (a) Show that $\liminf_n A_n \subset \limsup_n A_n$.
   (b) If the A_n form an expanding sequence whose union is A, or a contracting sequence whose intersection is A, show that $\liminf_n A_n = \limsup_n A_n = A$.
   (c) In general, if $\limsup_n A_n = \liminf_n A_n = A$, we say that A is the *limit* of the sequence $\{A_n\}$ (notation: $A = \lim_n A_n$). Give an example of a sequence that is not eventually expanding or contracting (i.e., that does not become an expanding or contracting sequence if an appropriate finite number of terms is omitted), but that has a limit.
   (d) If $A = \lim_n A_n$, show that $P(A) = \lim_{n \to \infty} P(A_n)$. HINT: $\bigcap_{k=n}^{\infty} A_k$ expands to $\liminf A_n$, $\bigcup_{k=n}^{\infty} A_k$ contracts to $\limsup A_n$, and $\bigcap_{k=n}^{\infty} A_k \subset A_n \subset \bigcup_{k=n}^{\infty} A_k$.
   (e) Show that $(\liminf_n A_n)^c = \limsup_n A_n^c$, and $(\limsup_n A_n)^c = \liminf_r A_n^c$.

5. Find $\limsup_n A_n$ and $\liminf_n A_n$ if $\Omega = E^1$ and

$$A_n = \left[0, 1 - \frac{1}{n}\right] \quad \text{if } n \text{ is even}$$

$$= \left[-1, \frac{1}{n}\right] \quad \text{if } n \text{ is odd}$$

6. Let $\Omega = E^2$ and take $A_n =$ the interior of the circle with center at $((-1)^n/n, 0)$ and radius 1. Find $\limsup_n A_n$ and $\liminf_n A_n$.

7. Let $x_1, x_2, \ldots$ be a sequence of real numbers, and let $A_n = (-\infty, x_n)$. What is the connection between $\limsup_n x_n$ and $\limsup_n A_n$ (similarly for $\liminf$)?

8. (Second Borel-Cantelli lemma) If $A_1, A_2, \ldots$ are *independent* events and $\sum_{n=1}^{\infty} P(A_n) = \infty$, show that $P(\limsup_n A_n) = 1$. HINT: Show that $P(\limsup_n A_n) = \lim_{n \to \infty} \lim_{m \to \infty} P(\bigcup_{k=n}^{m} A_k)$, and consider $(\bigcup_{k=n}^{m} A_k)^c$; use the fact that $e^{-x} \geq 1 - x$.

9. Let $R_1, R_2, \ldots$ be a sequence of independent random variables, all defined on

the same probability space. Let c be any real number. Show that $R_n \xrightarrow{\text{a .s.}} c$ if and only if for every $\varepsilon > 0$, $\sum_{n=1}^{\infty} P\{|R_n - c| \geq \varepsilon\} < \infty$.

10. Let $R_1, R_2, \ldots$ be a sequence of independent random variables, and let c be any real number. Show that either $R_n \xrightarrow{\text{a.s.}} c$ or R_n "diverges almost surely" from c; that is, $P\{x: \lim_{n\to\infty} R_n(x) = c\} = 0$. Thus, for example, it is impossible that $P\{R_n \to c\} = 1/3$.

11. Let $R_1, R_2, \ldots$ be independent random variables, with $R_n = 1$ with probability p_n, $R_n = 0$ with probability $1 - p_n$.

   (a) What conditions on the p_n are equivalent to the statement that $R_n \xrightarrow{P} 0$?

   (b) What conditions on the p_n are equivalent to the statement that $R_n \xrightarrow{\text{a.s.}} 0$?

12. Let $R_1, R_2, \ldots$ be independent random variables, with $E(R_n) \equiv 0$, $\mathrm{Var}\, R_n = \sigma_n^2 \leq M/n$, where M is some fixed positive constant. Show that $(R_1 + \cdots + R_n)/n \xrightarrow{\text{a.s.}} 0$.

13. Give an example of a particular sequence of random variables $R_1, R_2, \ldots$ and a random variable R such that $0 < P\{R_n \to R\} < 1$.

7

Markov Chains

7.1 INTRODUCTION

Suppose that a machine with two components is inspected every hour. A given component operating at $t = n$ hours has probability p of failing before the next inspection at $t = n + 1$; a component that is not operating at $t = n$ has probability r of being repaired at $t = n + 1$, regardless of how long it has been out of service. (Each repair crew works a 1-hour day and refuses to inform the next crew of any insights it may have gained.) The components are assumed to fail and be repaired independently of each other.

The situation may be summarized as follows. If R_n is the number of components in operation at $t = n$, then

$$P\{R_{n+1} = 0 \mid R_n = 0\} = (1 - r)^2$$
$$P\{R_{n+1} = 1 \mid R_n = 0\} = 2r(1 - r)$$
$$P\{R_{n+1} = 2 \mid R_n = 0\} = r^2$$

$$P\{R_{n+1} = 0 \mid R_n = 1\} = p(1 - r)$$
$$P\{R_{n+1} = 1 \mid R_n = 1\} = pr + (1 - p)(1 - r)$$
$$P\{R_{n+1} = 2 \mid R_n = 1\} = (1 - p)r$$

$$P\{R_{n+1} = 0 \mid R_n = 2\} = p^2$$
$$P\{R_{n+1} = 1 \mid R_n = 2\} = 2p(1 - p)$$
$$P\{R_{n+1} = 2 \mid R_n = 2\} = (1 - p)^2 \qquad (7.1.1)$$

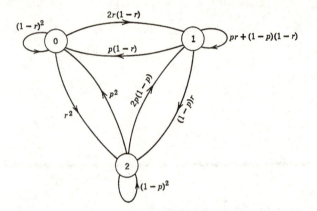

FIGURE 7.1.1   A Markov Chain.

For example, if component A is operating and component B is out of service at $t = n$, then in order to have one component in service at $t = n + 1$, either A fails and B is repaired between $t = n$ and $t = n + 1$, or A does not fail and B is not repaired between $t = n$ and $t = n + 1$. In order to have two components in service at $t = n + 1$, A must not fail and B must be repaired. The other entires of (7.1.1) are derived similarly.

Thus, at time $t = n$, there are three possible states, 0, 1, and 2; to be in state i means that i components are in operation; that is, $R_n = i$. There are various transition probabilities p_{ij} indicating the probability of moving to state j at $t = n + 1$, given that we are in state i at $t = n$; thus

$$p_{ij} = P\{R_{n+1} = j \mid R_n = i\}$$

(see Figure 7.1.1).

The p_{ij} may be arranged in the form of a matrix:

$$
\Pi = [p_{ij}] = \begin{array}{c} 0 \\ 1 \\ 2 \end{array}
\begin{array}{ccc}
0 & 1 & 2 \\
\left[\begin{array}{ccc}
(1 - r)^2 & 2r(1 - r) & r^2 \\
p(1 - r) & pr + (1 - p)(1 - r) & (1 - p)r \\
p^2 & 2p(1 - p) & (1 - p)^2
\end{array}\right]
\end{array}
$$

Notice that Π is *stochastic*; that is, the elements are nonnegative and the row sums $\sum_j p_{ij}$ are 1 for all i.

If R_n is the state of the process at time n, then, according to the way the problem is stated, if we know that $R_0 = i_0, R_1 = i_1, \ldots, R_n = i_n = $ (say) 2, we are in state 2 at $t = n$. Regardless of how we got there, once we know that $R_n = 2$, the probability that $R_{n+1} = j$ is $\binom{2}{j}(1 - p)^j p^{2-j}$, $j = 0, 1, 2$. In other words,

$$P\{R_{n+1} = i_{n+1} \mid R_0 = i_0, \ldots, R_n = i_n\} = P\{R_{n+1} = i_{n+1} \mid R_n = i_n\} \quad (7.1.2)$$

This is the *Markov property*.

What we intend to show is that, given a description of a process in terms of states and transition probabilities (formally, given a stochastic matrix), we can construct in a natural way an infinite sequence of random variables satisfying (7.1.2). Assume that we are given a stochastic matrix $\Pi = [p_{ij}]$, where i and j range over the *state space* S, which we take to be a finite or infinite subset of the integers. Let p_i, $i \in S$, be a set of nonnegative numbers adding to 1 (the *initial distribution*). We specify the joint probability function of $R_0, R_1, \ldots, R_n$ as follows.

$$P\{R_0 = i_0, R_1 = i_1, \ldots, R_n = i_n\}$$
$$= p_{i_0} p_{i_0 i_1} p_{i_1 i_2} \cdots p_{i_{n-1} i_n}, n = 1, 2, \ldots, P\{R_0 = i_0\} = p_{i_0} \quad (7.1.3)$$

If we sum the right side of (7.1.3) over $i_{k+1}, \ldots, i_n$, we obtain $p_{i_0} p_{i_0 i_1} \cdots p_{i_{k-1} i_k}$, since Π is stochastic. But this coincides with the original specification of $P\{R_0 = i_0, R_1 = i_1, \ldots, R_k = i_k\}$.

Thus the joint probability functions (7.1.3) are consistent, and therefore, by the discussion in Section 6.1, we can construct a sequence of random variables $R_0, R_1, \ldots$ such that for each n the joint probability function of $(R_0, \ldots, R_n)$ is given by (7.1.3).

Let us verify that the Markov property (7.1.2) is in fact satisfied. We have

$$P\{R_{n+1} = i_{n+1} \mid R_0 = i_0, \ldots, R_n = i_n\}$$
$$= \frac{P\{R_0 = i_0, \ldots, R_{n+1} = i_{n+1}\}}{P\{R_0 = i_0, \ldots, R_n = i_n\}}$$

(assuming the denominator is not zero)

$$= p_{i_n i_{n+1}} \quad \text{by (7.1.3)}$$

But

$$P\{R_{n+1} = i_{n+1} \mid R_n = i_n\} = \frac{P\{R_n = i_n, R_{n+1} = i_{n+1}\}}{P\{R_n = i_n\}}$$

$$= \frac{\displaystyle\sum_{i_0, \ldots, i_{n-1}} P\{R_0 = i_0, \ldots, R_{n-1} = i_{n-1}, R_n = i_n, R_{n+1} = i_{n+1}\}}{\displaystyle\sum_{i_0, \ldots, i_{n-1}} P\{R_0 = i_0, \ldots, R_n = i_n\}}$$

$$= \frac{\displaystyle\sum_{i_0, \ldots, i_{n-1}} p_{i_0} p_{i_0 i_1} \cdots p_{i_{n-1} i_n} p_{i_n i_{n+1}}}{\displaystyle\sum_{i_0, \ldots, i_{n-1}} p_{i_0} p_{i_0 i_1} \cdots p_{i_{n-1} i_n}}$$

$$= p_{i_n i_{n+1}}$$

establishing the Markov property.

The sequence $\{R_n\}$ is called the *Markov chain* corresponding to the matrix Π and the initial distribution $\{p_i\}$. Π is called the *transition matrix*, and the p_{ij} the *transition probabilities*, of the chain.

REMARK.   The basic construction given here can be carried out if we have "nonstationary transition probabilities," that is, if, instead of the "stationary transition probabilities" p_{ij}, we have, for each $n = 0$, $1, \ldots$, a stochastic matrix $[_np_{ij}]$. $_np_{ij}$ is interpreted as the probability of moving to state j at time $n + 1$ when the state at time n is i. We define $P\{R_0 = i_0, \ldots, R_n = i_n\} = p_{i_0}\, {_0p_{i_0i_1}}\, {_1p_{i_1i_2}} \cdots {_{n-1}p_{i_{n-1}i_n}}$. This yields a Markov process, a sequence of random variables satisfying the Markov property, with $P\{R_n = i_n \mid R_0 = i_0, \ldots, R_{n-1} = i_{n-1}\} = {_{n-1}p_{i_{n-1}i_n}}$.

We begin the analysis of Markov chains by calculating the probability function of R_n in terms of the initial probabilities p_i and the matrix Π. Let $p_j^{(n)} = P\{R_n = j\}$, $n = 0, 1, \ldots, j \in S$ (thus $p_j^{(0)} = p_j$). If we are in state i at time $n - 1$, we move to state j at time n with probability $P\{R_n = j \mid R_{n-1} = i\} = p_{ij}$; thus, by the theorem of total probability,

$$p_j^{(n)} = \sum_i P\{R_{n-1} = i\}P\{R_n = j \mid R_{n-1} = i\} = \sum_i p_i^{(n-1)}p_{ij} \quad (7.1.4)$$

If $v^{(n)} = (p_i^{(n)}, i \in S)$ is the "state distribution" or "state probability vector" at time n, then (7.1.4) may be written in matrix form as

$$v^{(n)} = v^{(n-1)}\Pi$$

Iterating this, we obtain

$$v^{(n)} = v^{(0)}\Pi^n \quad (7.1.5)$$

But suppose that we specify $R_0 = i$; that is, $p_i^{(0)} = 1$, $p_j^{(0)} = 0$, $i \neq j$. Then $v^{(0)}$ has a 1 in the ith coordinate and 0's elsewhere, so that by (7.1.5) $v^{(n)}$ is simply row i of Π^n. Thus the element $p_{ij}^{(n)}$ in row i and column j of Π^n is the probability that $R_n = j$ when the initial state is i. In other words,

$$p_{ij}^{(n)} = P\{R_n = j \mid R_0 = i\} \quad (7.1.6)$$

(A slight formal quibble lies behind the phrase "in other words"; see Problem 1.) Because of (7.1.6), Π^n is called the *n-step transition matrix*: it follows immediately that Π^n is stochastic.

We shall be interested in the behavior of Π^n for large n. As an example, suppose that

$$\Pi = \begin{bmatrix} \frac{1}{2} & \frac{1}{2} \\ \frac{3}{4} & \frac{1}{4} \end{bmatrix}$$

We compute

$$\Pi^2 = \begin{bmatrix} \frac{5}{8} & \frac{3}{8} \\ \frac{9}{16} & \frac{7}{16} \end{bmatrix}, \qquad \Pi^4 = \begin{bmatrix} \frac{154}{256} & \frac{102}{256} \\ \frac{153}{256} & \frac{103}{256} \end{bmatrix}$$

Thus $p_{11}^{(4)} \sim p_{21}^{(4)}$ and $p_{12}^{(4)} \sim p_{22}^{(4)}$, so that the probability of being in state j at time $t = 4$ is almost independent of the initial state. It appears as if, for large n, a "steady state" condition will be approached; the probability of being in a particular state j at $t = n$ will be almost independent of the initial state at $t = 0$. Mathematically, we express this condition by saying that

$$\lim_{n \to \infty} p_{ij}^{(n)} = v_j, i, j \in S \qquad (7.1.7)$$

where v_j does not depend on i.

In Sections 7.4 and 7.5 we investigate the conditions under which (7.1.7) holds; it is not true for an arbitrary Markov chain.

Note that (7.1.7) is equivalent to the statement that $\Pi^n \to$ a matrix with identical rows, the rows being $(v_j, j \in S)$.

▶ **Example 1.** Consider the simple random walk with no barriers. Then $S =$ the integers and $p_{i,i+1} = p, p_{i,i-1} = q = 1 - p, i = 0, \pm 1, \pm 2, \dots$.

If there is an absorbing varrier at 0 (gambler's ruin problem when the adversary has infinite capital), $S =$ the nonnegative integers and $p_{i,i+1} = p$, $p_{i,i-1} = q, i = 1, 2, \dots, p_{00} = 1$ (hence $p_{0j} = 0, j \neq 0$).

If there are absorbing barriers at 0 and b, then $S = \{0, 1, \dots, b\}$, $p_{i,i+1} = p, p_{i,i-1} = q, i = 1, 2, \dots, b - 1, p_{00} = p_{bb} = 1$. ◄

▶ **Example 2.** Consider an infinite sequence of Bernoulli trials. Let state 1 (at $t = n$) correspond to successes (S) at $t = n - 1$ and at $t = n$; state 2 to success at $t = n - 1$ and failure (F) at $t = n$; state 3 to failure at $t = n - 1$ and success at $t = n$; state 4 to failures at $t = n - 1$ and at $t = n$ (see Figure 7.1.2). We observe that Π^2 has identical rows, the rows being

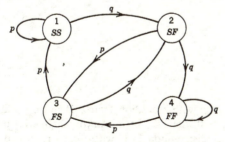

$$\Pi = \begin{bmatrix} p & q & 0 & 0 \\ 0 & 0 & p & q \\ p & q & 0 & 0 \\ 0 & 0 & p & q \end{bmatrix}$$

FIGURE 7.1.2

(p^2, pq, qp, q^2). Hence Π^n has identical rows for $n \geq 2$ (see Problem 2), so that Π^n approaches a limit. ◄

▶ **Example 3.**   (A queueing example) Assume that customers are to be served at discrete times $t = 0, 1, \ldots$, and at most one customer can be served at a given time. Say there are R_n customers before the completion of service at time n, and, in the interval $[n, n + 1)$, N_n new customers arrive, where $P\{N_n = k\} = p_k$, $k = 0, 1, \ldots$. The number of customers before completion of service at time $n + 1$ is

$$R_{n+1} = (R_n - 1)^+ + N_n$$

That is,

$$R_{n+1} = R_n - 1 + N_n \quad \text{if } R_n \geq 1$$
$$= N_n \quad \text{if } R_n = 0$$

If the number of customers at time $n + 1$ is $\geq M$, a new serving counter automatically opens and immediately serves all customers who are waiting and also those who arrive in the interval $[n + 1, n + 2)$; thus $R_{n+2} = 0$.

The queueing process may be represented as a Markov chain with $S = $ the nonnegative integers and transition matrix

$$
\Pi = \quad
\begin{array}{c|ccccccccc}
 & 0 & 1 & 2 & 3 & \cdots & & & & \\
\hline
0 & p_0 & p_1 & p_2 & \cdots & & & & & \\
1 & p_0 & p_1 & p_2 & \cdots & & & & & \\
2 & 0 & p_0 & p_1 & p_2 & \cdots & & & & \\
3 & 0 & 0 & p_0 & p_1 & p_2 & \cdots & & & \\
\cdot & & & & & & & & & \\
\cdot & & & & & & & & & \\
\cdot & & & & & & & & & \\
M-1 & 0 & 0 & 0 & & 0 & 0 & \cdots & p_0 & p_1 & p_2 & \cdots \\
M & 1 & 0 & \cdots & & & & & & \\
M+1 & 1 & 0 & \cdots & & & & & & \\
\cdot & \cdot & & & & & & & & \\
\cdot & \cdot & & & & & & & & \\
\end{array}
$$

◄

PROBLEMS

1. Consider a Markov chain $\{R_n\}$ with transition matrix $\Pi = [p_{ij}]$ and initial distribution $\{p_i\}$; assume $p_r > 0$. Let $\{T_n\}$ be a Markov chain with the same transition matrix, and initial distribution $\{q_i\}$, where $q_r = 1$, $q_j = 0$, $j \neq r$. Show that $P\{R_n = j \mid R_0 = r\} = P\{T_n = j\} = p_{rj}^{(n)}$; this justifies (7.1.6).

2. If Π is a transition matrix of a Markov chain, and Π^k has identical rows, show that Π^n has identical rows for all $n \geq k$. Similarly, if Π^k has a column all of whose elements are $\geq \delta > 0$, show that Π^n has this property for all $n \geq k$.

3. Let $\{R_n\}$ be a Markov chain with state space S, and let g be a function from S to S. If g is one-to-one, $\{g(R_n)\}$ is also a Markov chain (this simply amounts to relabeling the states). Give an example to show that if g is not one-to-one, $\{g(R_n)\}$ need not have the Markov property.

4. If $\{R_n\}$ is a Markov chain, show that

$$P\{R_{n+1} = i_1, \ldots, R_{n+k} = i_k \mid R_n = i\} = p_{ii_1}p_{i_1 i_2} \cdots p_{i_{k-1} i_k}$$

7.2 STOPPING TIMES AND THE STRONG MARKOV PROPERTY

Let $\{R_n\}$ be a Markov chain, and let T be the first time at which a particular state i is reached; set $T = \infty$ if i is never reached. For example, if $i = 3$ and $R_0(\omega) = 4$, $R_1(\omega) = 2$, $R_2(\omega) = 2$, $R_3(\omega) = 5$, $R_4(\omega) = 3$, $R_5(\omega) = 1$, $R_6(\omega) = 3, \ldots$, then $T(\omega) = 4$. For our present purposes, the key feature of T is that if we examine $R_0, R_1, \ldots, R_k$, we can come to a definite decision as to whether or not $T = k$. Formally, for each $k = 0$, $1, 2, \ldots$, $I_{\{T=k\}}$ can be expressed as $g_k(R_0, R_1, \ldots, R_k)$, where g_k is a function from S^{k+1} to $\{0, 1\}$. A random variable T, whose possible values are the nonnegative integers together with ∞, that satisfies this condition for each $k = 0, 1, \ldots$ is said to be a *stopping time* for the chain $\{R_n\}$.

Now let T be the first time at which the state i is reached, as above. If we look at the sequence $\{R_n\}$ after we arrive at i, in other words, the sequence $R_T, R_{T+1}, \ldots$, it is reasonable to expect that we have a Markov chain with the same transition probabilities as the original chain. After all, if $T = k$, we are looking at the sequence $R_k, R_{k+1}, \ldots$. However, since T is a random variable rather than a constant, there is something to be proved. We first introduce a new concept.

If T is a stopping time, an event A is said to be *prior to* T iff, whenever $T = k$, we can tell by examination of $R_0, \ldots, R_k$ whether or not A has occurred. Formally, for each $k = 0, 1, \ldots$, $I_{A \cap \{T=k\}}$ can be expressed as $h_k(R_0, R_1, \ldots, R_k)$, where h_k is a function from S^{k+1} to $\{0, 1\}$

▶ **Example 1.** If T is a stopping time for the Markov chain $\{R_n\}$, define the random variable R_T as follows.

If $T(\omega) = k$, take $R_T(\omega) = R_k(\omega)$, $k = 0, 1, \ldots$.

If $T(\omega) = \infty$, take $R_T(\omega) = c$, where c is an arbitrary element not belonging to the state space S. If we like we can replace S by $S \cup \{c\}$ and define $p_{cc} = 1$, $p_{cj} = 0$, $j \in S$.

If $A = \{R_{T-j} = i\}$, $i \in S$, $j = 0, 1, \ldots$, then A is prior to T. (Set $R_{T-j} = R_0$ if $j > T$.)

To see this, note that $\{R_{T-j} = i\} \cap \{T = k\} = \{R_{k-j} = i\} \cap \{T = k\}$; examination of $R_0, \ldots, R_k$ determines the value of R_{k-j}, and also determines whether or not $T = k$. ◄

▶ **Example 2.** If T is a stopping time for the Markov chain $\{R_n\}$, then $\{T = r\}$ is prior to T for all $r = 0, 1, \ldots$. For

$$\{T = r\} \cap \{T = k\} = \varnothing \qquad \text{if } r \neq k$$
$$= \{T = k\} \qquad \text{if } r = k$$

In either case $I_{\{T=r\} \cap \{T=k\}}$ is a function of $R_0, \ldots, R_k$, since T is a stopping time. ◄

Theorem 1. *Let T be a stopping time for the Markov chain $\{R_n\}$. If A is prior to T, then*

$$P(A \cap \{R_T = i, R_{T+1} = i_1, \ldots, R_{T+k} = i_k\})$$
$$= P(A \cap \{R_T = i\}) p_{ii_1} p_{i_1 i_2} \cdots p_{i_{k-1} i_k} (i, i_1, \ldots, i_k \in S)$$

PROOF.   The probability of the set on the left is

$$\sum_{n=0}^{\infty} P(A \cap \{T = n, R_n = i, R_{n+1} = i_1, \ldots, R_{n+k} = i_k\})$$

$$= \sum_{n=0}^{\infty} P(A \cap \{T = n, R_n = i\})$$

$$\times P(R_{n+1} = i_1, \ldots, R_{n+k} = i_k \mid A \cap \{T = n, R_n = i\})$$

(Actually we sum only over those n for which $P(A \cap \{T = n, R_n = i\}) > 0$.) Now

$$P(R_{n+1} = i_1, \ldots, R_{n+k} = i_k \mid A \cap \{T = n, R_n = i\})$$
$$= P\{R_{n+1} = i_1, \ldots, R_{n+k} = i_k \mid R_n = i\}$$

since $I_{A \cap \{T=n\}}$ is a function of $R_0, R_1, \ldots, R_n$ (see Problem 1)

$$= p_{ii_1} p_{i_1 i_2} \cdots p_{i_{k-1} i_k}$$

(Problem 4, Section 7.1). Thus the summation becomes

$$\sum_{n=0}^{\infty} P(A \cap \{T = n, R_n = i\}) p_{ii_1} p_{i_1 i_2} \cdots p_{i_{k-1} i_k}$$

$$= P(A \cap \{R_T = i\}) p_{ii_1} p_{i_1 i_2} \cdots p_{i_{k-1} i_k}$$

and the result follows.

Theorem 2 (Strong Markov Property). *Let T be a stopping time for the Markov chain $\{R_n\}$. Then*

(a) $P\{R_{T+1} = i_1, \ldots, R_{T+k} = i_k \mid R_T = i\} = p_{ii_1} p_{i_1 i_2} \cdots p_{i_{k-1} i_k}$ *if*
    $P\{R_T = i\} > 0$ $(i, i_1, \ldots, i_k \in S)$

(b) *If A is prior to T, then*

$$P(R_{T+1} = i_1, \ldots, R_{T+k} = i_k \mid A \cap \{R_T = i\})$$
$$= P\{R_{T+1} = i_1, \ldots, R_{T+k} = i_k \mid R_T = i\}$$
$$\text{if } P(A \cap \{R_T = i\}) > 0 \quad (i, i_1, \ldots, i_k \in S)$$

PROOF.   (a) follows from Theorem 1 by taking $A = \Omega$. (b) follows upon dividing the equality of Theorem 1 by $P(A \cap \{R_T = i\})$ and using (a).

Thus the sequence $R_T, R_{T+1}, \ldots$ has essentially the same properties as the original sequence $R_0, R_1, \ldots$.

REMARK.   The strong Markov property reduces to the ordinary Markov property (7.1.2) if we set $k = 1$, $T \equiv n$, and $A = \{R_0 = j_0, \ldots, R_{n-1} = j_{n-1}\}$. For T is a stopping time since

$$I_{\{T=k\}} = g_k(R_0, \ldots, R_k) = 0 \qquad \text{if } k \neq n$$
$$= 1 \qquad \text{if } k = n$$

and A is prior to T since

$$I_{A \cap \{T=k\}} = h_k(R_0, \ldots, R_k)$$
$$= 1 \qquad \text{if } k = n \text{ and } R_0 = j_0, \ldots, R_{n-1} = j_{n-1}$$
$$= 0 \qquad \text{otherwise}$$

PROBLEMS

1. Let $\{R_n\}$ be a Markov chain. If D is an event whose occurrence or nonoccurrence is determined by examination of $R_0, \ldots, R_n$, that is, I_D is a function of $R_0, \ldots, R_n$, or, equivalently, D is of the form $\{(R_0, \ldots, R_n) \in B\}$ for some $B \subset S^{n+1}$, show that

$$P(R_{n+1} = i_1, \ldots, R_{n+k} = i_k \mid D \cap \{R_n = i\})$$
$$= P\{R_{n+1} = i_1, \ldots, R_{n+k} = i_k \mid R_n = i\} \quad \text{if } P(D \cap \{R_n = i\}) > 0$$

2. If $\{R_n\}$ is a Markov chain, show that the "reversed sequence" $\cdots R_n, R_{n-1}, R_{n-2}, \ldots$ also has the Markov property.

7.3   CLASSIFICATION OF STATES

In this section we examine various modes of behavior of Markov chains. A key to the analysis is the following result. We consider a fixed Markov chain $\{R_n\}$ throughout.

Theorem 1 (First Entrance Theorem). *Let $f_{ii}^{(n)}$ be the probability that the first return to i will occur at time n, when the initial state is i, that is,*

$$f_{ii}^{(n)} = P\{R_n = i, R_k \neq i \text{ for } 1 \leq k \leq n-1 \mid R_0 = i\}, \qquad n = 1, 2, \ldots$$

If $i \neq j$, let $f_{ij}^{(n)}$ be the probability that the first visit to state j will occur at time n, when the initial state is i; that is,

$$f_{ij}^{(n)} = P\{R_n = j, R_k \neq j \text{ for } 1 \leq k \leq n-1 \mid R_0 = i\}, \qquad n = 1, 2, \ldots$$

Then

$$p_{ij}^{(n)} = \sum_{k=1}^{n} f_{ij}^{(k)} p_{jj}^{(n-k)}, \qquad n = 1, 2, \ldots$$

PROOF.   Intuitively, if we are to be in state j after n steps, we must reach j for the first time at step k, $1 \leq k \leq n$. After this happens, we are in state j and must be in state j again after the $n - k$ remaining steps. For a formal proof, we use the strong Markov property. Assume that the initial state is i, and let T be the time of the first visit to j ($T = \min\{k \geq 1 : R_k = j\}$ if $R_k = j$ for some $k = 1, 2, \ldots$; $T = \infty$ if $R_k \neq j$ for all $k = 1, 2, \ldots$). Then

$$P\{R_n = j\} = \sum_{k=1}^{n} P\{R_n = j, T = k\}$$

$$= \sum_{k=1}^{n} P\{T = k, R_{T+n-k} = j\}$$

But

$$P\{T = k, R_{T+n-k} = j\} = P\{T = k\} P\{R_{T+n-k} = j \mid T = k\}$$

and since $\{T = k\} = \{T = k, R_T = j\}$,

$$P\{R_{T+n-k} = j \mid T = k\} = P\{R_{T+n-k} = j \mid R_T = j, T = k\}$$

$$= P\{R_{T+n-k} = j \mid R_T = j\}$$

by Theorem 2b of Section 7.2

$$= p_{jj}^{(n-k)} \qquad \text{by Theorem 2a of Section 7.2}$$

Since $P\{T = k\} = f_{ij}^{(k)}$, the result follows.

Now let

$$f_{ii} = \sum_{n=1}^{\infty} f_{ii}^{(n)}$$

f_{ii} is the probability of eventual return to state i when the initial state is i.

Theorem 2.   *If the initial state is i, the probability of returning to i at least r times is* $(f_{ii})^r$.

PROOF.   The result is immediate if $r = 1$. If it holds when $r = m - 1$, let T be the time of first return to i. Then, starting from i, $P\{$return to i at least m times$\} = \sum_{k=1}^{\infty} P\{T = k$, at least $m - 1$ returns after $T\}$. But

$P\{T = k$, at least $m - 1$ returns after $T\}$

$\quad = P\{T = k\} P\{R_{T+1}, R_{T+2}, \ldots$ returns to i at least $m - 1$ times $\mid T = k\}$

$\quad = P\{T = k\}$

$\quad\quad \times P\{R_{T+1}, R_{T+2}, \ldots$ returns to i at least $m - 1$ times $\mid R_T = i, T = k\}$

By the strong Markov property this may be written as

$\quad P\{T = k\}P\{R_{T+1}, R_{T+2}, \ldots$ returns to i at least $m - 1$ times $\mid R_T = i\}$

$\quad\quad = P\{T = k\}P\{R_1, R_2, \ldots$ returns to i at least $m - 1$ times $\mid R_0 = i\}$

$\quad\quad = f_{ii}^{(k)}(f_{ii})^{m-1} \qquad$ by the induction hypothesis

Thus the probability of returning to i at least m times is

$$\sum_{k=1}^{\infty} f_{ii}^{(k)}(f_{ii})^{m-1} = f_{ii}(f_{ii})^{m-1}$$

which is the desired result.

COROLLARY.   Let the initial state be i. If $f_{ii} = 1$, the probability of returning to i infinitely often is 1. If $f_{ii} < 1$, the probability of returning to i infinitely often is 0.

PROOF.   The events $\{$return to i at least r times$\}$, $r = 1, 2, \ldots$ form a contracting sequence whose intersection is $\{$return to i infinitely often$\}$. Thus the probability of returning to i infinitely often is $\lim_{r \to \infty} (f_{ii})^r$, and the result follows.

DEFINITION.   If $f_{ii} = 1$, we say that the state i is *recurrent* or *persistent*; if $f_{ii} = 1$, we say that i is *transient*.

It is useful to have a criterion for recurrence in terms of the probabilities $p_{ii}^{(n)}$, since these numbers are often easier to handle than the $f_{ii}^{(n)}$.

Theorem 3.   *The state i is recurrent iff* $\sum_{n=1}^{\infty} p_{ii}^{(n)} = \infty$.

PROOF.   By the first entrance theorem,

$$p_{ii}^{(n)} = \sum_{k=1}^{n} f_{ii}^{(k)} p_{ii}^{(n-k)}$$

so that

$$\sum_{n=1}^{\infty} p_{ii}^{(n)} = \sum_{n=1}^{\infty} \sum_{k=1}^{n} f_{ii}^{(k)} p_{ii}^{(n-k)} = \sum_{k=1}^{\infty} f_{ii}^{(k)} \sum_{n=k}^{\infty} p_{ii}^{(n-k)}$$

$$= f_{ii} \sum_{n=0}^{\infty} p_{ii}^{(n)}$$

Thus

$$\sum_{n=1}^{\infty} p_{ii}^{(n)} = f_{ii}\left(1 + \sum_{n=1}^{\infty} p_{ii}^{(n)}\right)$$

Hence, if

$$\sum_{n=1}^{\infty} p_{ii}^{(n)} < \infty$$

then $f_{ii} < 1$, so that i is transient. Now

$$\sum_{n=1}^{N} p_{ii}^{(n)} = \sum_{n=1}^{N} \sum_{k=1}^{n} f_{ii}^{(k)} p_{ii}^{(n-k)} = \sum_{k=1}^{N} f_{ii}^{(k)} \sum_{n=k}^{N} p_{ii}^{(n-k)} \le \sum_{k=1}^{N} f_{ii}^{(k)} \sum_{r=0}^{N} p_{ii}^{(r)}$$

Thus

$$f_{ii} = \sum_{k=1}^{\infty} f_{ii}^{(k)} \ge \sum_{k=1}^{N} f_{ii}^{(k)} \ge \frac{\sum_{n=1}^{N} p_{ii}^{(n)}}{\sum_{r=0}^{N} p_{ii}^{(r)}} \to 1 \quad \text{as} \quad N \to \infty \quad \text{if} \quad \sum_{n=1}^{\infty} p_{ii}^{(n)} = \infty$$

Therefore $\sum_{n=1}^{\infty} p_{ii}^{(n)} = \infty$ implies that $f_{ii} = 1$, so that i is recurrent.

We denote by f_{ij} the probability of ever visiting j at some future time, starting from i; that is,

$$f_{ij} = \sum_{k=1}^{\infty} f_{ij}^{(k)}$$

Theorem 4. *If j is a transient state and i an arbitrary state, then*

$$\sum_{n=1}^{\infty} p_{ij}^{(n)} < \infty$$

hence

$$p_{ij}^{(n)} \to 0 \quad \text{as} \quad n \to \infty$$

PROOF. By the first entrance theorem,

$$\sum_{n=1}^{\infty} p_{ij}^{(n)} = \sum_{n=1}^{\infty} \sum_{k=1}^{n} f_{ij}^{(k)} p_{jj}^{(n-k)} = \sum_{k=1}^{\infty} f_{ij}^{(k)} \sum_{n=0}^{\infty} p_{jj}^{(n)}$$

$$= f_{ij} \sum_{n=0}^{\infty} p_{jj}^{(n)}$$

But f_{ij}, being a probability, is ≤ 1, and $\sum_{n=0}^{\infty} p_{jj}^{(n)} < \infty$ by Theorem 3; the result follows.

REMARK. If j is a transient state and the initial state of the chain is i, then, by Theorem 4 above and the Borel-Cantelli lemma (Theorem 4, Section 6.6), with probability 1 the state j will be visited only finitely many times. Alternatively, we may use the argument of Theorem 2 (with initial state i and $T =$ the time of first visit to j) to show that the probability that j will be visited at least m times is $f_{ij}(f_{jj})^{m-1}$. Now $f_{jj} < 1$ since j is transient, and thus, if we let $m \to \infty$, we find (as in the corollary to Theorem 2) that the probability that j will be visited infinitely often is 0.

In fact this result holds for an arbitrary initial distribution. For

$P\{R_n = j \text{ for infinitely many } n\}$
$$= \sum_i P\{R_0 = i\}P\{R_n = j \text{ for infinitely many } n \mid R_0 = i\} = 0$$

It follows that if B is a finite set of transient states, then the probability of remaining in B forever is 0. For if $R_n \in B$ for all n, then, since B is finite, we have, for some $j \in B$, $R_n = j$ for infinitely many n; thus

$$P\{R_n \in B \text{ for all } n\} \leq \sum_{j \in B} P\{R_n = j \text{ for infinitely many } n\} = 0$$

One of the our main problems will be to classify the states of a given chain as to recurrence or nonrecurrence. The first step is to introduce an equivalence relation on the state space and show that within each equivalence class all states are of the same type.

DEFINITION. If i and j are distinct states, we say that i *leads to j* iff $f_{ij} > 0$; that is, it is possible to reach j, starting from i. Equivalently, i leads to j iff $p_{ij}^{(n)} > 0$ for some $n \geq 1$. By convention, i leads to itself. We say that i and j *communicate* iff i leads to j and j leads to i.

We define an *equivalence relation* on the state space S by taking i equivalent to j iff i and j communicate. (It is not difficult to verify that we have a legitimate equivalence relation.) The next theorem shows that *recurrence or nonrecurrence is a class property*: that is, if one state in a given equivalence class is recurrent, all states are recurrent.

Theorem 5. *If i is recurrent and i leads to j, then j is recurrent. Furthermore, $f_{ij} = f_{ji} = 1$. In fact, if f_{ij}' is the probability that j will be visited infinitely often when the initial state is i, then $f_{ij}' = f_{ji}' = 1$.*

PROOF.   Start in state i, and let T be the time of the first visit to j. Then

$$1 = \sum_{k=1}^{\infty} P\{T = k\} + P\{T = \infty\}$$

$$= \sum_{k=1}^{\infty} P\{T = k, \text{ infinitely many visits to } i \text{ after } T\} + P\{T = \infty\}$$

$$\text{since } i \text{ is recurrent}$$

$$= \sum_{k=1}^{\infty} P\{T = k\} P\{R_{T+1}, R_{T+2}, \dots \text{ visits } i \text{ infinitely often} \mid T = k, R_T = j\}$$

$$+ 1 - f_{ij}$$

$$= \sum_{k=1}^{\infty} f_{ij}^{(k)} P\{R_1, R_2, \dots \text{ visits } i \text{ infinitely often} \mid R_0 = j\} + 1 - f_{ij}$$

Thus

$$1 = f_{ij}f_{ji}' + 1 - f_{ij}, \quad \text{or} \quad f_{ij} = f_{ij}f_{ji}'$$

Since $f_{ij} > 0$ by hypothesis, $f_{ji}' = 1$.

Now if $p_{ij}^{(r)} > 0$, $p_{ji}^{(s)} > 0$, then

$$p_{jj}^{(n+r+s)} \geq p_{ji}^{(s)} p_{ii}^{(n)} p_{ij}^{(r)}$$

since one way of going from j to j in $n + r + s$ steps is to go from j to i in s steps, from i to i in n steps, and finally from i to j in r steps. It follows from Theorem 3 that $\sum_{n=1}^{\infty} p_{jj}^{(n)} = \infty$; hence j is recurrent.

Finally, we have j recurrent and $f_{ji} > 0$. By the above argument, with i and j interchanged, $f_{ij}' = 1$. Since $f_{ij} \geq f_{ij}'$, $f_{ji} \geq f_{ji}'$, it follows that $f_{ij} = f_{ji} = 1$ and the theorem is proved.

Theorem 6.   *If a finite chain (i.e., S a finite set), it is not possible for all states to be transient.*

In particular, if every state in a finite chain can be reached from every other state, so that there is only one equivalence class (namely S), then all states are recurrent.

PROOF.   If $S = \{1, 2, \dots, r\}$, then $\sum_{j=1}^{r} p_{ij}^{(n)} = 1$ for all n. Let $n \to \infty$. By Theorem 4 and the fact that the limit of a finite sum is the sum of the limits, we have $0 = \sum_{j=1}^{r} \lim_{n \to \infty} p_{ij}^{(n)} = \lim_{n \to \infty} \sum_{j=1}^{r} p_{ij}^{(n)} = 1$, a contradiction.

In the case of a finite chain, it is easy to decide whether or not a given class is recurrent; we shall see how to do this in a moment.

DEFINITION.   A nonempty subset C of the state space S is said to be *closed* iff it is not possible to leave C; that is, $\sum_{j \in C} p_{ij} = 1$ for all $i \in C$. Notice that if C is closed, then the submatrix $[p_{ij}]$, $i, j, \in C$, is stochastic; hence so is $[p_{ij}^{(n)}]$, $i, j \in C$.

Theorem 7. *C is closed iff for all $i \in C$ (i leads to j implies $j \in C$).*

PROOF. Let C be closed. If $i \in C$ and i leads to j, then $p_{ij}^{(n)} > 0$ for some n. If $j \notin C$, then $\sum_{k \in C} p_{ik}^{(n)} < 1$, a contradiction. Conversely, if the condition is satisfied and C is not closed, then $\sum_{j \in C} p_{ij} < 1$ for some $i \in C$; hence $p_{ij} > 0$ for some $i \in C, j \notin C$, a contradiction.

Theorem 8.
(a) *Let C be a recurrent class. Then C is closed.*
(b) *If C is any equivalence class, no proper subset of C is closed.*
(c) *In a finite chain, every closed equivalence class C is recurrent.*
Thus, in a finite chain, the recurrent classes are simply those classes that are closed.

PROOF.
(a) Let C be a recurrent class. If C is not closed, then by Theorem 7 we have some $i \in C$ leading to a $j \notin C$. But by Theorem 5, i and j communicate, and so i is equivalent to j. This contradicts $i \in C, j \notin C$.
(b) Let D be a (nonempty) proper subset of the arbitrary equivalence class C. Pick $i \in D$ and $j \in C, j \notin D$. Then i leads to j, since both states belong to the same equivalence class. Thus D cannot be closed.
(c) Consider C itself as a chain; this is possible since $\sum_{j \in C} p_{ij} = 1$, $i \in C$. (We are simply restricting the original transition matrix to C.) By Theorem 6 and the fact that recurrence is a class property, C is recurrent.

▶ **Example 1.** Consider the chain of Figure 7.3.1. (An arrow from i to j indicates that $p_{ij} > 0$.) There are three equivalence classes, $C_1 = \{1, 2\}$, $C_2 = \{3, 4, 5\}$, and $C_3 = \{6\}$. By Theorem 8, C_2 is recurrent and C_1 and C_3 are transient.

There is no foolproof method for classifying the states of an infinite chain, but in some cases an analysis can be done quickly. Consider the chain of Example 3, Section 7.1, and assume that all $p_k > 0$. Then every state is reachable from every other state, so that the entire state space forms a

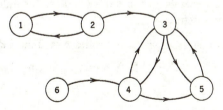

FIGURE 7.3.1 A Finite Markov Chain.

single equivalence class. We claim that the class is recurrent. For assume the contrary; then all states are transient. Let 0 be the initial state; by the remark after Theorem 4, the set $B = \{0, 1, \ldots, M - 1\}$ will be visited only finitely many times; that is, eventually $R_n \geq M$ (with probability 1). But by definition of the transition matrix, if $R_n \geq M$ then $R_{n+1} = 0$, a contradiction. ◄

We now describe another basic class property, that of periodicity.

If $p_{ii}^{(n)} > 0$ for some $n \geq 1$, that is, if starting from i it is possible to return to i, we define the *period* of i (notation: d_i) as the greatest common divisor of the set of positive integers n such that $p_{ii}^{(n)} > 0$. Equivalently, the period of i is the greatest common divisor of the set of positive integers n such that $f_{ii}^{(n)} > 0$ (see Problem 1c).

Theorem 9. *If the distinct states i and j are in the same equivalence class, they have the same period.*

PROOF. Since i and j communicate, each has a period. If $p_{ij}^{(r)} > 0$, $p_{ji}^{(s)} > 0$, then $p_{jj}^{(n+r+s)} \geq p_{ji}^{(s)} p_{ii}^{(n)} p_{ij}^{(r)}$ (see the argument of Theorem 5). Set $n = 0$ to obtain $p_{jj}^{(r+s)} > 0$, so that $r + s$ is a multiple of d_j. Thus if n is not a multiple of d_j (so neither is $n + r + s$) we have $p_{jj}^{(n+r+s)} = 0$; hence $p_{ii}^{(n)} = 0$. But this says that if $p_{ii}^{(n)} > 0$ then n is a multiple of d_j; hence $d_j \leq d_i$. By a symmetrical argument, $d_i \leq d_j$.

The transitions from state to state within a closed equivalence class C of period $d > 1$, although random, have a certain cyclic pattern, which we now describe.

Let $i, j \in C$; if $p_{ij}^{(r)} > 0$ and $p_{ij}^{(s)} > 0$, let t be such that $p_{ji}^{(t)} > 0$. Then $p_{ii}^{(r+t)} \geq p_{ij}^{(r)} p_{ji}^{(t)} > 0$; hence d divides $r + t$. Similarly, d divides $s + t$, and so d divides $s - r$.

Thus, if $r = ad + b$, a and b integers, $0 \leq b \leq d - 1$, then $s = cd + b$ for some integer c. Consequently, if i leads to j in n steps, then n is of the form $ed + b$, that is,

$$n \equiv b \bmod d$$

where the integer b depends on the states i and j but is independent of n.

Now fix $i \in C$ and define

$$C_0 = \{ j \in C \colon p_{ij}^{(n)} > 0 \text{ implies } n \equiv 0 \bmod d \}$$

$$C_1 = \{ j \in C \colon p_{ij}^{(n)} > 0 \text{ implies } n \equiv 1 \bmod d \}$$

.

.

.

$$C_{d-1} = \{ j \in C \colon p_{ij}^{(n)} > 0 \text{ implies } n \equiv d - 1 \bmod d \}$$

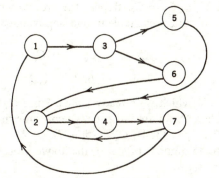

FIGURE 7.3.2 A Periodic Chain.

Then

$$C = \bigcup_{j=0}^{d-1} C_j$$

Theorem 10. *If $k \in C_t$ and $p_{kj} > 0$, then $j \in C_{t+1}$ (with indices reduced mod d; i.e., $C_d = C_0$, $C_{d+1} = C_1$, etc.). Thus, starting from i, the chain moves from C_0 to C_1 to . . . to C_{d-1} back to C_0, and so on. The C_j are called the cyclically moving subclasses of C.*

PROOF. Choose an n such that $p_{ik}^{(n)} > 0$. Then n is of the form $ad + t$. Now $p_{ij}^{(n+1)} \geq p_{ik}^{(n)} p_{kj} > 0$; hence i leads to j, and therefore $j \in C$, since C is closed. But $n \equiv t$ mod d; hence $n + 1 \equiv t + 1$ mod d, so that $j \in C_{t+1}$.

▶ **Example 2.** Consider the chain of Figure 7.3.2. Since every state leads to every other state, the entire state space forms a closed equivalence class C (necessarily recurrent by Theorem 6). We now describe an effective procedure that can be used to find the period of any finite closed equivalence class. Start with any state, say 1, and let C_0 be the subclass containing 1. Then all states reachable in one step from 1 belong to C_1; in this case $3 \in C_1$. All states reachable in one step from 3 belong to C_2; in this case $5, 6 \in C_2$. Continue in this fashion to obtain the following table, constructed according to the rule that all states reachable in one step from at least one state in C_i belong to C_{i+1}.

C_0	C_1	C_2	C_3	C_4	C_5	C_6
1	3	5, 6	2	4	7	1, 2

Stop the construction when a class C_k is reached that contains a state belonging to some C_j, $j < k$. Here we have $2 \in C_3 \cap C_6$; hence $C_3 = C_6$. Also,

$1 \in C_0 \cap C_6$; hence $C_0 = C_3 = C_6$. Repeat the process with $C_0 =$ the class containing 1 and 2; all states reachable in one step from either 1 or 2 belong to C_1. We obtain

C_0	C_1	C_2	C_3
1, 2	3, 4	5, 6, 7	1, 2

We find that $C_0 = C_3$ (which we already knew). Since $C_0 \cup C_1 \cup C_2$ is the entire equivalence class C, we are finished. We conclude that the period is 3, and that $C_0 = \{1, 2\}$, $C_1 = \{3, 4\}$, $C_2 = \{5, 6, 7\}$.

If C has only a finite number of states, the above process must terminate in a finite number of steps.

We have the following schematic representation of the powers of the transition matrix.

$$\Pi = \begin{array}{c} \\ C_0 \\ C_1 \\ C_2 \end{array} \begin{array}{ccc} C_0 & C_1 & C_2 \\ \left[\begin{array}{ccc} 0 & x & 0 \\ 0 & 0 & x \\ x & 0 & 0 \end{array}\right] \end{array} \qquad \Pi^2 = \begin{array}{c} \\ C_0 \\ C_1 \\ C_2 \end{array} \begin{array}{ccc} C_0 & C_1 & C_2 \\ \left[\begin{array}{ccc} 0 & 0 & x \\ x & 0 & 0 \\ 0 & x & 0 \end{array}\right] \end{array}$$

$$\Pi^3 = \begin{array}{c} \\ C_0 \\ C_1 \\ C_2 \end{array} \begin{array}{ccc} C_0 & C_1 & C_2 \\ \left[\begin{array}{ccc} x & 0 & 0 \\ 0 & x & 0 \\ 0 & 0 & x \end{array}\right] \end{array} \qquad (x \text{ stands for positive element})$$

Notice that Π^4 has the same form as Π but is not the same numerically; similarly, Π^5 has the same form as Π^2, Π^6 has the same form as Π^3, and so on. ◄

▶ **Example 3.**   Consider the simple random walk.
   (a) If there are no barriers, the entire state space forms a closed equivalence class with period 2. We have seen that $f_{00} = 1 - |p - q|$ [see (6.2.7)]; by symmetry, $f_{ii} = f_{00}$ for all i. Thus if $p = q$ the class is recurrent, and if $p \neq q$ the class is transient.
   (b) If there is an absorbing barrier at 0, then there are two classes, $C = \{0\}$ and $D = \{1, 2, \ldots\}$. C is clearly recurrent, and since D is not closed, it is transient by Theorem 8. C has period 1, and D has period 2.
   (c) If there are absorbing barriers at 0 and b, then there are three equivalence classes, $C = \{0\}$, $D = \{1, 2, \ldots, b - 1\}$, $E = \{b\}$. C and E have period 1 and are recurrent; D has period 2 and is transient. ◄

REMARK.   We have seen that if B is a finite set of transient states, the probability of remaining forever in B is 0. This is not true for an infinite set of transient states. For example, in the simple random walk with an absorbing barrier at 0, if the initial state is $x \geq 1$ and $p > q$, there is a probability $1 - (q/p)^x > 0$ of remaining forever in the transient class $D = \{1, 2, \ldots\}$ [see (6.2.6)].

TERMINOLOGY.   A state (or class) is said to be *aperiodic* iff its period d is 1, *periodic* iff $d > 1$.

PROBLEMS

1. (a) Let A be a (possibly infinite) set of positive integers with greatest common divisor d. Show that there is a finite subset of A with greatest common divisor d.
   (b) If A is a nonempty set of positive integers with greatest common divisor d, and A is closed under addition, show that all sufficiently large multiples of d belong to A.
   (c) If d_i is the period of the state i, show that d_i is the greatest common divisor of $\{n \geq 1 : f_{ii}^{(n)} > 0\}$.

2. A state i is said to be *essential* iff its equivalence class is closed. Show that i is essential iff, whenever i leads to j, it follows that j leads to i.

3. Prove directly (without using Theorem 5) that an equivalence class that is not closed must be transient.

4. Classify the states of the following Markov chains. [In (a) and (b) assume $0 < p < 1$.]
   (a) Simple random walk with reflecting barrier at 0 ($S = \{1, 2, \ldots\}$, $p_{11} = q$, $p_{i,i+1} = p$ for all i, $p_{i,i-1} = q$, $i = 2, 3, \ldots$)
   (b) Simple random walk with reflecting barriers at 0 and $l + 1$ ($S = \{1, 2, \ldots, l\}$, $p_{11} = q$, $p_{ll} = p$, $p_{i,i+1} = p$, $i = 1, 2, \ldots, l - 1$, $p_{i,i-1} = q$, $i = 2, 3, \ldots, l$)
   (c)

$$\Pi = \begin{bmatrix} .2 & .8 & 0 & 0 \\ 0 & 0 & .1 & .9 \\ 0 & 0 & .2 & .8 \\ .7 & .3 & 0 & 0 \end{bmatrix}$$

   (d)

$$\Pi = \begin{bmatrix} \frac{1}{2} & \frac{1}{2} & 0 \\ 0 & \frac{1}{3} & \frac{2}{3} \\ 0 & \frac{1}{2} & \frac{1}{2} \end{bmatrix}$$

5. Let $R_0, R_1, R_2, \ldots$ be independent random variables, all having the same distribution function, with values in the countable set S; assume $P\{R_i = j\} > 0$ for all $j \in S$.
   (a) Show that $\{R_n\}$ may be regarded as a Markov chain; what is Π?
   (b) Classify the states of the chain.

6. Let i be a state of a Markov chain, and let

$$H(z) = \sum_{n=0}^{\infty} f_{ii}^{(n)} z^n, \qquad U(z) = \sum_{n=0}^{\infty} p_{ii}^{(n)} z^n, \qquad |z| \leq 1$$

[take $f_{ii}^{(0)} = 0$]. Use the first entrance theorem to show that $U(z) - 1 = H(z)U(z)$.

7.4   LIMITING PROBABILITIES

In this section we investigate the limiting behavior of the n-step transition probability $p_{ij}^{(n)}$. The basic result is the following.

Theorem 1. *Let* $f_1, f_2, \ldots$ *be a sequence of nonnegative numbers with* $\sum_{n=1}^{\infty} f_n = 1$, *such that the greatest common divisor of* $\{j: f_j > 0\}$ *is 1. Set* $u_0 = 1$, $u_n = \sum_{k=1}^{n} f_k u_{n-k}$, $n = 1, 2, \ldots$ *. Define* $\mu = \sum_{n=1}^{\infty} n f_n$. *Then* $u_n \to 1/\mu$ *as* $n \to \infty$.

We shall apply the theorem to a Markov chain with $f_n = f_{ii}^{(n)}$, i a given recurrent state with period 1; then $u_n = p_{ii}^{(n)}$ by the first entrance theorem. Also, $\mu = \mu_i = \sum_{n=1}^{\infty} n f_{ii}^{(n)}$, so that if T is the time required to return to i when the initial state is i, then $\mu_i = E(T)$. If i is an arbitrary recurrent state of a Markov chain, μ_i is called the *mean recurrence time* of i.

Theorem 1 states that $p_{ii}^{(n)} \to 1/\mu_i$; thus, starting in i, there is a limiting probability for state i, namely, the reciprocal of the mean recurrence time. Intuitively, if $\mu_i = $ (say) 4, then for large n we should be in state i roughly one quarter of the time, and it is reasonable to expect that $p_{ii}^{(n)} \to 1/4$.

PROOF.   We first list three results from real analysis that will be needed. All numbers a_{kj}, c_j, a_j, b_j are assumed real.

1. *Fatou's Lemma:*   If $|a_{kj}| \leq c_j$, $k, j = 1, 2, \ldots$, and $\sum_j c_j < \infty$, then

$$\limsup_{k \to \infty} \sum_j a_{kj} \leq \sum_j \limsup_{k \to \infty} a_{kj}$$

and

$$\liminf_{k \to \infty} \sum_j a_{kj} \geq \sum_j \liminf_{k \to \infty} a_{kj}$$

The "lim inf" statement holds without the hypothesis that

$$|a_{kj}| \leq c_j, \sum_j c_j < \infty, \qquad \text{if all } a_{kj} \geq 0$$

2. *Dominated Convergence Theorem:* If $|a_{kj}| \leq c_j$, k, $j = 1, 2, \ldots$, $\sum_j c_j < \infty$, and $\lim_{k \to \infty} a_{kj} = a_j$, $j = 1, 2, \ldots$, then

$$\lim_{k \to \infty} \sum_j a_{kj} = \sum_j \lim_{k \to \infty} a_{kj} \left(= \sum_j a_j \right)$$

(The dominated convergence theorem follows from Fatou's lemma. Alternatively, a fairly short direct proof may be given.)

3. $\liminf_{k \to \infty} (a_k + b_k) \leq \liminf_{k \to \infty} a_k + \limsup_{k \to \infty} b_k$.
(This follows quickly from the definitions of lim inf and lim sup.)

We now prove Theorem 1. First notice that $0 \leq u_n \leq 1$ for all n (by induction). Define

$$r_n = \sum_{j=n+1}^{\infty} f_j, \qquad n = 0, 1, \ldots$$

Then

$$u_n = f_1 u_{n-1} + \cdots + f_n u_0 = (r_0 - r_1) u_{n-1} + \cdots + (r_{n-1} - r_n) u_0, \qquad n \geq 1 \tag{7.4.1}$$

Since $r_0 = \sum_{j=1}^{\infty} f_j = 1$, we have $u_n = r_0 u_n$, and thus we may rearrange terms in (7.4.1) to obtain

$$r_0 u_n + r_1 u_{n-1} + \cdots + r_n u_0 = r_0 u_{n-1} + \cdots + r_{n-1} u_0, \qquad n \geq 1$$

This indicates that $\sum_{k=0}^{n} r_k u_{n-k}$ is independent of n; hence

$$\sum_{k=0}^{n} r_k u_{n-k} = r_0 u_0 = 1, \qquad n = 0, 1, \ldots \tag{7.4.2}$$

[An alternative proof that $\sum_{k=0}^{n} r_k u_{n-k} = 1$: construct a Markov chain with $f_{ii}^{(n)} = f_n$, $p_{ii}^{(n)} = u_n$ (see Problem 1). Then

$$\sum_{k=0}^{n} r_k u_{n-k} = \sum_{k=0}^{n} u_k r_{n-k}$$

$$= \sum_{k=0}^{n} p_{ii}^{(k)} P\{T > n - k \mid R_0 = i\}$$

(where T is the time required to return to i when the initial state is i)

$$= \sum_{k=0}^{n} P\{R_k = i, R_{k+1} \neq i, \ldots, R_n \neq i \mid R_0 = i\}$$

$$= P\{R_k = i \text{ for some } k = 0, 1, \ldots, n \mid R_0 = i\}$$

$$= 1]$$

Now let $b = \limsup_n u_n$. Pick a subsequence $\{u_{n_k}\}$ converging to b. Then

$$b = \lim_k u_{n_k} = \liminf_k u_{n_k}$$

$$= \liminf_k \left[f_i u_{n_k-i} + \sum_{\substack{j=1 \\ j \neq i}}^{n_k} f_j u_{n_k-j} \right]$$

$$\leq \liminf_k (f_i u_{n_k-i}) + \left(\sum_{\substack{j=1 \\ j \neq i}}^{\infty} f_j \right) b$$

[We use here the fact that $\liminf_k (a_k + b_k) \leq \liminf_k a_k + \limsup_k b_k$. Furthermore, if we take $u_n = 0$ for $n < 0$, then

$$\limsup_k \left(\sum_{\substack{j=1 \\ j \neq i}}^{\infty} f_j u_{n_k-j} \right) \leq \sum_{\substack{j=1 \\ j \neq i}}^{\infty} f_j \limsup_k (u_{n_k-j}) \leq b \sum_{\substack{j=1 \\ j \neq i}}^{\infty} f_j$$

Notice that since $|f_j u_{n_k-j}| \leq f_j$ and $\sum_{j=1}^{\infty} f_j < \infty$, Fatou's lemma applies.]
   Therefore

$$b \leq f_i \liminf_k u_{n_k-i} + (1 - f_i)b$$

or

$$f_i \liminf_k u_{n_k-i} \geq f_i b$$

Thus $f_i > 0$ implies $u_{n_k-i} \to b$ as $k \to \infty$.

It follows that $u_{n_k-i} \to b$ for sufficiently large i. For if $f_i > 0$, we apply the above argument to the sequence u_{n_k-i}, $k = 1, 2, \ldots$, to show that $f_j > 0$ implies $u_{n_k-i-j} \to b$. Thus if $t = \sum_{r=1}^{m} a_r i_r$, where the a_r are positive integers and $f_{i_r} > 0$, then $u_{n_k-t} \to b$. The set S of all such t's is closed under addition and has greatest common divisor 1, since S is generated by the positive integers i for which $f_i > 0$. Thus (Problem 1b, Section 7.3) S contains all sufficiently large positive integers. Say $u_{n_k-i} \to b$ for $i \geq I$. By (7.4.2),

$$\sum_{j=0}^{\infty} r_j u_{n_k-I-j} = 1 \qquad \text{(with } u_n = 0 \text{ for } n < 0) \qquad (7.4.3)$$

If $\sum_{j=0}^{\infty} r_j < \infty$, the dominated convergence theorem shows that we may let $k \to \infty$ and take limits term by term in (7.4.3) to obtain $b \sum_{j=0}^{\infty} r_j = 1$. If $\sum_{j=0}^{\infty} r_j = \infty$, Fatou's lemma gives $1 \geq b \sum_{j=0}^{\infty} r_j$; hence $b = 0$. In either case, then,

$$b = \left[\sum_{j=0}^{\infty} r_j \right]^{-1}$$

But

$$r_0 = f_1 + f_2 + f_3 + \cdots$$
$$r_1 = \qquad f_2 + f_3 + \cdots$$
$$r_2 = \qquad\qquad f_3 + \cdots$$
.
.
.

Hence

$$\sum_{n=0}^{\infty} r_n = \sum_{n=1}^{\infty} nf_n = \mu$$

Consequently $b = \lim \sup_n u_n = 1/\mu$. By an entirely symmetric argument, $\lim \inf_n u_n = 1/\mu$, and the result follows.

We now apply Theorem 1 to gain complete information about the limiting behavior of the n-step transition probability $p_{ij}^{(n)}$. A recurrent state j is said to be *positive* iff its mean recurrence time μ_j is $< \infty$, *null* iff $\mu_j = \infty$.

Theorem 2.
(a) *If the state j is transient then $\sum_{n=1}^{\infty} p_{ij}^{(n)} < \infty$ for all i, hence*

$$p_{ij}^{(n)} \to 0 \qquad \text{as } n \to \infty$$

PROOF. This is Theorem 4 of Section 7.3.

(b) *If j is recurrent and aperiodic, and i belongs to the same equivalence class as j, then $p_{ij}^{(n)} \to 1/\mu_j$. Furthermore, μ_j is finite iff μ_i is finite. If i belongs to a different class, then $p_{ij}^{(n)} \to f_{ij}/\mu_j$.*

PROOF.

$$p_{ij}^{(n)} = \sum_{k=1}^{\infty} f_{ij}^{(k)} p_{jj}^{(n-k)}$$

by the first entrance theorem [take $p_{jj}^{(r)} = 0$, $r < 0$]. By the dominated convergence theorem, we may take limits term by term as $n \to \infty$; since $p_{jj}^{(n-k)} \to 1/\mu_j$ by Theorem 1, we have

$$p_{ij}^{(n)} \to \left(\sum_{k=1}^{\infty} f_{ij}^{(k)} \right) \frac{1}{\mu_j} = \frac{f_{ij}}{\mu_j}$$

If i and j belong to the same recurrent class, $f_{ij} = 1$.

Now assume that μ_j is finite. If $p_{ij}^{(r)}, p_{ji}^{(s)} > 0$, then $p_{ii}^{(n+r+s)} \geq p_{ij}^{(r)} p_{jj}^{(n)} p_{ji}^{(s)}$; this is bounded away from 0 for large n, since $p_{jj}^{(n)} \to 1/\mu_j > 0$. But if $\mu_i = \infty$, then $p_{ii}^{(n)} \to 0$ as $n \to \infty$, a contradiction. This proves (b).

(c) *Let j be recurrent with period $d > 1$. Let i be in the same class as j, with $i \in$ the cyclically moving subclass C_r, $j \in C_{r+a}$. Then $p_{ij}^{(nd+a)} \to d/\mu_j$. Also, μ_j is finite iff μ_i is finite, so that the property of being recurrent positive (or recurrent null) is a class property.*

PROOF. First assume $a = 0$. Then j is recurrent and aperiodic relative to the chain with transition matrix Π^d. (If A has greatest common divisor d, then the greatest common divisor of $\{x/d: x \in A\}$ is 1.) By (b),

$$p_{ij}^{(nd)} \to \frac{1}{\displaystyle\sum_{k=1}^{\infty} k f_{jj}^{(kd)}} = \frac{d}{\displaystyle\sum_{k=1}^{\infty} kd f_{jj}^{(kd)}} = \frac{d}{\mu_j}$$

Now, having established the result for $a = r$, assume $a = r + 1$ and write

$$p_{ij}^{(nd+r+1)} = \sum_k p_{ik} p_{kj}^{(nd+r)} \to \sum_k p_{ik} \frac{d}{\mu_j} = \frac{d}{\mu_j}$$

as asserted.

The argument that μ_j is finite iff μ_i is finite is the same as in (b), with nd replacing n.

(d) *If j is recurrent with period $d > 1$, and i is an arbitrary state, then*

$$p_{ij}^{(nd+a)} \to \left[\sum_{k=0}^{\infty} f_{ij}^{(kd+a)} \right] \frac{d}{\mu_j}, \qquad a = 0, 1, \ldots, d - 1$$

The expression in brackets is the probability of reaching j from i in a number of steps that is $\equiv a \bmod d$. Thus, if j is recurrent null, $p_{ij}^{(n)} \to 0$ as $n \to \infty$ for all i.

PROOF.

$$p_{ij}^{(nd+a)} = \sum_{k=1}^{nd+a} f_{ij}^{(k)} p_{jj}^{(nd+a-k)}, \qquad a = 0, 1, \ldots, d - 1$$

Since j has period d, $p_{jj}^{(nd+a-k)} = 0$ unless $k - a$ is of the form rd (necessarily $r \le n$); hence

$$p_{ij}^{(nd+a)} = \sum_{r=0}^{n} f_{ij}^{(rd+a)} p_{jj}^{((n-r)d)}$$

Let $n \to \infty$ and use (c) to finish the proof.

(e) *A finite chain has no recurrent null states.*

PROOF.   Let C be a finite recurrent null class, say, $C = \{1, 2, \ldots, r\}$. Then

$$\sum_{j=1}^{r} p_{ij}^{(n)} = 1, \qquad i \in C$$

Let $n \to \infty$; by (d) we obtain $0 = 1$, a contradiction.

PROBLEMS

1. With f_n and u_n as in Theorem 1, show how to construct a Markov chain with a state i such that $f_{ii}^{(n)} = f_n$ and $p_{ii}^{(n)} = u_n$ for all n.

2. (The renewal theorem) Let $T_1, T_2, \ldots$ be independent random variables, all with the same distribution function, taking values on the positive integers. (Think of the T_i as waiting times for customers to arrive, or as lifetimes of a succession of products such as light bulbs. If $T_1 + \cdots + T_n = x$, bulb n has burned out at time x, and the light must be renewed by placing bulb $n + 1$ in position.) Assume that the greatest common divisor of $\{x : P\{T_k = x\} > 0\}$ is d and let $G(n) = \sum_{k=1}^{\infty} P\{T_1 + \cdots + T_k = n\}, n = 1, 2, \ldots$. If $\mu = E(T_i)$, show that $\lim_{n \to \infty} G(nd) = d/\mu$; interpret the result intuitively.

3. Show that in any Markov chain, $(1/n) \sum_{k=1}^{n} p_{ij}^{(k)}$ approaches a limit as $n \to \infty$, namely, f_{ij}/μ_j. (Define $\mu_j = \infty$ if j is transient.) HINT:

$$\frac{1}{n} \sum_{k=1}^{n} p_{ij}^{(k)} = \sum_{r=1}^{d} \frac{1}{n} \sum_{\substack{k=1 \\ k \equiv r \bmod d}}^{n} p_{ij}^{(k)}, \qquad d = \text{period of } j$$

4. Let V_{ij} be the number of visits to the state j, starting at i. (If $i = j$, $t = 0$ counts as a visit.)
   (a) Show that $E(V_{ij}) = \sum_{n=0}^{\infty} p_{ij}^{(n)}$. Thus i is recurrent iff $E(V_{ii}) = \infty$, and if j is transient, $E(V_{ij}) < \infty$ for all i.
   (b) Let C be a transient class, $N_{ij} = E(V_{ij})$, $i, j \in C$. Show that

$$N_{ij} = \delta_{ij} + \sum_{k \in C} p_{ik} N_{kj} \qquad (\delta_{ij} = 1, i = j$$
$$= 0, i \neq j)$$

   In matrix form, $N = I + QN$, $Q = \Pi$ restricted to C.
   (c) Show that $(I - Q)N = N(I - Q) = I$ so that $N = (I - Q)^{-1}$ in the case of a finite chain (the inverse of an *infinite* matrix need not be unique).

REMARK.   (a) implies that in the gambler's ruin problem with finite capital, the average duration of the game is finite. For if the initial capital is i and D is the duration of the game, then $D = \sum_{j=1}^{b-1} V_{ij}$, so that $E(D) < \infty$.

7.5 STATIONARY AND STEADY-STATE DISTRIBUTIONS

A *stationary distribution* for a Markov chain with state space S is a set of numbers v_i, $i \in S$, such that $v_i \geq 0$, $\sum_{i \in S} v_i = 1$, and

$$\sum_{i \in S} v_i p_{ij} = v_j, \qquad j \in S$$

Thus, if $V = (v_i, i \in S)$, then $V\Pi = V$. By induction, $V\Pi^n = V\Pi (\Pi^{n-1}) = V\Pi^{n-1} = \cdots = V\Pi = V$, so that $V\Pi^n = V$ for all $n = 0, 1, \ldots$. Therefore, if the initial state distribution is V, the state distribution at all future times is still V. Furthermore, since

$$P\{R_n = i, R_{n+1} = i_1, \ldots, R_{n+k} = i_k\}$$

$$= P\{R_n = i\} p_{ii_1} p_{i_1 i_2} \cdots p_{i_{k-1} i_k} \qquad \text{(Problem 4, Section 7.1)}$$

$$= v_i p_{ii_1} p_{i_1 i_2} \cdots p_{i_{k-1} i_k}$$

the sequence $\{R_n\}$ is stationary; that is, the joint probability function of $R_n, R_{n+1}, \ldots, R_{n+k}$ does not depend on n.

Stationary distributions are closely related to limiting probabilities. The main result is the following.

Theorem 1. *Consider a Markov chain with transition matrix $[p_{ij}]$. Assume*

$$\lim_{n \to \infty} p_{ij}^{(n)} = q_j$$

for all states i, j (where q_j does not depend on i). Then

(a) *$\sum_{j \in S} q_j \leq 1$ and $\sum_{i \in S} q_i p_{ij} = q_j, j \in S$.*
(b) *Either all $q_j = 0$, or else $\sum_{j \in S} q_j = 1$.*
(c) *If all $q_j = 0$, there is no stationary distribution. If $\sum_{j \in S} q_j = 1$, then $\{q_j\}$ is the unique stationary distribution.*

PROOF.

$$\sum_j q_j = \sum_j \lim_n p_{ij}^{(n)} \leq \liminf_n \sum_j p_{ij}^{(n)}$$

by Fatou's lemma; hence

$$\sum_j q_j \leq 1$$

Now

$$\sum_i q_i p_{ij} = \sum_i (\lim_n p_{ki}^{(n)}) p_{ij} \leq \liminf_n \sum_i p_{ki}^{(n)} p_{ij} = \liminf_n p_{kj}^{(n+1)} = q_j$$

But if $\sum_i q_i p_{ij_0} < q_{j_0}$ for some j_0, then

$$\sum_j q_j > \sum_j \sum_i q_i p_{ij} = \sum_i q_i \sum_j p_{ij} = \sum_i q_i$$

which is a contradiction. This proves (a).

Now if $Q = (q_i, i \in S)$, then by (a), $Q\Pi = Q$; hence, by induction, $Q\Pi^n = Q$, that is, $\sum_i q_i p_{ij}^{(n)} = q_j$. Thus

$$q_j = \lim_n \sum_i q_i p_{ij}^{(n)} = \sum_i q_i \lim_n p_{ij}^{(n)}$$

by the dominated convergence theorem. Hence $q_j = (\sum_i q_i)q_j$, proving (b).

Finally, if $\{v_i\}$ is a stationary distribution, then $\sum_i v_i p_{ij}^{(n)} = v_j$. Let $n \to \infty$ to obtain $\sum_i v_i q_j = v_j$, so that $q_j = v_j$. Consequently, if a stationary distribution exists, it is unique and coincides with $\{q_j\}$. Therefore no stationary distribution can exist if all $q_j = 0$; if $\sum_j q_j = 1$, then, by (a), $\{q_j\}$ is stationary and the result is established.

The numbers v_i, $i \in S$, are said to form a *steady-state distribution* iff $\lim_{n \to \infty} p_{ij}^{(n)} = v_j$ for all $i, j \in S$, and $\sum_{j \in S} v_j = 1$. Thus we require that limiting probabilities exist (independent of the initial state) and form a probability distribution.

In the case of a finite chain, a set of limiting probabilities that are independent of the initial state must form a steady-state distribution, that is, the case in which all $q_j = 0$ cannot occur in Theorem 1. For $\sum_{j \in S} p_{ij}^{(n)} = 1$ for all $i \in S$; let $n \to \infty$ to obtain, since S is finite, $\sum_{j \in S} q_j = 1$. If the chain is infinite, this result is no longer valid. For example, if all states are transient, then $p_{ij}^{(n)} \to 0$ for all i, j.

If $\{q_j\}$ is a steady-state distribution, $\{q_j\}$ is the unique stationary distribution, by Theorem 1. However, a chain can have a unique stationary distribution without having a steady-state distribution, in fact without having limiting probabilities. We give examples later in the section.

We shall establish conditions under which a steady-state distribution exists after we discuss the existence and uniqueness of stationary distributions.

Let N be the number of positive recurrent classes.

CASE 1. $N = 0$. Then all states are transient or recurrent null. Hence $p_{ij}^{(n)} \to 0$ for all i, j by Theorem 2 of Section 7.4, so that, by Theorem 1 of this section, there is no stationary distribution.

CASE 2. $N = 1$. Let C be the unique positive recurrent class. If C is aperiodic, then, by Theorem 2 of Section 7.4, $p_{ij}^{(n)} \to 1/\mu_j$, $i, j \in C$. If $j \notin C$, then j is transient or recurrent null, so that $p_{ij}^{(n)} \to 0$ for all i. By Theorem 1, if we assign $v_j = 1/\mu_j, j \in C, v_j = 0, j \notin C$, then $\{v_j\}$ is the unique stationary distribution, and $p_{ij}^{(n)} \to v_j$ for all i, j.

Now assume C periodic, with period $d > 1$. Let D be a cyclically moving subclass of C. The states of D are recurrent and aperiodic relative to the transition matrix Π^d. By Theorem 2 of Section 7.4, $p_{ij}^{(nd)} \to d/\mu_j$, $i, j \in D$; hence $\{d/\mu_j, j \in D\}$ is the unique stationary distribution for D relative to Π^d (in particular, $\sum_{j \in D} 1/\mu_j = 1/d$). It follows that $v_j = 1/\mu_j, j \in C, v_j = 0, j \notin C$, gives the unique stationary distribution for the original chain (see Problem 1).

CASE 3. $N \geq 2$. There is a unique stationary distribution for each positive recurrent class, hence uncountably many stationary distributions for the original chain. For if $V_1 \Pi = V_1, V_2 \Pi = V_2$, then, if $a_1, a_2 \geq 0$, $a_1 + a_2 = 1$, we have

$$(a_1 V_1 + a_2 V_2)\Pi = a_1 V_1 + a_2 V_2$$

In summary, *there is a unique stationary distribution if and only if there is exactly one positive recurrent class.*

Finally, we have the basic theorem concerning steady-state distributions.

Theorem 2.
(a) *If there is a steady-state distribution, there is exactly one positive recurrent class C, and this class is aperiodic; also, $f_{ij} = 1$ for all $j \in C$ and all $i \in S$.*
(b) *Conversely, if there is exactly one positive recurrent class C, which is aperiodic, and, in addition, $f_{ij} = 1$ for all $j \in C$ and all $i \in S$, then a steady-state distribution exists.*

PROOF.
(a) Let $\{v_j\}$ be a steady-state distribution. By Theorem 1, $\{v_j\}$ is the unique stationary distribution; hence there must be exactly one positive recurrent class C. Suppose that C has period $d > 1$, and let $i \in$ a cyclically moving subclass $C_0, j \in C_1$. Then $p_{ij}^{(nd+1)} \to d/\mu_j$ by Theorem 2 of Section 7.4, and $p_{ij}^{(nd)} = 0$ for all n. Since $d/\mu_j > 0$, $p_{ij}^{(n)}$ has no limit as $n \to \infty$, contradicting the hypothesis. If $j \in C$ and $i \in S$, then by Theorem 2(b) of Section 7.4, $p_{ij}^{(n)} \to f_{ij}/\mu_j$, hence $v_j = f_{ij}/\mu_j$. Since v_j does not depend on i, we have $f_{ij} = f_{jj} = 1$.
(b) By Theorem 2(b) of Section 7.4,

$$p_{ij}^{(n)} \to \frac{f_{ij}}{\mu_j} \qquad \text{for all } i, \text{ if } j \in C$$

$$\to 0 \qquad \text{for all } i \text{ if } j \notin C$$

since in this case j is transient or recurrent null. Therefore, if $f_{ij} = 1$ for all $i \in S$ and $j \in C$, the limit v_j is independent of i. Since C is positive, $v_j > 0$ for $j \in C$; hence, by Theorem 1, $\sum_j v_j = 1$ and the result follows.

[Note that if a steady state distribution exists, there are no recurrent null classes (or closed transient classes). For if D is such a class and $i \in D$, then since D is closed, $f_{ij} = 0$ for all $j \in C$, a contradiction. Thus in Theorem 2, the statement "there is exactly one positive recurrent class, which is aperiodic" may be replaced by "there is exactly one recurrent class, which is positive and aperiodic".]

COROLLARY.   Consider a finite chain.
(a) A steady-state distribution exists iff there is exactly one closed equivalence class C, and C is aperiodic.
(b) There is a unique stationary distribution iff there is exactly one closed equivalence class.

PROOF.   The result follows from Theorem 2, with the aid of Theorem 8c of Section 7.3, Theorem 2e of Section 7.4, and the fact that if B is a finite set of transient states, the probability of remaining forever in B is 0 (see the remark after Theorem 4 of Section 7.3).

[It is not difficult to verify that a finite chain has at least one closed equivalence class. Thus a finite chain always has at least one stationary distribution.]

REMARK.   Consider a finite chain with exactly one closed equivalence class, which is periodic. Then, by the above corollary, there is a unique stationary distribution but no steady-state distribution, in fact no limiting probabilities (see the argument of Theorem 2a). For example, consider the chain with transition matrix

$$\Pi = \begin{bmatrix} 0 & 1 \\ 1 & 0 \end{bmatrix}$$

The unique stationary distribution is $(1/2, 1/2)$, but

$$\Pi^n = \begin{bmatrix} 1 & 0 \\ 0 & 1 \end{bmatrix}, \qquad n \text{ even}$$

$$\Pi^n = \begin{bmatrix} 0 & 1 \\ 1 & 0 \end{bmatrix}, \qquad n \text{ odd}$$

and therefore Π^n does not approach a limit.

Usually the easiest way to find a steady-state distribution $\{v_j\}$, if it exists, is to use the fact that a steady-state distribution must be the unique stationary distribution. Thus we solve the equations

$$\sum_{i \in S} v_i p_{ij} = v_j, \qquad j \in S$$

under the conditions that all $v_i \geq 0$ and $\sum_{i \in S} v_i = 1$.

PROBLEMS

1. Show that if there is a single positive recurrent class C, then $\{1/\mu_j, j \in C\}$, with probability 0 assigned to states outside C, gives the unique stationary distribution for the chain. HINT: $p_{ii}^{(nd)} = \sum_{k \in C} p_{ik}^{(nd-1)} p_{ki}$, $i \in C$. Use Fatou's lemma to show that $1/\mu_i \geq \sum_{k \in C}(1/\mu_k) p_{ki}$. Then use the fact that $\sum_{j \in C} 1/\mu_j = 1$.

2. (a) If, for some N, Π^N has a column bounded away from 0, that is, if for some j_0 and some $\delta > 0$ we have $p_{ij_0}^{(N)} \geq \delta > 0$ for all i, show that there is exactly one recurrent class (namely, the class of j_0); this class is positive and aperiodic.

   (b) In the case of a finite chain, show that a steady-state distribution exists iff Π^N has a positive column for some N.

3. Classify the states of the following Markov chains. Discuss the limiting behavior of the transition probabilities and the existence of steady-state and stationary distributions.

   1. Simple random walk with no barriers.
   2. Simple random walk with absorbing barrier at 0.
   3. Simple random walk with absorbing barriers at 0 and b.
   4. Simple random walk with reflecting barrier at 0.
   5. Simple random walk with reflecting barriers at 0 and $l + 1$.
   6. The chain of Example 2, Section 7.1.
   7. The chain of Problem 4c, Section 7.3.
   8. The chain of Problem 4d, Section 7.3.
   9. A sequence of independent random variables (Problem 5, Section 7.3).
   10. The chain with transition matrix

$$
\Pi =
\begin{array}{c c}
 & \begin{array}{c c c c c c c} 1 & 2 & 3 & 4 & 5 & 6 & 7 \end{array} \\
\begin{array}{c} 1 \\ 2 \\ 3 \\ 4 \\ 5 \\ 6 \\ 7 \end{array} &
\left[
\begin{array}{c c c c c c c}
0 & 0 & 1 & 0 & 0 & 0 & 0 \\
0 & 0 & 0 & 1 & 0 & 0 & 0 \\
0 & 0 & 0 & 0 & \frac{3}{4} & \frac{1}{4} & 0 \\
0 & 0 & 0 & 0 & 0 & 0 & 1 \\
0 & 1 & 0 & 0 & 0 & 0 & 0 \\
0 & 1 & 0 & 0 & 0 & 0 & 0 \\
\frac{1}{2} & \frac{1}{2} & 0 & 0 & 0 & 0 & 0
\end{array}
\right]
\end{array}
$$

8

Introduction to Statistics

8.1 STATISTICAL DECISIONS

Suppose that the number of telephone calls made per day at a given exchange is known to have a Poisson distribution with parameter θ, but θ itself is unknown. In order to obtain some information about θ, we observe the number of calls over a certain period of time, and then try to come to a decision about θ. The nature of the decision will depend on the type of information desired. For example, it may be that extra equipment will be needed if $\theta > \theta_0$, but not if $\theta \leq \theta_0$. In this case we make one of two possible decisions: we decide either that $\theta > \theta_0$ or that $\theta \leq \theta_0$. Alternatively, we may want to estimate the actual value of θ in order to know how much equipment to install. In this case the decision results in a number $\hat{\theta}$, which we hope is as close to θ as possible. In general, an incorrect decision will result in a loss, which may be measurable in precise terms, as in the case of the cost of unnecessary equipment, but which also may have intangible components. For example, it may be difficult to assign a numerical value to losses due to customer complaints, unfavorable publicity, or government investigations.

Decision problems such as the one just discussed may be formulated mathematically by means of a *statistical decision model*. The ingredients of the model are as follows.

1. N, the set of *states of nature*.

2. A random variable (or random vector) R, the *observable*, whose distribution function F_θ depends on the particular $\theta \in N$. We may imagine that "nature" chooses the parameter $\theta \in N$ (without revealing the result to us); we then observe the value of a random variable R with distribution function F_θ. In the above example, N is the set of positive real numbers, and F_θ is the distribution function of a Poisson random variable with parameter θ.

3. A, the set of possible *actions*. In the above example, since we are trying to determine the value of θ, $A = N = (0, \infty)$.

4. A *loss function* (or *cost function*) $L(\theta, a)$, $\theta \in N$, $a \in A$; $L(\theta, a)$ represents our loss when the true state of nature is θ and we take action a.

The process by which we arrive at a decision may be described by means of a *decision function*, defined as follows.

Let E be the range of the observable R (e.g., E^1 if R is a random variable, E^n if R is an n-dimensional random vector). A *nonrandomized decision function* is a function φ from E to A. Thus, if R takes the value x, we take action $\varphi(x)$. φ is to be chosen so as to minimize the loss, in some sense.

Nonrandomized decision functions are not adequate to describe all aspects of the decision-making process. For example, under certain conditions we may flip a coin or use some other chance device to determine the appropriate action. (If you are a statistician employed by a company, it is best to do this out of sight of the customer.) The general concept of a *decision function* is that of a mapping assigning to each $x \in E$ a probability measure P_x on an appropriate sigma field of subsets of A. Thus $P_x(B)$ is the probability of taking an action in the set B when $R = x$ is observed. A nonrandomized decision function may be regarded as a decision function with each P_x concentrated on a single point; that is, for each x we have $P_x\{a\} = 1$ for some a $(=\varphi(x))$ in A.

We shall concentrate on the two most important special cases of the statistical decision problem, hypothesis testing and estimation.

A typical physical situation in which decisions of this type occur is the problem of signal detection. The input to a radar receiver at a particular instant of time may be regarded as a random variable R with density f_θ, where θ is related to the signal strength. In the simplest model, $R = \theta + R'$, where R' (the noise) is a random variable with a specified density, and θ is a fixed but unknown constant determined by the strength of the signal. We may be interested in distinguishing between two conditions: the absence of a target ($\theta = \theta_0$) versus its presence ($\theta = \theta_1$); this is an example of a hypothesis-testing problem. Alternatively, we may know that a signal is present and wish to estimate its strength. Thus, after observing R, we record a number that we hope is close to the true value of θ; this is an example of a problem in estimation.

As another example, suppose that θ is the (unknown) percentage of defective components produced on an assembly line. We inspect n components (i.e., we observe $R_1, \ldots, R_n$, where $R_i = 1$ if component i is defective, $R_i = 0$ if component i is accceptable) and then try to say something about θ. We may be trying to distinguish between the two conditions $\theta \leq \theta_0$ and $\theta > \theta_0$ (hypothesis testing), or we may be trying to come as close as possible to the true value of θ (estimation).

In working the specific examples in the chapter, the table of common density and probability functions and their properties given at the end of the book may be helpful.

8.2 HYPOTHESIS TESTING

Consider again the statistical decision model of the preceding section. Suppose that H_0 and H_1 are disjoint nonempty subsets of N whose union is N, and our objective is to determine whether the true state of nature θ belongs to H_0 or to H_1. (In the example on the telephone exchange, H_0 might correspond to $\theta \leq \theta_0$, and H_1 to $\theta > \theta_0$.) Thus our ultimate decision must be either "$\theta \in H_0$" or "$\theta \in H_1$," so that the action space A contains only two points, labeled 0 and 1 for convenience.

The above decision problem is called a *hypothesis-testing problem*; H_0 is called the *null hypothesis*, and H_1 the *alternative*. H_0 is said to be *simple* iff it contains only one element; otherwise H_0 is said to be *composite*, and similarly for H_1. To take action 1 is to *reject the null hypothesis H_0*; to take action 0 is to *accept H_0*.

We first consider the case of *simple hypothesis versus simple alternative*. Here H_0 and H_1 each contain one element, say θ_0 and θ_1. For the sake of definiteness, we assume that under H_0, R is absolutely continuous with density f_0, and under H_1, R is absolutely continuous with density f_1. (The results of this section will also apply to the discrete case upon replacing integrals by sums.) Thus the problem essentially comes down to deciding, after observing R, whether R has density f_0 or f_1.

A decision function may be specified by giving a (Borel measurable) function φ from E to $[0, 1]$, with $\varphi(x)$ interpreted as the probability of rejecting H_0 when x is observed. Thus, if $\varphi(x) = 1$, we reject H_0; if $\varphi(x) = 0$, we accept H_0; and if $\varphi(x) = a$, $0 < a < 1$, we toss a coin with probability a of heads: if the coin comes up heads, we reject H_0; if tails, we accept H_0. The set $\{x: \varphi(x) = 1\}$ is called the *rejection region* or the *critical region*; the function φ is called a *test*. The decision we arrive at may be in error in two possible ways. A *type 1 error* occurs if we reject H_0 when it is in fact true, and a *type 2 error* occurs if H_0 is accepted when it is false, that is, when H_1 is true. Now if H_0 is true and we observe $R = x$, an error will be made if H_0 is rejected; this happens with probability $\varphi(x)$. Thus the probability of a type 1 error is

$$\alpha = \int_{-\infty}^{\infty} \varphi(x) f_0(x) \, dx \qquad (8.2.1)$$

Similarly, the probability of a type 2 error is

$$\beta = \int_{-\infty}^{\infty} (1 - \varphi(x)) f_1(x) \, dx \qquad (8.2.2)$$

Note that α is the expectation of $\varphi(R)$ under H_0, sometimes written $E_{\theta_0} \varphi$; similarly, $\beta = 1 - E_\theta \varphi$.

It would be desirable to choose φ so that both α and β will be small, but, as we shall see, a decrease in one of the two error probabilities usually results in an increase in the other. For example, if we ignore the observed data and always accept H_0, then $\alpha = 0$ but $\beta = 1$.

There is no unique answer to the question of what is a good test; we shall consider several possibilities First, suppose that there is a nonnegative cost c_i associated with a type i error, $i = 1, 2$. (For simplicity, assume that the cost of a correct decision is 0.) Suppose also that we know the probability p that the null hypothesis will be true. (p is called the *a priori probability* of H_0. In many situations it will be difficult to estimate; for example, in a radar reception problem, H_0 might correspond to no signal being present.)

Let φ be a test with error probabilities $\alpha(\varphi)$ and $\beta(\varphi)$ The over-all average cost associated with φ is

$$B(\varphi) = p c_1 \alpha(\varphi) + (1 - p) c_2 \beta(\varphi) \qquad (8.2.3)$$

$B(\varphi)$ is called the *Bayes risk* associated with φ; a test that minimizes $B(\varphi)$ is called a *Bayes test* corresponding to the given p, c_1, c_2, f_0, and f_1.

The Bayes solution can be computed in a straightforward way. We have, from (8.2.1–8.2.3),

$$B(\varphi) = \int_{-\infty}^{\infty} [p c_1 \varphi(x) f_0(x) + (1 - p) c_2 (1 - \varphi(x)) f_1(x)] \, dx$$

$$= \int_{-\infty}^{\infty} \varphi(x)[p c_1 f_0(x) - (1 - p) c_2 f_1(x)] \, dx + (1 - p) c_2 \quad (8.2.4)$$

Now if we wish to minimize $\int_S \varphi(x) g(x) \, dx$ and $g(x) < 0$ on S, we can do no better than to take $\varphi(x) = 1$ for all x in S; if $g(x) > 0$ on S, we should take $\varphi(x) = 0$ for all x in S; if $g(x) = 0$ on S, $\varphi(x)$ may be chosen arbitrarily. In this case $g(x) = p c_1 f_0(x) - (1 - p) c_2 f_1(x)$, and the Bayes solution may therefore be given as follows.

Let $L(x) = f_1(x)/f_0(x)$.

If $L(x) > pc_1/(1 - p)c_2$, take $\varphi(x) = 1$; that is, reject H_0.

If $L(x) < pc_1/(1 - p)c_2$, take $\varphi(x) = 0$; that is, accept H_0.

If $L(x) = pc_1/(1 - p)c_2$, take $\varphi(x) = $ anything.

L is called the *likelihood ratio*, and a test φ such that for some constant λ, $0 \le \lambda \le \infty$, $\varphi(x) = 1$ when $L(x) > \lambda$ and $\varphi(x) = 0$ when $L(x) < \lambda$, is called a *likelihood ratio test*, abbreviated LRT.

To avoid ambiguity, if $f_1(x) > 0$ and $f_0(x) = 0$, we take $L(x) = \infty$. The set on which $f_1(x) = f_0(x) = 0$ may be ignored, since it will have probability 0 under both H_0 and H_1. Also, if we observe an x for which $f_1(x) > 0$ and $f_0(x) = 0$, it must be associated with H_1, so that we should take $\varphi(x) = 1$. It will be convenient to build this requirement into the definition of a likelihood ratio test: *if $L(x) = \infty$ we assume that $\varphi(x) = 1$.*

In fact, likelihood ratio tests are completely adequate to describe the problem of testing a simple hypothesis versus a simple alternative. This assertion will be justified by the sequence of theorems to follow.

From now on, the notation $P_\theta(B)$ will indicate the probability that the value of R will belong to the set B when the true state of nature is θ.

Theorem 1. *For any α, $0 \le \alpha \le 1$, there is a likelihood ratio test whose probability of type 1 error is α.*

PROOF. If $\alpha = 0$, the test given by $\varphi(x) = 1$ if $L(x) = \infty$; $\varphi(x) = 0$ if $L(x) < \infty$, is the desired LRT, so assume $\alpha > 0$. Now $G(y) = P_{\theta_0}\{x: L(x) \le y\}$, $-\infty < y < \infty$, is a distribution function [of the random variable $L(R)$; notice that $L(R) \ge 0$, and $L(R)$ cannot be infinite under H_0]. Thus either we can find λ, $0 \le \lambda < \infty$, such that $G(\lambda) = 1 - \alpha$, or else G jumps through $1 - \alpha$; that is, for some λ we have $G(\lambda^-) \le 1 - \alpha \le G(\lambda)$ (see Figure 8.2.1). Define

$$\varphi(x) = 1 \quad \text{if } L(x) > \lambda$$
$$= 0 \quad \text{if } L(x) < \lambda$$
$$= a \quad \text{if } L(x) = \lambda$$

where $a = [G(\lambda) - (1 - \alpha)]/[G(\lambda) - G(\lambda^-)]$ if $G(\lambda) > G(\lambda^-)$, $a = $ an arbitrary number in $[0, 1]$ if $G(\lambda) = G(\lambda^-)$. Then the probability of a type 1 error is

$$P_{\theta_0}\{x: L(x) > \lambda\} + aP_{\theta_0}\{x: L(x) = \lambda\} = 1 - G(\lambda) + a[G(\lambda) - G(\lambda^-)] = \alpha$$

as desired.

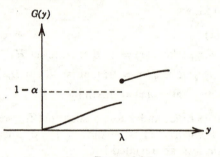

FIGURE 8.2.1

A test is said to be *at level* α_0 if its probability α of type 1 error is $\le \alpha_0$. α itself is called the *size* of the test, and $1 - \beta$, the probability of rejecting the null hypothesis when it is false, is called the *power* of the test.

The following result, known as the *Neyman-Pearson lemma*, is the fundamental theorem of hypothesis testing.

Theorem 2. *Let φ_λ be a LRT with parameter λ and error probabilities α_λ and β_λ. Let φ be an arbitrary test with error probabilities α and β; if $\alpha \le \alpha_\lambda$ then $\beta \ge \beta_\lambda$. In other words, the LRT has maximum power among all tests at level α_λ.*

We give two proofs

FIRST PROOF. Consider the Bayes problem with costs $c_1 = c_2 = 1$, and set $\lambda = pc_1/(1 - p)c_2 = p/(1 - p)$. Assuming first that $\lambda < \infty$, we have $p = \lambda/(1 + \lambda)$. Thus φ_λ is the Bayes solution when the a priori probability is $p = \lambda/(1 + \lambda)$.

If $\beta < \beta_\lambda$, we compute the Bayes risk [see (8.2.3)] for $p = \lambda/(1 + \lambda)$, using the test φ.

$$B(\varphi) = p\alpha + (1 - p)\beta$$

But $\alpha \le \alpha_\lambda$ by hypothesis, while $\beta < \beta_\lambda$ and $p < 1$ by assumption. Thus $B(\varphi) < B(\varphi_\lambda)$, contradicting the fact that φ_λ is the Bayes solution.

It remains to consider the case $\lambda = \infty$. Then we must have $\varphi_\lambda(x) = 1$ if $L(x) = \infty$, $\varphi_\lambda(x) = 0$ if $L(x) < \infty$. Then $\alpha_\lambda = 0$, since $L(R)$ is never infinite under H_0; consequently $\alpha = 0$, so that, by (8.2.1), $\varphi(x)f_0(x) \equiv 0$ [strictly speaking, $\varphi(x)f_0(x) = 0$ except possibly on a set of Lebesgue measure 0]. By (8.2.2),

$$\beta = \int_{\{x:L(x) < \infty\}} (1 - \varphi(x))f_1(x)\,dx + \int_{\{x:L(x) = \infty\}} (1 - \varphi(x))f_1(x)\,dx$$

If $L(x) < \infty$, then $f_0(x) > 0$; hence $\varphi(x) = 0$. Thus, in order to minimize β, we must take $\varphi(x) = 1$ when $L(x) = \infty$. But this says that $\beta \geq \beta_\lambda$, completing the proof.

SECOND PROOF. First assume $\lambda < \infty$. We claim that $[\varphi_\lambda(x) - \varphi(x)] \times [f_1(x) - \lambda f_0(x)] \geq 0$ for all x. For if $f_1(x) > \lambda f_0(x)$, then $\varphi_\lambda(x) = 1 \geq \varphi(x)$, and if $f_1(x) < \lambda f_0(x)$, then $\varphi_\lambda(x) = 0 \leq \varphi(x)$. Thus

$$\int_{-\infty}^{\infty} [\varphi_\lambda(x) - \varphi(x)][f_1(x) - \lambda f_0(x)] \geq 0$$

By (8.2.1) and (8.2.2),

$$1 - \beta_\lambda - (1 - \beta) - \lambda\alpha_\lambda + \lambda\alpha \geq 0$$

or

$$\beta - \beta_\lambda \geq \lambda(\alpha_\lambda - \alpha) \geq 0$$

The case $\lambda = \infty$ is handled just as in the first proof.

If we wish to construct a test that is best at level α in the sense of maximum power, we find, by Theorem 1, a LRT of size α. By Theorem 2, the test has maximum power among all tests at level α. We shall illustrate the procedure with examples and problems later in the section.

Finally, we show that no matter what criterion the statistician adopts in defining a good test, he can restrict himself to the class of likelihood ratio tests.

A test φ with error probabilities α and β is said to be *inadmissible* iff there is a test φ' with error probabilities α' and β', with $\alpha' \leq \alpha$, $\beta' \leq \beta$, and either $\alpha' < \alpha$ or $\beta' < \beta$. (In this case we say that φ' is *better than* φ.) Of course, φ is *admissible* iff it is not inadmissible.

Theorem 3.   *Every LRT is admissible.*

PROOF. Let φ_λ be a LRT with parameter λ and error probabilities α_λ and β_λ, and φ an arbitrary test with error probabilities α and β. We have seen that if $\alpha \leq \alpha_\lambda$, then $\beta \geq \beta_\lambda$. But *the Neyman-Pearson lemma is symmetric in H_0 and H_1*. In other words, if we relabel H_1 as the null hypothesis and H_0 as the alternative, Theorem 2 states that if $\beta \leq \beta_\lambda$, then $\alpha \geq \alpha_\lambda$; the result follows.

Thus no test can be better than a LRT. In fact, if φ is any test, then there is a LRT φ_λ that is *as good as* φ; that is, $\alpha_\lambda \leq \alpha$ and $\beta_\lambda \leq \beta$. For by Theorem 1 there is a LRT φ_λ with $\alpha_\lambda = \alpha$, and by Theorem 2 $\beta_\lambda \leq \beta$. This argument establishes the following result, essentially a converse to Theorem 3.

Theorem 4. *If φ is an admissible test, there is a LRT with exactly the same error probabilities.*

PROOF. As above, we find a LRT φ_λ with $\alpha_\lambda = \alpha$ and $\beta_\lambda \leq \beta$; since φ is admissible, we must have $\beta_\lambda = \beta$. •

▶ **Example 1.** Suppose that under H_0, R is uniformly distributed between 0 and 1, and under H_1, R has density $3x^2$, $0 \leq x \leq 1$. For short we write

$$H_0: f_0(x) = 1, \qquad 0 \leq x \leq 1$$
$$H_1: f_1(x) = 3x^2, \qquad 0 \leq x \leq 1$$

We are going to find the *risk set* S, that is, the set of points $(\alpha(\varphi), \beta(\varphi))$ where φ ranges over all possible tests. [The individual points $(\alpha(\varphi), \beta(\varphi))$ are called *risk points*.] We are also going to find the set S_A of *admissible risk points*, that is, the set of risk points corresponding to admissible tests. By Theorems 3 and 4, S_A is the set of risk points corresponding to LRTs.

First we notice two general properties of S.

1. *S is convex*; that is, if Q_1 and Q_2 belong to S, so do all points on the line segment joining Q_1 to Q_2. In other words, $(1 - a)Q_1 + aQ_2 \in S$ for all $a \in [0, 1]$.

For if $Q_1 = (\alpha(\varphi_1), \beta(\varphi_1))$, $Q_2 = (\alpha(\varphi_2), \beta(\varphi_2))$ and $0 \leq a \leq 1$, let $\varphi = (1 - a)\varphi_1 + a\varphi_2$. Then φ is a test, and by (8.2.1) and (8.2.2), $\alpha(\varphi) = (1 - a)\alpha(\varphi_1) + a\alpha(\varphi_2)$, $\beta(\varphi) = (1 - a)\beta(\varphi_1) + a\beta(\varphi_2)$. If $Q = (\alpha(\varphi), \beta(\varphi))$, then $Q \in S$, since φ is a test, and $Q = (1 - a)Q_1 + aQ_2$.

2. *S is symmetric about $(1/2, 1/2)$*; that is, if $|\varepsilon|, |\delta| \leq 1/2$ and $(1/2 - \varepsilon, 1/2 - \delta) \in S$, then $(1/2 + \varepsilon, 1/2 + \delta) \in S$. Equivalently, $(\alpha, \beta) \in S$ implies $(1 - \alpha, 1 - \beta) \in S$.

For if $(\alpha(\varphi), \beta(\varphi)) \in S$, let $\varphi' = 1 - \varphi$; then φ' is a test and $\alpha(\varphi') = 1 - \alpha(\varphi)$, $\beta(\varphi') = 1 - \beta(\varphi)$.

To return to the present example, we have $L(x) = 3x^2$, $0 \leq x \leq 1$. Thus the error probabilities for a LRT with parameter $\lambda \leq 3$ are

$$\alpha = P_{\theta_0}\{x: L(x) > \lambda\} = P_{\theta_0}\left\{x: x > \left(\frac{\lambda}{3}\right)^{1/2}\right\} = 1 - \left(\frac{\lambda}{3}\right)^{1/2}$$

$$\beta = P_{\theta_1}\{x: L(x) < \lambda\} = P_{\theta_1}\left\{x: x < \left(\frac{\lambda}{3}\right)^{1/2}\right\}$$

$$= \int_0^{(\lambda/3)^{1/2}} 3x^2\, dx = \left(\frac{\lambda}{3}\right)^{3/2} = (1 - \alpha)^3$$

(If $\lambda > 3$, then $\alpha = 0$, $\beta = 1$.) Thus $S_A = \{(\alpha, (1 - \alpha)^3): 0 \leq \alpha \leq 1\}$. Since no test can be better than a LRT, S_A is the lower boundary of the

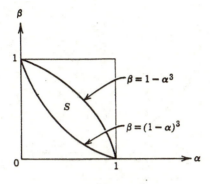

FIGURE 8.2.2

set S; hence, by symmetry, $\{(1 - \alpha, 1 - (1 - \alpha)^3), 0 \le \alpha \le 1\} = \{(\alpha, 1 - \alpha^3): 0 \le \alpha \le 1\}$ is the upper boundary of S. Thus S must be $\{(\alpha, \beta): 0 \le \alpha \le 1, (1 - \alpha)^3 \le \beta \le 1 - \alpha^3\}$ (see Figure 8.2.2).

Various tests may now be computed without difficulty. We give some typical illustrations.

(a) Find a most powerful test at level .15. Set $\alpha = .15 = 1 - (\lambda/3)^{1/2}$. Since $L(x) > \lambda$ iff $x > (\lambda/3)^{1/2}$, the test is given by

$$\varphi(x) = 1 \quad \text{if } x > .85$$
$$= 0 \quad \text{if } x < .85$$
$$= \text{anything} \quad \text{if } x = .85$$

We have $\beta = (1 - \alpha)^3 = (.85)^3 = .614$.

(b) Find a Bayes test corresponding to $c_1 = 3/2$, $c_2 = 3$, $p = 3/4$. This is a LRT with $\lambda = pc_1/(1 - p)c_2 = 3/2$; that is,

$$\varphi(x) = 1 \quad \text{if } x > \left(\frac{\lambda}{3}\right)^{1/2} = \frac{\sqrt{2}}{2} = .707$$
$$= 0 \quad \text{if } x < \frac{\sqrt{2}}{2}$$
$$= \text{anything} \quad \text{if } x = \frac{\sqrt{2}}{2}$$

Thus $\alpha = 1 - (\lambda/3)^{1/2} = .293$, $\beta = (1 - \alpha)^3$, and the Bayes risk may be computed using (8.2.3).

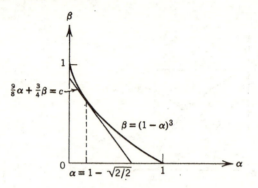

FIGURE 8.2.3   Geometric Interpretation of Bayes Solution.

The Bayes solution may be interpreted geometrically as follows. We are trying to find a test that minimizes the Bayes risk $pc_1\alpha + (1 - p)c_2\beta = (9/8)\alpha + (3/4)\beta$. If we vary c until the line $(9/8)\alpha + (3/4)\beta = c$ intersects S_A, we find the desired test (see Figure 8.2.3).

Notice also that to find the Bayes solution we may differentiate $(9/8)\alpha + (3/4)(1 - \alpha)^3$ and set the result equal to zero to obtain $\alpha = 1 - \sqrt{2}/2$, as before.

(c) Find a *minimax test*, that is, a test that minimizes max (α, β). It is immediate from the definition of admissibility that *an admissible test with constant risk (i.e., $\alpha = \beta$) is minimax*. Thus we set $\alpha = \beta = (1 - \alpha)^3$, which yields $\alpha = .318$ (approximately). Therefore $(\lambda/3)^{1/2} = 1 - \alpha = .682$, and so we reject H_0 if $x > .682$ and accept H_0 if $x < .682$. ◀

▶ **Example 2.** Let R be a discrete random variable taking on only the values 0, 1, 2, 3. Let the probability function of R under H_i be p_i, $i = 0, 1$, where the p_i are as follows.

x	0	1	2	3
$p_0(x)$	.1	.2	.3	.4
$p_1(x)$	.2	.1	.4	.3

The appropriate likelihood ratio here is $L(x) = p_1(x)/p_0(x)$. Arranging the values of $L(x)$ in increasing order, we have the following table.

x	1	3	2	0
$L(x)$	$\frac{1}{2}$	$\frac{3}{4}$	$\frac{4}{3}$	2

We may therefore describe the LRT with parameter λ as follows.

LRT	Rejection Region	Acceptance Region	α	β
$0 \le \lambda < \frac{1}{2}$	All x	Empty	1	0
$1/2 < \lambda < \frac{3}{4}$	$x = 0, 2, 3$	$x = 1$	.8	.1
$3/4 < \lambda < \frac{4}{3}$	$x = 0, 2$	$x = 1, 3$	.4	.4
$4/3 < \lambda < 2$	$x = 0$	$x = 1, 2, 3$	.1	.8
$2 < \lambda \le \infty$	Empty	All x	0	1

Now assume $\lambda = 3/4$. Then we reject H_0 if $x = 0$ or 2, accept H_0 if $x = 1$, and if $x = 3$ we *randomize*, that is, reject H_0 with probability a, $0 \le a \le 1$. Thus

$$\alpha = p_0(0) + p_0(2) + ap_0(3) = .4 + .4a$$
$$\beta = p_1(1) + (1 - a)p_1(3) = .1 + .3(1 - a)$$

As a ranges over $[0, 1]$, (α, β) traces out the line segment joining $(.4, .4)$ to $(.8, .1)$. In a similar fashion we calculate the error probabilities for $\lambda = 1/2$, $4/3$, and 2. The admissible risk points are shown in Figure 8.2.4.

We compute several tests.

(a) Find a most powerful test at level .25. Since $.1 < .25 < .4$, we have $\lambda = 4/3$. Thus we reject H_0 if $x = 0$, accept H_0 if $x = 1$ or 3, and reject H_0 with probability a if $x = 2$, where $.1(1 - a) + .4a = .25$, so that $a = 1/2$. Notice that $\beta = .8(1 - a) + .4a = .6$.

(b) Find a Bayes test with $c_1 = c_2 = 1$, $p = .6$. We have $\lambda = pc_1/(1 - p)c_2 = 3/2$. Thus we reject H_0 if $x = 0$ and accept H_0 otherwise. The error probabilities are $\alpha = .1$, $\beta = .8$, and the Bayes risk is $pc_1\alpha + (1 - p)c_2\beta = .38$.

(c) Find a minimax test. The only admissible test with $\alpha = \beta$ has $\alpha = \beta = .4$, so that $3/4 < \lambda < 4/3$. We reject H_0 when $x = 0$ or 2 and accept H_0 if $x = 1$ or 3. ◀

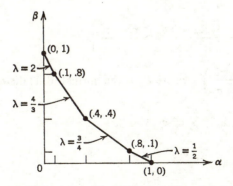

FIGURE 8.2.4 Admissible Risk Points When R is Discrete.

▶ **Example 3.** Let R be normally distributed with mean θ and variance σ^2, where σ^2 is known. We wish to test the null hypothesis that $\theta = \theta_0$ against the alternative that $\theta = \theta_1$, and the test is to be based on n independent observations $R_1, \ldots, R_n$ of R. (Assume $\theta_0 < \theta_1$.)

The appropriate likelihood ratio is

$$L(x_1, \ldots, x_n) = \frac{f_1(x_1, \ldots, x_n)}{f_0(x_1, \ldots, x_n)} = \frac{(2\pi\sigma^2)^{-n/2} \exp\left[-\sum_{k=1}^{n}(x_k - \theta_1)^2/2\sigma^2\right]}{(2\pi\sigma^2)^{-n/2} \exp\left[-\sum_{k=1}^{n}(x_k - \theta_0)^2/2\sigma^2\right]}$$

The condition $L(x) > \lambda$ is equivalent to $\ln L(x) > \ln \lambda$; that is,

$$\sum_{k=1}^{n} 2(\theta_1 - \theta_0)x_k + n(\theta_0^2 - \theta_1^2) > 2\sigma^2 \ln \lambda \qquad (8.2.5)$$

This is of the form $\sum_{k=1}^{n} x_k > c$. Thus a LRT must be of the form

$$\varphi(x_1, \ldots, x_n) = 1 \qquad \text{if } \sum_{k=1}^{n} x_k > c$$

$$= 0 \qquad \text{if } \sum_{k=1}^{n} x_k < c$$

$$= \text{anything} \qquad \text{if } \sum_{k=1}^{n} x_k = c$$

Now $R_1 + \cdots + R_n$ is normal with mean $n\theta$ and variance $n\sigma^2$, so that the error probabilities are

$$\alpha = P_{\theta_0}\left\{(x_1, \ldots, x_n): \sum_{k=1}^{n} x_k > c\right\}$$

$$= P_{\theta_0}\{R_1 + \cdots + R_n > c\}$$

$$= P_{\theta_0}\left\{\frac{R_1 + \cdots + R_n - n\theta_0}{\sqrt{n}\,\sigma} > \frac{c - n\theta_0}{\sqrt{n}\,\sigma}\right\}$$

$$= 1 - F^*\left(\frac{c - n\theta_0}{\sqrt{n}\,\sigma}\right) \qquad \text{where } F^* \text{ is the normal } (0, 1) \text{ distribution function}$$

$$\beta = P_{\theta_1}\left\{(x_1, \ldots, x_n): \sum_{k=1}^{n} x_k < c\right\}$$

$$= P_{\theta_1}\{R_1 + \cdots + R_n < c\}$$

$$= F^*\left(\frac{c - n\theta_1}{\sqrt{n}\,\sigma}\right)$$

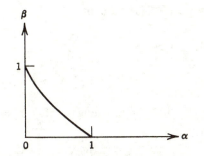

FIGURE 8.2.5 Admissible Risk Points
When R is Normal.

Thus we have parametric equations for α and β with c as parameter, $-\infty \leq c \leq \infty$. The admissible risk points are sketched in Figure 8.2.5.

Suppose that we want a LRT of size α. If N_α is the number such that $1 - F^*(N_\alpha) = \alpha$, then $(c - n\theta_0)/\sqrt{n}\,\sigma = N_\alpha$, so that $c = n\theta_0 + \sqrt{n}\,\sigma N_\alpha$.

We now apply the results to a problem in testing a simple hypothesis versus a composite alternative. Again let R be normal (θ, σ^2), and take $H_0: \theta = \theta_0$, $H_1: \theta > \theta_0$.

If we choose any particular $\theta_1 > \theta_0$ and test $\theta = \theta_0$ against $\theta = \theta_1$, the test described above is most powerful at level α. However, the test is completely specified by c, and c does not depend on θ_1. Thus, for any $\theta_1 > \theta_0$, the test has the highest power of any test at level α of $\theta = \theta_0$ versus $\theta = \theta_1$. Such a test is called a *uniformly most powerful* (UMP) level α test of $\theta = \theta_0$ versus $\theta > \theta_0$.

We expect intuitively that the larger the separation between θ_0 and θ_1, the better the performance of the test in distinguishing between the two possibilities. This may be verified by considering the *power function* Q, defined by

$$Q(\theta) = E_\theta \varphi$$

= the probability of rejecting H_0 when the true state of nature is θ

$$= P_\theta\{R_1 + \cdots + R_n > c\}$$

$$= 1 - F^*\left(\frac{c - n\theta}{\sqrt{n}\,c}\right)$$

Thus $Q(\theta)$ increases with θ.

Now if $H_0: \theta = \theta_0$, $H_1: \theta = \theta_1$, where $\theta_1 < \theta_0$, the same technique as above shows that a size α LRT is of the form

$$\varphi(x_1, \ldots, x_n) = 1 \quad \text{if } \sum_{k=1}^{n} x_k < c$$

$$= 0 \quad \text{if } \sum_{k=1}^{n} x_k > c$$

$$= \text{anything} \quad \text{if } \sum_{k=1}^{n} x_k = c$$

where

$$\alpha = F^*\left(\frac{c - n\theta_0}{\sqrt{n}\,\sigma}\right)$$

$$\beta = 1 - F^*\left(\frac{c - n\theta_1}{\sqrt{n}\,c}\right)$$

$$c = n\theta_0 + \sqrt{n}\,\sigma N_{1-\alpha}$$

Again, the test is UMP at level α for $\theta = \theta_0$ versus $\theta < \theta_0$, with power function

$$Q'(\theta) = F^*\left(\frac{c - n\theta}{\sqrt{n}\,\sigma}\right)$$

which increases as θ decreases (see Figure 8.2.6).

The above discussion suggests that there can be no UMP level α test of $\theta = \theta_0$ versus $\theta \neq \theta_0$. For any such test φ must have power function $Q(\theta)$ for $\theta > \theta_0$, and $Q'(\theta)$ for $\theta < \theta_0$. But the power function of φ is given by

$$E_\theta \varphi = \int_{-\infty}^{\infty} \cdots \int_{-\infty}^{\infty} \varphi(x_1, \ldots, x_n) f_\theta(x_1, \ldots, x_n)\, dx_1 \cdots dx_n$$

where f_θ is the joint density of n independent normal random variables with mean θ and variance σ^2. It can be shown that this is differentiable for all θ (the derivative can be taken under the integral sign). But a function that is $Q(\theta)$ for $\theta > \theta_0$ and $Q'(\theta)$ for $\theta < \theta_0$ cannot be differentiable at θ_0.

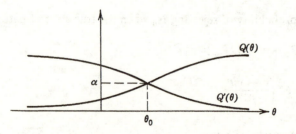

FIGURE 8.2.6   Power Functions.

In fact, the test φ with power function $Q(\theta)$ is UMP at level α for the composite hypothesis $H_0 \colon \theta \leq \theta_0$ versus the composite alternative $H_1 \colon \theta > \theta_0$. Let us explain what this means.

φ is said to be at level α for H_0 versus H_1 iff $E_\theta \varphi \leq \alpha$ for all $\theta \leq \theta_0$; φ is UMP at level α if for any test φ' at level α for H_0 versus H_1 we have $E_\theta \varphi' \leq E_\theta \varphi$ for all $\theta > \theta_0$.

In the present case $E_\theta \varphi = Q(\theta) \leq \alpha$ for $\theta \leq \theta_0$ by monotonicity of $Q(\theta)$, and $E_\theta \varphi' \leq E_\theta \varphi$ for $\theta > \theta_0$, since φ is UMP at level α for $\theta = \theta_0$ versus $\theta > \theta_0$.

The underlying reason for the existence of uniformly most powerful tests is the following. If $\theta < \theta'$, the likelihood ratio $f_{\theta'}(x)/f_\theta(x)$ can be expressed as a nondecreasing function of $t(x)$ [where, in this case, $t(x) = x_1 + \cdots + x_n$; see (8.2.5)]. Whenever this happens, the family of densities f_θ is said to have the *monotone likelihood ratio* (MLR) property.

Suppose that the f_θ have the MLR property. Consider the following test of $\theta = \theta_0$ versus $\theta = \theta_1$, $\theta_1 > \theta_0$.

$$\begin{aligned}
\varphi(x) &= 1 && \text{if } t(x) > c \\
&= 0 && \text{if } t(x) < c \\
&= a && \text{if } t(x) = c
\end{aligned}$$

where $P_{\theta_0}\{x \colon t(x) > c\} + a P_{\theta_0}\{x \colon t(x) = c\} = \alpha$ (notice that c does not depend on θ_1). Let λ be the value of the likelihood ratio when $t(x) = c$; then $L(x) > \lambda$ implies $t(x) > c$; hence $\varphi(x) = 1$. Also $L(x) < \lambda$ implies $t(x) < c$, so that $\varphi(x) = 0$. Thus φ is a LRT and hence is most powerful at level α. We may make the following observations.

1. φ is UMP at level α for $\theta = \theta_0$ versus $\theta > \theta_0$.

This is immediate from the Neyman-Pearson lemma and the fact that c does not depend on the particular $\theta > \theta_0$.

2. If $\theta_1 < \theta_2$, φ is the most powerful test at level $\alpha_1 = E_{\theta_1} \varphi$ for $\theta = \theta_1$ versus $\theta = \theta_2$.

Since φ is a LRT, the Neyman-Pearson lemma yields this result immediately.

3. If $\theta_1 < \theta_2$, then $E_{\theta_1} \varphi \leq E_{\theta_2} \varphi$; that is, φ has a monotone nondecreasing power function. It follows, as in the earlier discussion, that φ is UMP at level α for $\theta \leq \theta_0$ versus $\theta > \theta_0$.

By property 2, φ is most powerful at level $\alpha_1 = E_{\theta_1} \varphi$ for $\theta = \theta_1$ versus $\theta = \theta_2$. But the test $\varphi'(x) \equiv \alpha_1$ is also at level α_1; hence $E_{\theta_2} \varphi' \leq E_{\theta_2} \varphi$, that is, $\alpha_1 = E_{\theta_1} \varphi \leq E_{\theta_2} \varphi$.

REMARK. Since the Neyman-Pearson lemma is symmetric in H_0 and H_1, if $\theta_1 < \theta_2$, then for all tests φ' with $\beta(\varphi') \leq \beta(\varphi)$, we have $E_{\theta_1} \varphi \leq E_{\theta_1} \varphi'$.

We might say that φ is *uniformly least powerful* for $\theta < \theta_0$ among all tests whose type 2 error is $\leq \beta$ whenever $\theta \geq \theta_0$. ◄

PROBLEMS

1. Let $H_0 : f_0(x) = e^{-x}, x \geq 0; H_1 : f_1(x) = 2e^{-2x}, x \geq 0$.
   (a) Find the risk set and the admissible risk points.
   (b) Find a most powerful test at level .05.
   (c) Find a minimax test.

2. Show that the following families have the MLR property, and thus UMP tests may be constructed as in the discussion of Example 3.
   (a) p_θ = the joint probability function of n independent random variables, each Poisson with parameter θ.
   (b) p_θ = the joint probability function of n independent random variables R_i, where R_i is *Bernoulli with parameter* θ; that is, $P\{R_i = 1\} = \theta, P\{R_i = 0\} = 1 - \theta, 0 \leq \theta \leq 1$; notice that $R_1 + \cdots + R_n$ has the binomial distribution with parameters n and θ.
   (c) Suppose that of N objects, θ are defective. If n objects are drawn without replacement, the probability that exactly x defective objects will be found in the sample is

   $$p_\theta(x) = \frac{\binom{\theta}{x}\binom{N-\theta}{n-x}}{\binom{N}{n}}, \qquad x = 0, 1, \ldots, \theta \ (\theta = 0, 1, \ldots, N)$$

   This is the hypergeometric probability function; see Problem 7, Section 1.5.
   (d) f_θ = the joint density of n independent normally distributed random variables with mean 0 and variance $\theta > 0$.

3. It is desired to test the null hypothesis that a die is unbiased versus the alternative that the die is loaded, with faces 1 and 2 having probability 1/4 and faces 3, 4, 5, and 6 having probability 1/8.
   (a) Sketch the set of admissible risk points.
   (b) Find a most powerful test at level .1.
   (c) Find a Bayes solution if the cost of a type 1 error is c_1, the cost of a type 2 error is $2c_1$, and the null hypothesis has probability 3/4.

4. It is desired to test the null hypothesis that R is normal with mean θ_0 and known variance σ^2 versus the alternative that R is normal with mean $\theta_1 = \theta_0 + \sigma$ and variance σ^2, on the basis of n independent observations of R. Find the minimum value of n such that $\alpha \leq .05$ and $\beta \leq .03$.

5. Consider the problem of testing the null hypothesis that R is normal $(0, \theta_0)$ versus the alternative that R is normal $(0, \theta_1), \theta_1 > \theta_0$ (notice that in this case a UMP test of $\theta \leq \theta_0$ versus $\theta > \theta_0$ exists; see Problem 2d). Describe a most

powerful test at level α and indicate how to find the minimum number of independent observations of R necessary to reduce the probability of a type 2 error below a given figure.

6. Let $R_1, \ldots, R_n$ be independent random variables, each uniformly distributed between 0 and θ, $\theta > 0$. Show that the following test is UMP at level α for $H_0: \theta = \theta_0$ versus $H_1: \theta \neq \theta_0$.

$$\varphi(x_1, \ldots, x_n) = 1 \quad \text{if } \max_{1 \leq i \leq n} x_i \leq \theta_0 \alpha^{1/n} \quad \text{or if } \max_{1 \leq i \leq n} x_i > \theta_0$$
$$= 0 \quad \text{otherwise}$$

Find the sketch the power function of the test.

7. Consider the test of Problem 6 with $H_0: \theta = 1$; $H_1: \theta = 2$. Find the risk set and the set of admissible risk points.

8. Let R_1, R_2, and R_3 be independent, each Bernoulli with parameter θ, $0 \leq \theta \leq 1$. Find the UMP test of size $\alpha = .1$ of $\theta \leq 1/4$ versus $\theta > 1/4$, and find the power function of the test.

9. Show that every admissible test is a Bayes test for some choice of costs c_1 and c_2 and a priori probability p. Conversely, show that every Bayes test with $c_1 > 0$, $c_2 > 0$, $0 < p < 1$ is admissible. Give an example of an inadmissible Bayes test with $c_1 > 0$, $c_2 > 0$.

10. If φ is most powerful at level α_0 and $\beta(\varphi) > 0$, show that φ is actually of size α_0. Give a counterexample to the assertion if $\beta(\varphi) = 0$.

*11. Let φ be a most powerful test at level α. Show that for some constant λ we have $\varphi(x) = 1$ if $x > \lambda$; $\varphi(x) = 0$ if $x < \lambda$, except possibly for x in a set of Lebesgue measure 0.

12. A class C of tests is said to be *essentially complete* iff for any test φ_1 there is a test $\varphi_2 \in C$ such that φ_2 is as good as φ_1. Show that the following classes are essentially complete.
    (a) The likelihood ratio tests.
    (b) The admissible tests.
    (c) The Bayes tests (i.e., considering all possible c_1, c_2, and p).

13. Give an example of tests φ_1 and φ_2 such that the statements "φ_1 is as good as φ_2" and "φ_2 is as good as φ_1" are both false.

14. Let $R_1, R_2, \ldots$ be independent random variables, each with density h_θ, and let $H_0: \theta = \theta_0$, $H_1: \theta = \theta_1$.
    (a) If φ_n is a test based on n observations that minimizes the sum of the error probabilities, show that $\varphi_n(x) = 1$ if $g_n(x) = \prod_{i=1}^n [h_{\theta_1}(x_i)/h_{\theta_0}(x_i)] > 1$, $\varphi_n(x) = 0$ if $g_n(x) < 1$. Thus

$$\alpha_n + \beta_n = P_{\theta_0}\{x: g_n(x) > 1\} + P_{\theta_1}\left\{x: \frac{1}{g_n(x)} > 1\right\}$$

    (b) Let $t(x_i) = [h_{\theta_1}(x_i)/h_{\theta_0}(x_i)]^{1/2}$. Show that

$$P_{\theta_0}\{x: g_n(x) > 1\} \leq \prod_{i=1}^n E_{\theta_0} t(R_i) = [E_{\theta_0} t(R_1)]^n$$

(c) Show that $E_{\theta_0} t(R_1) < 1$; hence $\alpha_n \to 0$ as $n \to \infty$. A similar argument with θ_0 and θ_1 interchanged shows that $\beta_n \to 0$ as $n \to \infty$, so that if enough observations are taken, both error probabilities can be made arbitrarily small.

8.3   ESTIMATION

Consider the statistical decision model of Section 8.1. Suppose that γ is a real-valued function on the set N of states of nature, and we wish to estimate $\gamma(\theta)$. If we observe $R = x$ we must produce a number $\psi(x)$ that we hope will be close to $\gamma(\theta)$. Thus the action space A is the set of reals E^1, and a decision function may be specified by giving a (Borel measurable) function ψ from the range of R to E^1; such a ψ is called an *estimate*, and the above decision problem is called a problem of *point estimation of a real parameter*.

Although the estimate ψ appears intrinsically nonrandomized, it is possible to introduce randomization without an essential change in the model. If R_1 is the observable, we let R_2 be a random variable independent of R_1 and θ, with an arbitrary distribution function F. Formally, assume $P_\theta\{R_1 \in B_1, R_2 \in B_2\} = P_\theta\{R_1 \in B_1\}P\{R_2 \in B_2\}$, where $P\{R_2 \in B_2\}$ is determined by the distribution function F and is unaffected by θ. If $R_1 = x$ and $R_2 = y$, we estimate $\gamma(\theta)$ by a number $\psi(x, y)$. Thus we introduce randomization by enlarging the observable.

There is no unique way of specifying a good estimate; we shall discuss several classes of estimates that have desirable properties.

We first consider *maximum likelihood* estimates. Let f_θ be the density (or probability) function corresponding to the state of nature θ, and assume for simplicity that $\gamma(\theta) = \theta$. If $R = x$, the maximum likelihood estimate of θ is given by $\gamma(x) = \hat{\theta} =$ the value of θ that maximizes $f_\theta(x)$. Thus (at least in the discrete case) the estimate is the state of nature that makes the particular observation most likely. In many cases the maximum likelihood estimate is easily computable.

▶ **Example 1.**   Let R have the binomial distribution with parameters n and θ, $0 \leq \theta \leq 1$, so that $p_\theta(x) = \binom{n}{x}\theta^x(1 - \theta)^{n-x}$, $x = 0, 1, \ldots, n$. To find $\hat{\theta}$ we may set

$$\frac{\partial}{\partial \theta} \ln p_\theta(x) = 0$$

to obtain

$$\frac{x}{\theta} - \frac{n - x}{1 - \theta} = 0 \quad \text{or} \quad \theta = \frac{x}{n}$$

Notice that R may be regarded as a sum of independent random variables

$R_1, \ldots, R_n$, where R_i is 1 with probability θ and 0 with probability $1 - \theta$. In terms of the R_i we have $\hat{\theta}(R) = (R_1 + \cdots + R_n)/n$, which converges in probability to $E(R_i) = \theta$ by the weak law of large numbers. Convergence in probability of the maximum likelihood estimate to the true parameter can be established under rather general conditions. ◄

▶ **Example 2.** Let $R_1, \ldots, R_n$ be independent, normally distributed random variables with mean μ and variance σ^2. Find the maximum likelihood estimate of $\theta = (\mu, \sigma^2)$. (Here θ is a point in E^2 rather than a real number, but the maximum likelihood estimate is defined as before.)

We have

$$f_\theta(x) = (2\pi\sigma^2)^{-n/2} \exp\left[-\frac{1}{2\sigma^2}\sum_{i=1}^n (x_i - \mu)^2\right]$$

so that

$$\ln f_\theta(x) = -\frac{n}{2}\ln 2\pi - n\ln\sigma - \frac{1}{2\sigma^2}\sum_{i=1}^n (x_i - \mu)^2$$

Thus

$$\frac{\partial}{\partial\mu}\ln f_\theta(x) = \frac{1}{\sigma^2}\sum_{i=1}^n (x_i - \mu) = \frac{n}{\sigma^2}(\bar{x} - \mu)$$

where

$$\bar{x} = \frac{1}{n}\sum_{i=1}^n x_i$$

and

$$\frac{\partial}{\partial\sigma}\ln f_\theta(x) = -\frac{n}{\sigma} + \frac{1}{\sigma^3}\sum_{i=1}^n (x_i - \mu)^2 = \frac{n}{\sigma^3}\left(-\sigma^2 + \frac{1}{n}\sum_{i=1}^n (x_i - \mu)^2\right)$$

Setting the partial derivatives equal to zero, we obtain

$$\theta = (\bar{x}, s^2)$$

where

$$s^2 = \frac{1}{n}\sum_{i=1}^n (x_i - \bar{x})^2$$

(A standard calculus argument shows that this is actually a maximum.) In terms of the R_i, we have

$$\hat{\theta}(R_1, \ldots, R_n) = (\bar{R}, V^2)$$

where $\bar{R}$ is the *sample mean* $(R_1 + \cdots + R_n)/n$ and V^2 is the *sample variance* $(1/n)\sum_{i=1}^n (R_i - \bar{R})^2$.

If the problem is changed so that $\theta = \mu$ (i.e., σ^2 is known), we obtain $\hat{\theta} = \bar{R}$ as above. However, if $\theta = \sigma^2$, then we find $\hat{\theta} = (1/n)\sum_{i=1}^n (x_i - \mu)^2$, since the equation $\partial\ln f_\theta(x)/\partial\mu = 0$ is no longer present. ◄

We now discuss *Bayes* estimates. For the sake of definiteness we consider the absolutely continuous case. Assume $N = E^1$, and let f_θ be the density of R when the state of nature is θ. Assume that there is an *a priori density g* for θ; that is, the probability that the state of nature will lie in the set B is given by $\int_B g(\theta) \, d\theta$. Finally, assume that we are given a (nonnegative) loss function $L(\gamma(\theta), a)$, $\theta \in N$, $a \in A$; $L(\gamma(\theta), a)$ is the cost when our estimate of $\gamma(\theta)$ turns out to be a. If ψ is an estimate, the over-all average cost associated with ψ is

$$B(\psi) = \int_{-\infty}^{\infty} \int_{-\infty}^{\infty} g(\theta) f_\theta(x) L(\gamma(\theta), \psi(x)) \, d\theta \, dx$$

$B(\psi)$ is called the *Bayes risk* of ψ, and an estimate that minimizes $B(\psi)$ is called a *Bayes estimate*. If we write

$$B(\psi) = \int_{-\infty}^{\infty} \left[\int_{-\infty}^{\infty} g(\theta) f_\theta(x) L(\gamma(\theta), \psi(x)) \, d\theta \right] dx \qquad (8.3.1)$$

it follows that in order to minimize $B(\psi)$ it is sufficient to minimize the expression in brackets for each x.

Often this is computationally feasible. In particular, let $L(\gamma(\theta), a) = (\gamma(\theta) - a)^2$. Thus we are trying to minimize

$$\int_{-\infty}^{\infty} g(\theta) f_\theta(x)(\gamma(\theta) - \psi(x))^2 \, d\theta$$

This is of the form $A\psi^2(x) - 2B\psi(x) + C$, which is a minimum when $\psi(x) = B/A$; that is,

$$\psi(x) = \frac{\displaystyle\int_{-\infty}^{\infty} g(\theta) f_\theta(x) \gamma(\theta) \, d\theta}{\displaystyle\int_{-\infty}^{\infty} g(\theta) f_\theta(x) \, d\theta} \qquad (8.3.2)$$

But the conditional density of θ given $R = x$ is $g(\theta) f_\theta(x) / \int_{-\infty}^{\infty} g(\theta) f_\theta(x) \, d\theta$, so that $\psi(x)$ is simply the conditional expectation of $\gamma(\theta)$ given $R = x$.

To summarize: To find a Bayes estimate with quadratic loss function, set $\psi(x)$ = the conditional expectation of the parameter to be estimated, given that the observable takes the value x.

▶ **Example 3.**   Let R have the binomial distribution with parameters n and θ, $0 \leq \theta \leq 1$, and let $\gamma(\theta) = \theta$. Take g as the *beta density* with parameters r and s; that is,

$$g(\theta) = \frac{\theta^{r-1}(1 - \theta)^{s-1}}{\beta(r, s)}, \qquad 0 \leq \theta \leq 1, r, s > 0$$

where $\beta(r, s)$ is the beta function (see Section 2 of Chapter 4). First we find a Bayes estimate of θ with quadratic loss function.

The discussion leading to (8.3.2) applies, with $f_\theta(x)$ replaced by $p_\theta(x) = \binom{n}{x}\theta^x(1 - \theta)^{n-x}$, $x = 0, 1, \ldots, n$. Thus

$$
\begin{aligned}
\psi(x) &= \frac{\displaystyle\int_0^1 \binom{n}{x}\theta^{r-1+x+1}(1 - \theta)^{s-1+n-x}\, d\theta}{\displaystyle\int_0^1 \binom{n}{x}\theta^{r-1+x}(1 - \theta)^{s-1+n-x}\, d\theta} \\[2mm]
&= \frac{\beta(r + x + 1, n - x + s)}{\beta(r + x, n - x + s)} \\[2mm]
&= \frac{\Gamma(r + x + 1)\Gamma(n - x + s)}{\Gamma(r + x)\Gamma(n - x + s)} \frac{\Gamma(r + s + n)}{\Gamma(r + s + n + 1)} \\[2mm]
&= \frac{r + x}{r + s + n}
\end{aligned}
$$

Now, for a given θ, the average loss $\rho_\psi(\theta)$, using ψ, may be computed as follows.

$$
\begin{aligned}
\rho_\psi(\theta) &= E_\theta\!\left[\left(\frac{r + R}{r + s + n} - \theta\right)^2\right] \\[2mm]
&= \frac{1}{(r + s + n)^2} E_\theta[(R - n\theta + r - r\theta - s\theta)^2]
\end{aligned}
$$

Since $E_\theta[(R - n\theta)^2] = \mathrm{Var}_\theta\, R = n\theta(1 - \theta)$ and $E_\theta R = n\theta$, we have

$$
\begin{aligned}
\rho_\psi(\theta) &= \frac{1}{(r + s + n)^2}\,[n\theta(1 - \theta) + (r - r\theta - s\theta)^2] \\[2mm]
&= \frac{1}{(r + s + n)^2}\,[((r + s)^2 - n)\theta^2 + (n - 2r(r + s))\theta + r^2]
\end{aligned}
$$

ρ_ψ is called the *risk function* of ψ; notice that

$$
B(\psi) = \int_0^1 g(\theta)\rho_\psi(\theta)\, d\theta \tag{8.3.3}
$$

It is possible to choose r and s so that ρ_ψ will be constant for all θ. For this to happen,

$$
n = (r + s)^2 = 2r(r + s)
$$

which is satisfied if $r = s = \sqrt{n}/2$. We then have

$$\psi(x) = \frac{x + \sqrt{n}/2}{n + \sqrt{n}} = \frac{\sqrt{n}}{1 + \sqrt{n}} \frac{x}{n} + \frac{1/2}{1 + \sqrt{n}}$$

$$\rho_\psi(\theta) = \frac{n/4}{(n + \sqrt{n})^2} = \frac{1}{4(1 + \sqrt{n})^2} = B(\psi)$$

Thus in this case ψ is a *Bayes estimate with constant risk*; we claim that ψ must be *minimax*, that is, ψ minimizes $\max_\theta \rho_\psi(\theta)$. For if ψ' had a maximum risk smaller than that of ψ, (8.3.3) shows that $B(\psi') < B(\psi)$, contradicting the fact that ψ is Bayes.

Notice that if $\hat\theta(x) = x/n$ is the maximum likelihood estimate, then $\psi(x) = a_n\hat\theta(x) + b_n$, where $a_n \to 1$, $b_n \to 0$ as $n \to \infty$. ◀

We have not yet discussed randomized estimates; in fact, in a wide variety of situations, including the case of quadratic loss functions, randomization can be ignored. In order to justify this, we first consider a basic theorem concerning convex functions.

A function f from the reals to the reals is said to be *convex* iff $f[(1 - a)x + ay] \leq (1 - a)f(x) + af(y)$ for all real x, y and all $a \in [0, 1]$. A sufficient condition for f to be convex is that it have a nonnegative second derivative ("concave upward" is the phrase used in calculus books). The geometric interpretation is that f lies on or above any of its tangents.

Theorem 1 (*Jensen's Inequality*).   *If R is a random variable, f is a convex function, and $E(R)$ is finite, then $E[f(R)] \geq f[E(R)]$. (For example, $E[R^{2n}] \geq [E(R)]^{2n}$, $n = 1, 2, \ldots$.)*

PROOF.   Consider a tangent to f at the point $E(R)$ (see Figure 8.3.1); let the equation of the tangent be $y = ax + b$. Since f is convex, $f(x) \geq ax + b$ for all x; hence $f(R) \geq aR + b$. Thus $E[f(R)] \geq aE(R) + b = f(E(R))$.

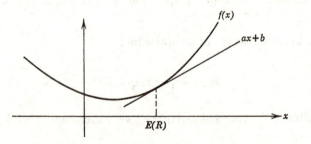

FIGURE 8.3.1   Proof of Jensen's Inequality.

We may now prove the theorem that allows us to ignore randomized estimates.

Theorem 2 (Rao-Blackwell). *Let R_1 be an observable, and let R_2 be independent of R_1 and θ, as indicated in the discussion of randomized estimates at the beginning of this section. Let $\psi = \psi(x, y)$ be any estimate of $\gamma(\theta)$ based on observation of R_1 and R_2. Assume that the loss function $L(\gamma(\theta), a)$ is a convex function of a for each θ (this includes the case of quadratic loss). Define*

$$\psi^*(x) = E_\theta[\psi(R_1, R_2) \mid R_1 = x]$$
$$= E[\psi(x, R_2)]$$

$$(E_\theta\psi(R_1, R_2) \text{ is assumed finite.})$$

Let ρ_ψ be the risk function of ψ, defined by $\rho_\psi(\theta) = E_\theta[L(\gamma(\theta), \psi(R_1, R_2))] =$ the average loss, using ψ, when the state of nature is θ. Similarly, let $\rho_{\psi^*}(\theta) = E_\theta[L(\gamma(\theta), \psi^*(R_1))]$. Then $\rho_{\psi^*}(\theta) \leq \rho_\psi(\theta)$ for all θ; hence the nonrandomized estimate ψ^* is at least as good as the randomized estimate ψ.

PROOF.

$$L(\gamma(\theta), E_\theta[\psi(R_1, R_2) \mid R_1 = x]) \leq E_\theta[L(\gamma(\theta), \psi(R_1, R_2)) \mid R_1 = x]$$

by the argument of Jensen's inequality applied to conditional expectations. Therefore

$$L(\gamma(\theta), \psi^*(R_1)) \leq E_\theta[L(\gamma(\theta), \psi(R_1, R_2)) \mid R_1]$$

Take expectations on both sides to obtain

$$\rho_{\psi^*}(\theta) \leq E_\theta[L(\gamma(\theta), \psi(R_1, R_2))] = \rho_\psi(\theta)$$

as desired.

PROBLEMS

1. Let $R_1, \ldots, R_n$ be independent random variables, all having the same density h_θ; thus $f_\theta(x_1, \ldots, x_n) = \prod_{i=1}^{n} h_\theta(x_i)$. In each case find the maximum likelihood estimate of θ.

(a) $h_\theta(x) = \theta x^{\theta-1}, \quad 0 \leq x \leq 1, \theta > 0$

(b) $h_\theta(x) = \dfrac{1}{\theta} e^{-x/\theta}, \quad x \geq 0, \theta > 0$

(c) $h_\theta(x) = \dfrac{1}{\theta}, \quad 0 \leq x \leq \theta, \theta > 0$

2. Let R have the Cauchy density with parameter θ; that is,

$$f_\theta(x) = \frac{\theta}{\pi(x^2 + \theta^2)}, \qquad \theta > 0$$

Find the maximum likelihood estimate of θ.

3. Let R have the negative binomial distribution; that is, (see Problem 6, Section 6.4),

$$p_\theta(x) = P\{R = x\} = \binom{x-1}{r-1}\theta^r(1 - \theta)^{x-r}, \qquad x = r, r + 1, \ldots, 0 < \theta \leq 1$$

Find the maximum likelihood estimate of θ.

4. Find the risk function in Example 3, using the maximum likelihood estimate $\hat{\theta} = x/n$.

5. In Example 3, find the Bayes estimate if θ is uniformly distributed between 0 and 1.

6. In Example 3, change the loss function to $L(\theta, a) = (\theta - a)^2/\theta(1 - \theta)$, and let θ be uniformly distributed between 0 and 1. Find the Bayes estimate and show that it has constant risk and is therefore minimax.

7. Let R have the Poisson distribution with parameter $\theta > 0$. Find the Bayes estimate ψ of θ with quadratic loss function if the a priori density is $g(\theta) = e^{-\theta}$. Compute the risk function and the Bayes risk using ψ, and compare with the results using the maximum likelihood estimate.

8.4 SUFFICIENT STATISTICS

In many situations the statistician is concerned with reduction of data. For example, if a sequence of observations results in numbers $x_1, \ldots, x_n$, it is easier to store the single number $x_1 + \cdots + x_n$ than to record the entire set of observations. Under certain conditions no essential information is lost in reducing the data; let us illustrate this by an example.

Let $R_1, \ldots, R_n$ be independent, Bernoulli random variables with parameter θ; that is, $P\{R_i = 1\} = \theta$, $P\{R_i = 0\} = 1 - \theta$, $0 \leq \theta \leq 1$. Let $T = t(R_1, \ldots, R_n) = R_1 + \cdots + R_n$, which has the binomial distribution with parameters n and θ. We claim that $P_\theta\{R_1 = x_1, \ldots, R_n = x_n \mid T = y\}$ actually does not depend on θ. We compute, for $x_i = 0$ or 1, $i = 1, \ldots, n$,

$$P_\theta\{R_1 = x_1, \ldots, R_n = x_n \mid T = y\} = \frac{P_\theta\{R_1 = x_1, \ldots, R_n = x_n, T = y\}}{P_\theta\{T = y\}}$$

This is 0 unless $y = x_1 + \cdots + x_n$, in which case we obtain

$$\frac{P_\theta\{R_1 = x_1, \ldots, R_n = x_n\}}{P_\theta\{T = y\}} = \frac{\theta^y(1 - \theta)^{n-y}}{\binom{n}{y}\theta^y(1 - \theta)^{n-y}} = \frac{1}{\binom{n}{y}}$$

The significance of this result is that for the purpose of making a statistical decision based on observation of $R_1, \ldots, R_n$, we may ignore the individual R_i and base the decision entirely on $R_1 + \cdots + R_n$. To justify this, consider two statisticians, A and B. Statistician A observes $R_1, \ldots, R_n$ and then makes his decision. Statistician B, on the other hand, is only given $T = R_1 + \cdots + R_n$. He then constructs random variables $R'_1, \ldots, R'_n$ as follows. If $T = y$, let $R'_1, \ldots, R'_n$ be chosen according to the conditional probability function of $R_1, \ldots, R_n$ given $T = y$. Explicitly,

$$P\{R'_1 = x_1, \ldots, R'_n = x_n \mid T = y\} = \frac{1}{\dbinom{n}{y}}$$

where $x_i = 0$ or 1, $i = 1, \ldots, n$, $x_1 + \cdots + x_n = y$. B then follows A's decision procedure, using $R'_1, \ldots, R'_n$. Note that since the conditional probability function of $R_1, \ldots, R_n$ given $T = y$ does not depend on the unknown parameter θ, B's procedure is sensible. Now if $x_1 + \cdots + x_n = y$,

$$
\begin{aligned}
P_\theta\{R'_1 = x_1, \ldots, R'_n = x_n\} &= P_\theta\{R'_1 = x_1, \ldots, R'_n = x_n, T = y\} \\
&= P_\theta\{T = y\}P_\theta\{R'_1 = x_1, \ldots, R'_n = x_n \mid T = y\} \\
&= \binom{n}{y}\theta^y(1 - \theta)^{n-y}\frac{1}{\dbinom{n}{y}} \\
&= \theta^y(1 - \theta)^{n-y} \\
&= P_\theta\{R_1 = x_1, \ldots, R_n = x_n\}
\end{aligned}
$$

Thus $(R'_1, \ldots, R'_n)$ has exactly the same probability function as $(R_1, \ldots, R_n)$, so that the procedures of A and B are equivalent. In other words, anything A can do, B can do at least as well, even though B starts with less information.

We now give the formal definitions. For simplicity, we restrict ourselves to the discrete case. However, the definition of sufficiency in the absolutely continuous case is the same, with probability functions replaced by densities. Also, the basic factorization theorem, to be proved below, holds in the absolutely continuous case (admittedly with a more difficult proof).

Let R be a discrete random variable (or random vector) whose probability function under the state of nature θ is p_θ. Let T be a *statistic* for R, that is, a function of R that is also a random variable. T is said to be *sufficient* for R (or for the family p_θ, $\theta \in N$) iff the conditional probability function of R given T does not depend on θ.

The definition is often unwieldy, and the following criterion for sufficiency is useful.

Theorem 1 (Factorization Theorem). *Let $T = t(R)$ be a statistic for R. T is sufficient for R if and only if the probability function p_θ can be factored in the form $p_\theta(x) = g(\theta, t(x))h(x)$.*

PROOF. Assume a factorization of this form. Then

$$P_\theta\{R = x \mid T = y\} = \frac{p_\theta\{R = x, T = y\}}{P_\theta\{T = y\}}$$

This is 0 unless $t(x) = y$, in which case we obtain

$$\frac{P_\theta\{R = x\}}{P_\theta\{T = y\}} = \frac{g(\theta, t(x))h(x)}{\sum\limits_{\{z:t(z)=y\}} g(\theta, t(z))h(z)}$$

$$= \frac{g(\theta, y)h(x)}{\sum\limits_{\{z:t(z)=y\}} g(\theta, y)h(z)}$$

$$= \frac{h(x)}{\sum\limits_{\{z:t(z)=y\}} h(z)}, \quad \text{which is free of } \theta$$

Conversely, if T is sufficient, then

$$p_\theta(x) = P_\theta\{R = x\} = P_\theta\{R = x, T = t(x)\}$$
$$= P_\theta\{T = t(x)\}P_\theta\{R = x \mid T = t(x)\}$$
$$= g(\theta, t(x))h(x) \quad \text{by definition of sufficiency}$$

▶ **Example 1.** Let $R_1, \ldots, R_n$ be independent, each Bernoulli with parameter θ. Show that $R_1 + \cdots + R_n$ is sufficient for $(R_1, \ldots, R_n)$.

We have done this in the introductory discussion, using the definition of sufficiency. Let us check the result using the factorization theorem. If $x = x_1 + \cdots + x_n$, $x_i = 0, 1$, and $t(x) = x_1 + \cdots + x_n$, then

$$p_\theta(x_1, \ldots, x_n) = \theta^{t(x)}(1 - \theta)^{n-t(x)}$$

which is of the form specified in the factorization theorem [with $h(x) = 1$]. ◀

▶ **Example 2.** Let $R_1, \ldots, R_n$ be independent, each Poisson with parameter θ. Again $R_1 + \cdots + R_n$ is sufficient for $(R_1, \ldots, R_n)$. (Notice that $R_1 + \cdots + R_n$ is Poisson with parameter $n\theta$.)

For

$$p_\theta(x_1, \ldots, x_n) = P_\theta\{R_1 = x_1, \ldots, R_n = x_n\}, \quad x_1, \ldots, x_n = 0, 1, \ldots$$

$$= \prod_{i=1}^{n} P_\theta\{R_i = x_i\}$$

$$= \frac{e^{-n\theta}\theta^{x_1+\cdots+x_n}}{x_1! \cdots x_n!}$$

The factorization theorem applies, with $g(\theta, t(x)) = e^{-n\theta}\theta^{t(x)}$, $h(x) = 1/x_1! \ldots x_n!$, $t(x) = x_1 + \cdots + x_n$. ◄

▶ **Example 3.** Let $R_1, \ldots, R_n$ be independent, each normally distributed with mean μ and variance σ^2. Find a sufficient statistic for $(R_1, \ldots, R_n)$ assuming

(a) μ and σ^2 both unknown; that is, $\theta = (\mu, \sigma^2)$.

(b) σ^2 known; that is, $\theta = \mu$.

(c) μ known; that is, $\theta = \sigma^2$.

[Of course $(R_1, \ldots, R_n)$ is always sufficient for itself, but we hope to reduce the data a bit more.] We compute

$$f_\theta(x) = (2\pi\sigma^2)^{-n/2} \exp\left[-\frac{1}{2\sigma^2} \sum_{i=1}^{n} (x_i - \mu)^2 \right] \qquad (8.4.1)$$

Let

$$\bar{x} = \frac{1}{n} \sum_{i=1}^{n} x_i, \qquad s^2 = \frac{1}{n} \sum_{i=1}^{n} (x_i - \bar{x})^2$$

Since $x_i - \bar{x} = x_i - \mu - (\bar{x} - \mu)$, we have

$$s^2 = \frac{1}{n} \sum_{i=1}^{n} (x_i - \mu)^2 - (\bar{x} - \mu)^2$$

Thus

$$f_\theta(x) = (2\pi\sigma^2)^{-n/2} e^{-ns^2/2\sigma^2} e^{-n(\bar{x}-\mu)^2/2\sigma^2} \qquad (8.4.2)$$

By (8.4.2), if μ and σ^2 are unknown, then [take $h(x) = 1$] $(\bar{R}, V^2)$ is sufficient, where $\bar{R}$ is the sample mean $(1/n) \sum_{i=1}^{n} R_i$ and V^2 is the sample variance $(1/n) \sum_{i=1}^{n} (R_i - \bar{R})^2$. If σ^2 is known, then the term $(2\pi\sigma^2)^{-n/2} e^{-ns^2/2\sigma^2}$ can be taken as $h(x)$ in the factorization theorem; hence $\bar{R}$ is sufficient. If μ is known, then, by (8.4.1), $\sum_{i=1}^{n} (R_i - \mu)^2$ is sufficient. ◄

PROBLEMS

1. Let $R_1, \ldots, R_n$ be independent, each uniformly distributed on the interval $[\theta_1, \theta_2]$. Find a sufficient statistic for $(R_1, \ldots, R_n)$, assuming

(a) θ_1, θ_2 both unknown

(b) θ_1 known

(c) θ_2 known

2. Repeat Problem 1 if each R_i has the gamma density with parameters θ_1 and θ_2, that is,

$$f(x) = \frac{x^{\theta_1-1} e^{-x/\theta_2}}{\Gamma(\theta_1)\theta_2^{\theta_1}}, \qquad x \geq 0, \theta_1, \theta_2 > 0$$

3. Repeat Problem 1 if each R_i has the beta density with parameters θ_1 and θ_2, that is,

$$f(x) = \frac{x^{\theta_1-1}(1-x)^{\theta_2-1}}{\beta(\theta_1,\theta_2)}, \qquad 0 \leq x \leq 1, \theta_1, \theta_2 > 0$$

4. Let R_1 and R_2 be independent, with R_1 normal (θ, σ^2), R_2 normal (θ, τ^2), where σ^2 and τ^2 are known. Show that $R_1/\sigma^2 + R_2/\tau^2$ is sufficient for (R_1, R_2).

5. An *exponential family* of densities is a family of the form

$$f_\theta(x) = a(\theta)b(x) \exp\left[\sum_{i=1}^{k} c_i(\theta)t_i(x) \right], \qquad x \text{ real}, \theta \in N$$

(a) Verify that the following density (or probability) functions can be put into the above form.

(i) Binomial (n, θ): $p_\theta(x) = \binom{n}{x}\theta^x(1-\theta)^{n-x}$, $x = 0, 1, \ldots, n, 0 < \theta < 1$

(ii) Poisson (θ): $p_\theta(x) = \dfrac{e^{-\theta}\theta^x}{x!}$, $x = 0, 1, \ldots, \theta > 0$

(iii) Normal (μ, σ^2): $f_\theta(x) = \dfrac{1}{\sqrt{2\pi}\,\sigma} e^{-(x-u)^2/2\theta^2}$, $\theta = (\mu, \sigma^2)$

(iv) Gamma (θ_1, θ_2): $f_\theta(x) = \dfrac{x^{\theta_1-1}e^{-x/\theta_2}}{\Gamma(\theta_1)\theta_2^{\theta_1}}$, $x > 0, \theta = (\theta_1, \theta_2), \theta_1, \theta_2 > 0$

(v) Beta (θ_1, θ_2): $f_\theta(x) = \dfrac{x^{\theta_1-1}(1-x)^{\theta_2-1}}{\beta(\theta_1,\theta_2)}$, $0 < x < 1, \theta = (\theta_1, \theta_2),$

$$\theta_1, \theta_2 > 0$$

(vi) Negative binomial (r, θ): $p_\theta(x) = \binom{x-1}{r-1}\theta^r(1-\theta)^{x-r}$, $x = r, r+1, \ldots,$

$$0 < \theta < 1, r \text{ a known positive integer}$$

(b) If $R_1, \ldots, R_n$ are independent, each R_i having the density f_θ of part (a), find a sufficient statistic for $(R_1, \ldots, R_n)$.

6. Let T be sufficient for the family of densities f_θ, $\theta \in N$. Consider the problem of testing the null hypothesis that $\theta \in H_0$ versus the alternative that $\theta \in H_1$. Show that all possible risk points can be obtained from tests based on T [i.e., $\varphi(x)$ expressible as a function of $t(x)$].

8.5 UNBIASED ESTIMATES BASED ON A COMPLETE SUFFICIENT STATISTIC

In this section we require our estimates ψ of $\gamma(\theta)$ to be *unbiased*; that is, $E_\theta\psi(R) = \gamma(\theta)$ for all $\theta \in N$. Our objective is to show that in a wide class of situations it is possible to construct unbiased estimates ψ that have *uniformly minimum risk*; that is, if ψ' is any unbiased estimate of $\gamma(\theta)$, then $\rho_\psi(\theta) \leq \rho_{\psi'}(\theta)$ for all θ. We need a technical definition first. If T is a statistic for R,

T is said to be *complete* iff there are no unbiased estimates of 0 based on T, that is, iff whenever $E_\theta g(T) = 0$ for all $\theta \in N$, we have $P_\theta\{g(T) = 0\} = 1$ for all $\theta \in N$.

Theorem 1. *Let* $T = t(R)$ *be a complete sufficient statistic for* R, *and let* ψ *be an unbiased estimate of* $\gamma(\theta)$ *based on* T [*i.e.,* $\psi(x)$ *can be expressed as a function of* $t(x)$]. *Assume that the loss function* $L(\gamma(\theta), a)$ *is convex in* a *for each fixed* θ. *Then* ψ *has uniformly minimum risk among all unbiased estimates of* $\gamma(\theta)$.

PROOF.   Let ψ' be any unbiased estimate of $\gamma(\theta)$, and define $\psi''(x) = E[\psi'(R) \mid T = t(x)]$. (Since T is sufficient, ψ'' does not depend on θ and hence is a legitimate estimate.) ψ'' is an unbiased estimate of $\gamma(\theta)$ based on T, and so is ψ, and therefore $E_\theta[\psi''(R) - \psi(R)] = 0$ for all θ. But $\psi''(R)$ and $\psi(R)$ can be expressed as functions of T; hence, by completeness,

$$P_\theta\{\psi''(R) = \psi(R)\} = 1 \qquad \text{for all } \theta$$

It follows that $\rho_{\psi''}(\theta) = \rho_\psi(\theta)$ for all θ. But the proof of the Rao-Blackwell theorem, with R_1 replaced by T, x by $t(x)$, $\psi(R_1, R_2)$ by $\psi'(R)$, and $\psi^*(R_1)$ by $\psi''(R)$, shows that $\rho_{\psi''}(\theta) \leq \rho_{\psi'}(\theta)$ for all θ, as desired.

If $L(\gamma(\theta), a) = (\gamma(\theta) - a)^2$, then $\rho_\psi(\theta) = E_\theta[(\gamma(\theta) - \psi(R))^2] = \text{Var}_\theta\, \psi(R)$. Thus ψ has the smallest variance of all unbiased estimates of $\gamma(\theta)$, regardless of the state of nature. In this case ψ is said to be a *uniformly minimum variance unbiased estimate* (UMVUE).

▶ **Example 1.**   Let $R_1, \ldots, R_n$ be independent, each Bernoulli with parameter θ, $0 \leq \theta \leq 1$. By Example 1, Section 8.4, $T = R_1 + \cdots + R_n$ is sufficient for $(R_1, \ldots, R_n)$; let us show that it is complete.

Now T is binomial with parameters n and θ; hence

$$E_\theta g(T) = \sum_{k=0}^{n} g(k) \binom{n}{k} \theta^k (1 - \theta)^{n-k}$$

$$= \left[\sum_{k=0}^{n} g(k) \binom{n}{k} \left(\frac{\theta}{1 - \theta} \right)^k \right] (1 - \theta)^n \qquad \text{if } \theta < 1$$

If $E_\theta g(T) = 0$ for all $\theta \in [0, 1]$, then $\sum_{k=0}^{n} g(k) \binom{n}{k} z^k = 0$ for all $z \in [0, \infty)$; hence $g(k) = 0$ for $k = 0, 1, \ldots, n$.

We now look for unbiased estimates of $\gamma(\theta)$ based on T. If $\psi(x_1, \ldots, x_n) = g(t(x_1, \ldots, x_n))$, $t(x_1, \ldots, x_n) = x_1 + \cdots + x_n$, is such an estimate, the above argument shows that $E_\theta \psi(R_1, \ldots, R_n) = E_\theta g(T)$ is a polynomial in

θ of degree $\leq n$. Thus $\gamma(\theta)$ must be of the form $a_0 + a_1\theta + \cdots + a_n\theta^n$; furthermore, an unbiased estimate of such an expression is easily found. If $T^{(r)} = T(T-1) \cdots (T-r+1)$, $n^{(r)} = (n-1) \cdots (n-r+1)$, then

$$E\left[\frac{T^{(r)}}{n^{(r)}}\right] = \sum_{k=0}^{n} \frac{k(k-1) \cdots (k-r+1)}{n(n-1) \cdots (n-r+1)} \frac{n!}{k!\,(n-k)!} \theta^k(1-\theta)^{n-k}$$

$$= \theta^r \sum_{k=r}^{n} \frac{(n-r)!}{(k-r)!\,(n-k)!} \theta^{k-r}(1-\theta)^{n-k} = \theta^r(\theta+1-\theta)^{n-r}$$

$$= \theta^r$$

Thus $\sum_{k=0}^{n} a_k[T^{(k)}/n^{(k)}]$ is a UMVUE of $\gamma(\theta) = \sum_{k=0}^{n} a_k\theta^k$; in particular the sample mean T/n is a UMVUE of θ. ◄

▶ **Example 2.** Let $R_1, \ldots, R_n$ be independent, each Poisson with parameter θ. By Example 2, Section 8.4, $T = R_1 + \cdots + R_n$ is sufficient for $(R_1, \ldots, R_n)$; T is also complete. For T is Poisson with parameter $n\theta$; hence

$$E_\theta g(T) = \sum_{k=0}^{\infty} g(k)e^{-n\theta}\frac{(n\theta)^k}{k!}$$

If $E_\theta g(T) = 0$ for all $\theta > 0$, then

$$\sum_{k=0}^{\infty}\left[\frac{g(k)n^k}{k!}\right]\theta^k = 0 \qquad \text{for all } \theta > 0$$

Since this is a power series in θ, we must have $g \equiv 0$.

If $\psi(x_1, \ldots, x_n) = g(t(x_1, \ldots, x_n))$ is an unbiased estimate of $\gamma(\theta)$, then

$$\gamma(\theta) = E_\theta\psi(R_1, \ldots, R_n) = E_\theta g(T) = e^{-n\theta}\sum_{k=0}^{\infty} g(k)\frac{(n\theta)^k}{k!}$$

Thus $\gamma(\theta)$ must be expressible as a power series in θ. If $\gamma(\theta) = \sum_{k=0}^{\infty} a_k\theta^k$, then

$$\gamma(\theta)e^{n\theta} = \sum_{j=0}^{\infty} a_j\theta^j \sum_{k=0}^{\infty}\frac{(n\theta)^k}{k!}$$

$$= \sum_{k=0}^{\infty} c_k\theta^k \qquad \text{where } c_k = \sum_{i=0}^{k}\frac{n^i}{i!} a_{k-i}$$

But

$$\gamma(\theta)e^{n\theta} = \sum_{k=0}^{\infty}\frac{g(k)n^k}{k!}\theta^k$$

hence

$$g(k) = \frac{k! \, c_k}{n^k}$$

$$= \sum_{i=0}^{k} \frac{k!}{i!} \frac{a_{k-i}}{n^{k-i}}$$

We conclude that

$$\sum_{i=0}^{T} \frac{T!}{i!} \frac{a_{T-i}}{n^{T-i}} \quad \text{is a UMVUE of} \quad \gamma(\theta) = \sum_{k=0}^{\infty} a_k \theta^k$$

For example, if $\gamma(\theta) = \theta^r$, $r = 1, 2, \ldots$, the UMVUE is

$$\frac{T!}{(T-r)! \, n^r} \frac{1}{} = \frac{T^{(r)}}{n^r} \left(= \frac{T}{n} = \text{the sample mean when } r = 1 \right)$$

[In this particular case the above computation could have been avoided, since we know that $E_\theta(T^{(r)}) = (n\theta)^r$ (Problem 8, Section 3.2). Since $T^{(r)}/n^r$ is an unbiased estimate of θ^r based on T, it is a UMVUE.]

As another example, a UMVUE of $1/(1 - \theta) = \sum_{k=0}^{\infty} \theta^k$, $0 < \theta < 1$, is

$$\sum_{i=0}^{T} \frac{T!}{i!} \frac{1}{n^{T-i}} \quad \blacktriangleleft$$

PROBLEMS

1. Find a UMVUE of $e^{-\theta}$ in Example 2.

2. Let $R_1, \ldots, R_n$ be independent, each uniformly distributed between 0 and $\theta > 0$. By Problem 1, Section 8.4, $T = \max R_i$ is sufficient for $(R_1, \ldots, R_n)$.
   (a) Show that T is complete.
   (b) Find a UMVUE of $\gamma(\theta)$, assuming that γ extends to a function with a continuous derivative on $[0, \infty)$, and $\theta^n \gamma(\theta) \to 0$ as $\theta \to 0$. [In part (a), use without proof the fact that if $\int_0^\theta h(y) \, dy = 0$ for all $\theta > 0$, then $h(y) = 0$ except on a set of Lebesgue measure 0. Notice that if it is known that h is continuous, then $h \equiv 0$ by the fundamental theorem of calculus.]

3. Let $R_1, \ldots, R_n$ be independent, each normal with mean θ and known variance σ^2.
   (a) Show that the sample mean $\bar{R}$ is a UMVUE of θ.
   (b) Show that $(\bar{R})^2 - (\sigma^2/n)$ is a UMVUE of θ^2.
   [Use without proof the fact that if $\int_{-\infty}^{\infty} h(y) e^{\theta y} \, dy = 0$ for all $\theta > 0$, then $h(y) = 0$ except on a set of Lebesgue measure 0.]

4. Let $R_1, \ldots, R_n$ be independent, with $P\{R_i = k\} = 1/N, k = 1, \ldots, N$; take $\theta = N, N = 1, 2, \ldots$.
   (a) Show that $\max_{1 \leq i \leq n} R_i$ is a complete sufficient statistic.
   (b) Find a UMVUE of $\gamma(N)$.

5. Let R have the negative binomial distribution:

$$P\{R = k\} = \binom{k-1}{r-1} p^r (1 - p)^{k-r}, \qquad k = r, r + 1, \ldots, 0 < p \leq 1$$

   Take $\theta = 1 - p, 0 \leq \theta < 1$. Show that $\gamma(\theta)$ has a UMVUE if and only if it is expressible as a power series in θ; find the form of the UMVUE.

6. Let

$$P\{R = k\} = \frac{e^{-\theta}}{1 - e^{-\theta}} \frac{\theta^k}{k!}, \qquad k = 1, 2, \ldots, \theta > 0$$

   (This is the conditional probability function of a Poisson random variable R', given that $R' \geq 1$.) R is clearly sufficient for itself, and is complete by an argument similar to that of Example 2.
   (a) Find a UMVUE of $e^{-\theta}$.
   (b) Show that (assuming quadratic loss function) the estimate ψ found in part (a) is *inadmissible*; that is, there is another estimate ψ' such that $\rho_{\psi'}(\theta) \leq \rho_{\psi}(\theta)$ for all θ, and $\rho_{\psi'}(\theta) < \rho_{\psi}(\theta)$ for some θ. This shows that unbiased estimates, while often easy to find, are not necessarily desirable.

7. The following is another method for obtaining a UMVUE. Let $R_1, \ldots, R_n$ be independent, each Bernoulli with parameter $\theta, 0 \leq \theta \leq 1$, as in Example 1. If $j = 1, \ldots, n$, then

$$E\left[\prod_{i=1}^{j} R_i \right] = P\{R_1 = \cdots = R_j = 1\} = \theta^j$$

   Thus $R_1 R_2 \cdots R_j$ is an unbiased estimate of θ^j. But then $\psi(k) = E[R_1 \cdots R_j \mid \sum_{i=1}^{n} R_i = k\}$ is an unbiased estimate of θ^j based on the complete sufficient statistic $\sum_{i=1}^{n} R_i$, so that ψ is a UMVUE. Compute ψ directly and show that the result agrees with Example 1.

8. Let $R_1, \ldots, R_n$ be independent, each Poisson with parameter $\theta > 0$. Show, using the analysis in Problem 7, that

$$E\left(R_1 R_2 \mid \sum_{i=1}^{n} R_i = k \right) = \frac{k(k - 1)}{n^2}, \qquad k = 0, 1, \ldots$$

9. Let $R_1, \ldots, R_n$ be independent, each uniformly distributed between 0 and θ; if $T = \max R_i$, then $[(n + 1)/n]T$ is a UMVUE of θ (see Problem 2). Compare the risk function $E_\theta[((1 + 1/n)T - \theta)^2]$ using $[(n + 1)/n]T$ with the risk function $E_\theta[((2/n) \sum_{i=1}^{n} R_i - \theta)^2]$ using the unbiased estimate $(2/n) \sum_{i=1}^{n} R_i$.

10. Let $R_1, \ldots, R_n$ be independent, each Bernoulli with parameter $\theta \in [0, 1]$. Show that (assuming quadratic loss function) there is no best estimate of θ based on $R_1, \ldots, R_n$; that is, there is no estimate ψ such that $\rho_{\psi}(\theta) \leq \rho_{\psi'}(\theta)$ for all θ and all estimates ψ' of θ.

11. If $E_\theta \psi_1(R) = E_\theta \psi_2(R) = \gamma(\theta)$ and ψ_1, ψ_2 both minimize $\rho_\psi(\theta) = E_\theta[(\psi(R) - \gamma(\theta))^2]$, θ fixed, show that $P_\theta\{\psi_1(R) = \psi_2(R)\} = 1$. Consequently, if ψ_1 and ψ_2 are UMVUEs of $\gamma(\theta)$, then, for each θ, $\psi_1(R) = \psi_2(R)$ with probability 1.

12. Let f_θ, $\theta \in N = $ an open interval of reals, be a family of densities. Assume that $\partial f_\theta(x)/\partial\theta$ exists and is continuous everywhere, and that $\int_{-\infty}^{\infty} f_\theta(x)\, dx$ can be differentiated under the integral sign with respect to θ.

(a) If R has density f_θ when the state of nature is θ, show that

$$E_\theta\left[\frac{\partial}{\partial\theta}\ln f_\theta(R)\right] = 0$$

(b) If $E_\theta \psi(R) = \gamma(\theta)$ and $\int_{-\infty}^{\infty} \psi(x) f_\theta(x)\, dx$ can be differentiated under the integral sign with respect to θ, show that

$$\frac{d}{d\theta}\gamma(\theta) = E_\theta\left[\psi(R)\frac{\partial}{\partial\theta}\ln f_\theta(R)\right]$$

(c) Under the assumptions of part (b), show that

$$\mathrm{Var}_\theta\, \psi(R) \geq \frac{[(d/d\theta)\,\gamma(\theta)]^2}{[E_\theta(\partial \ln f_\theta(R)/\partial\theta)^2]}$$

if the denominator is >0. In particular, if $f_\theta(x) = f_\theta(x_1, \ldots, x_n) = \prod_{i=1}^{n} h_\theta(x_i)$, then

$$E_\theta\left(\frac{\partial}{\partial\theta}\ln f_\theta(R)\right)^2 = \mathrm{Var}_\theta\, \frac{\partial}{\partial\theta}\ln f_\theta(R)$$

$$= n\,\mathrm{Var}_\theta\, \frac{\partial}{\partial\theta}\ln h_\theta(R_i)$$

$$= nE_\theta\left(\frac{\partial}{\partial\theta}\ln h_\theta(R_i)\right)^2$$

where $R = (R_1, \ldots, R_n)$.
The above result is called the *Cramer-Rao inequality* (an analogous theorem may be proved with densities replaced by probability functions). If ψ is an estimate that satisfies the Cramer-Rao lower bound with equality for all θ, then ψ is a UMVUE of $\gamma(\theta)$. This idea may be used to give an alternative proof that the sample mean is a UMVUE of the true mean in the Bernoulli, Poisson, and normal cases (see Examples 1 and 2 and Problem 3 of this section).

13. If $R_1, \ldots, R_n$ are independent, each with mean μ and variance σ^2, and V^2 is the sample variance, show that V^2 is a biased estimate of σ^2; specifically,

$$E(V^2) = \frac{(n-1)}{n}\sigma^2$$

8.6 SAMPLING FROM A NORMAL POPULATION

If $R_1, \ldots, R_n$ are independent, each normally distributed with mean μ and variance σ^2, we have seen that $(\bar{R}, V^2)$, where $\bar{R}$ is the sample mean $(1/n)(R_1 + \cdots + R_n)$ and $V^2 = (1/n)\sum_{i=1}^{n}(R_i - \bar{R})^2$ is the sample variance, is a sufficient statistic for $(R_1, \ldots, R_n)$. $\bar{R}$ and V^2 have some special properties that are often useful. First, $\bar{R}$ is a sum of the independent normal random variables R_i/n, each of which has mean μ/n and variance σ^2/n^2; hence $\bar{R}$ *is normal with mean μ and variance σ^2/n. We* now prove that $\bar{R}$ and V^2 are independent.

Theorem 1. If $R_1, \ldots, R_n$ are independent, each normal (μ, σ^2), the associated sample mean and variance are independent random variables.

*PROOF. Define random variables $W_1, \ldots, W_n$ by

$$W_1 = \frac{1}{\sqrt{n}} R_1 + \cdots + \frac{1}{\sqrt{n}} R_n$$

$$W_2 = c_{21}R_1 + \cdots + c_{2n}R_n$$

$$\vdots$$

$$W_n = c_{n1}R_1 + \cdots + c_{nn}R_n$$

where the c_{ij} are chosen so as to make the transformation orthogonal. [This may be accomplished by extending the vector $(1/\sqrt{n}, \ldots, 1/\sqrt{n})$ to an orthonormal basis for E^n.] The Jacobian J of the transformation is the determinant of the orthogonal matrix $A = [c_{ij}]$ (with $c_{1j} = 1/\sqrt{n}$, $j = 1, \ldots, n$), namely, ± 1. Thus (see Problem 12, Section 2.8) the density of $(W_1, \ldots, W_n)$ is given by

$$f^*(y_1, \ldots, y_n) = \frac{f(x_1, \ldots, x_n)}{|J|}$$

$$= (2\pi\sigma^2)^{-n/2} \exp\left[-\frac{1}{2\sigma^2}\sum_{i=1}^{n}(x_i - \mu)^2\right]$$

where

$$\begin{pmatrix} x_1 \\ \cdot \\ \cdot \\ \cdot \\ x_n \end{pmatrix} = A^{-1} \begin{pmatrix} y_1 \\ \cdot \\ \cdot \\ \cdot \\ y_n \end{pmatrix}$$

Since $\sum_{i=1}^{n} x_i^2 = \sum_{i=1}^{n} y_i^2$ by orthogonality, and $\sum_{i=1}^{n} x_i = \sqrt{n} \, y_1$,

$$f^*(y_1, \ldots, y_n) = (2\pi\sigma^2)^{-n/2} \exp\left[-\frac{1}{2\sigma^2}\left(\sum_{i=1}^{n} y_i^2 - 2\mu\sqrt{n}\,y_1 + n\mu^2\right)\right]$$

$$= \frac{1}{\sqrt{2\pi}\,\sigma} \exp\left[-\frac{(y_1 - \sqrt{n}\,\mu)^2}{2\sigma^2}\right] \prod_{i=2}^{n} \frac{1}{\sqrt{2\pi}\,\sigma} \exp\left(-\frac{y_i^2}{2\sigma^2}\right)$$

It follows that $W_1, \ldots, W_n$ are independent, with $W_2, \ldots, W_n$ each normal $(0, \sigma^2)$ and W_1 normal $(\sqrt{n}\,\mu, \sigma^2)$. But

$$nV^2 = \sum_{i=1}^{n}(R_i - \bar{R})^2 = \sum_{i=1}^{n} R_i^2 - 2\bar{R}\sum_{i=1}^{n} R_i + n(\bar{R})^2$$

$$= \sum_{i=1}^{n} R_i^2 - n(\bar{R})^2$$

$$= \sum_{i=1}^{n} W_i^2 - \left(\sum_{i=1}^{n} \frac{R_i}{\sqrt{n}}\right)^2$$

$$= \sum_{i=1}^{n} W_i^2 - W_1^2$$

$$= \sum_{i=2}^{n} W_i^2$$

Since $\sqrt{n}\,\bar{R} = W_1$, it follows that $\bar{R}$ and V^2 are independent, completing the proof.

The above argument also gives us the distribution of the sample variance. For

$$\frac{nV^2}{\sigma^2} = \sum_{i=2}^{n}\left(\frac{W_i}{\sigma}\right)^2$$

where the W_i/σ are independent, each normal $(0, 1)$. Thus nV^2/σ^2 *has the chi-square distribution with $n - 1$ degrees of freedom*; that is, the density of nV^2/σ^2 is

$$\frac{1}{2^{(n-1)/2}\Gamma((n-1)/2)}\, x^{(n-3)/2}e^{-x/2}, \qquad x \geq 0$$

(see Problem 3, Section 5.2).

Now since $\bar{R}$ is normal $(\mu, \sigma^2/n)$, $\sqrt{n}\,(\bar{R} - \mu)/\sigma$ is normal $(0, 1)$; hence

$$P\left\{-b \leq \sqrt{n}\,\frac{(\bar{R} - \mu)}{\sigma} \leq b\right\} = F^*(b) - F^*(-b)$$

$$= 2F^*(b) - 1$$

where F^* is the normal $(0, 1)$ distribution function. If b is chosen so that $2F^*(b) - 1 = 1 - \alpha$, that is, $F^*(b) = 1 - \alpha/2$ ($b = N_{\alpha/2}$ in the terminology of Example 3, Section 8.2),

$$P\left\{-N_{\alpha/2} \leq \sqrt{n}\,\frac{(\bar{R} - \mu)}{\sigma} \leq N_{\alpha/2}\right\} = 1 - \alpha$$

or

$$P\left\{\bar{R} - \frac{\sigma N_{\alpha/2}}{\sqrt{n}} \leq \mu \leq \bar{R} + \frac{\sigma N_{\alpha/2}}{\sqrt{n}}\right\} = 1 - \alpha$$

Thus, with probability $1 - \alpha$, the true mean μ lies in the random interval

$$I = \left[\bar{R} - \frac{\sigma N_{\alpha/2}}{\sqrt{n}}, \quad \bar{R} + \frac{\sigma N_{\alpha/2}}{\sqrt{n}}\right]$$

I is called a *confidence interval* for μ with *confidence coefficient* $1 - \alpha$.

The interval I is computable from the given observations of $R_1, \ldots, R_n$, provided that σ^2 is known. If σ^2 is unknown, it is natural to replace the true variance σ^2 by the sample variance V^2. However, we then must know something about the random variable $(\bar{R} - \mu)/V$. In order to provide the necessary information, we do the following computation.

Let R_1 be normal $(0, 1)$, and let R_2 have the chi square distribution with m degrees of freedom; assume that R_1 and R_2 are independent. We compute the density of $\sqrt{m}\,R_1/\sqrt{R_2}$, as follows.

Let

$$W_1 = \frac{\sqrt{m}\,R_1}{\sqrt{R_2}}$$

$$W_2 = R_2$$

so that

$$R_1 = \frac{W_1\sqrt{W_2}}{\sqrt{m}}$$

$$R_2 = W_2$$

Thus we have a transformation of the form

$$(y_1, y_2) = g(x_1, x_2) = \left(\frac{\sqrt{m}\,x_1}{\sqrt{x_2}}, x_2\right)$$

with inverse given by

$$(x_1, x_2) = h(y_1, y_2) = \left(\frac{y_1\sqrt{y_2}}{\sqrt{m}}, y_2\right)$$

g is defined on $\{(x_1, x_2) \in E^2 : x_2 > 0\}$ and h on $\{(y_1, y_2) \in E^2 : y_2 > 0\}$.

By Problem 12, Section 2.8, the density of (W_1, W_2) is given by

$$f_{12}^*(y_1, y_2) = f_{12}(h(y_1, y_2)) |J_h(y_1, y_2)|, \qquad y_2 > 0$$

where

$$J_h(y_1, y_2) = \begin{vmatrix} \dfrac{\partial x_1}{\partial y_1} & \dfrac{\partial x_1}{\partial y_2} \\[2mm] \dfrac{\partial x_2}{\partial y_1} & \dfrac{\partial x_2}{\partial y_2} \end{vmatrix} = \begin{vmatrix} \dfrac{\sqrt{y_2}}{\sqrt{m}} & \dfrac{y_1}{2\sqrt{my_2}} \\[2mm] 0 & 1 \end{vmatrix} = \sqrt{\dfrac{y_2}{m}}$$

Thus

$$f_{12}^*(y_1, y_2) = \frac{1}{\sqrt{2\pi}} e^{-y_1^2 y_2/2m} \frac{1}{2^{m/2}\Gamma(m/2)} y_2^{m/2-1} e^{-y_2/2} \sqrt{\frac{y_2}{m}}$$

Therefore the density of W_1 is

$$f_{W_1}(y_1) = \int_0^\infty f_{12}^*(y_1, y_2)\, dy_2$$

But

$$\int_0^\infty y_2^{(m+1)/2-1} e^{-(1+y_1^2/m)y_2/2} dy_2 = \frac{\Gamma((m+1)/2)2^{(m+1)/2}}{(1+y_1^2/m)^{(m+1)/2}}$$

Hence

$$f_{W_1}(y_1) = \frac{\Gamma((m+1)/2)}{\sqrt{m\pi}\,\Gamma(m/2)} \frac{1}{(1+y_1^2/m)^{(m+1)/2}}$$

A random variable with this density is said to have the t *distribution* with m degrees of freedom.

An application of Stirling's formula shows that the t density approaches the normal $(0, 1)$ density as $m \to \infty$.

Now we know that $\sqrt{n}\,(\bar{R} - \mu)/\sigma$ is normal $(0, 1)$, and nV^2/σ^2 has the chi-square distribution with $n - 1$ degrees of freedom. Thus

$$\frac{\sqrt{n-1}\sqrt{n}\,(\bar{R} - \mu)/\sigma}{\sqrt{n}\,V/\sigma} = \sqrt{n-1}\,\frac{(\bar{R} - \mu)}{V} = \frac{\bar{R} - \mu}{\left[(1/n(n-1))\sum\limits_{i=1}^n (R_i - \bar{R})^2\right]^{1/2}}$$

has the t distribution with $n - 1$ degrees of freedom.

If $t_{\beta,m}$ is such that $\int_{t_{\beta,m}}^\infty h_m(t)\, dt = \beta$, where h_m is the t density with m degrees of freedom, then

$$P\left\{-t_{\alpha/2,n-1} \leq \sqrt{n-1}\,\frac{(\bar{R} - \mu)}{V} \leq t_{\alpha/2,n-1}\right\} = 1 - \alpha$$

Thus

$$\left[\bar{R} - \frac{Vt_{\alpha/2,n-1}}{\sqrt{n-1}}, \quad \bar{R} + \frac{Vt_{\alpha/2,n-1}}{\sqrt{n-1}}\right]$$

is a confidence interval for μ with confidence coefficient $1 - \alpha$.

PROBLEMS

1. Let R_1 and R_2 be independent, chi-square random variables with m and n degrees of freedom, respectively. Show that $(R_1/m)/(R_2/n)$ has density

$$f_{mn}(x) = \frac{(m/n)^{m/2}}{\beta(m/2, n/2)} \frac{x^{(m/2)-1}}{(1 + mx/n)^{(m+n)/2}}, \quad x \geq 0$$

$(R_1/m)/(R_2/n)$ is said to have the *F distribution with m and n degrees of freedom*, abbreviated $F(m, n)$.

2. Calculate the mean and variance of the chi-square, t, and F distributions.

3. (a) If T has the t distribution with n degrees of freedom, show that T^2 has the $F(1, n)$ distribution.
   (b) If R has the $F(m, n)$ distribution, show that $1/R$ has the $F(n, m)$ distribution.
   (c) If R_1 is chi-square (m) and R_2 is chi-square (n), show that $R_1 + R_2$ is chi-square $(m + n)$.

4. Discuss the problem of obtaining confidence intervals for the variance σ^2 of a normally distributed random variable, assuming that
   (a) The mean μ is known
   (b) μ is unknown

5. (A two-sample problem) Let $R_{11}, R_{12}, \ldots, R_{1n_1}, R_{21}, R_{22}, \ldots, R_{2n_2}$ be independent, with the R_{1j} normal (μ_1, σ^2) and the R_{2j} normal (μ_2, σ^2) (μ_1, μ_2 and σ^2 unknown). Thus we are taking independent samples from two different normal populations. Show that if $\bar{R}_i$ and V_i^2, $i = 1, 2$, are the sample mean and variance of the two samples, and

$$k = (n_1 V_1^2 + n_2 V_2^2)^{1/2} \left[\frac{n_1 + n_2}{n_1 n_2 (n_1 + n_2 - 2)} \right]^{1/2}$$

then $[\bar{R}_1 - \bar{R}_2 - kt_{\alpha/2, n_1+n_2-2}, \bar{R}_1 - \bar{R}_2 + kt_{\alpha/2,n_1+n_2-2}]$ is a confidence interval for $\mu_1 - \mu_2$ with confidence coefficient $1 - \alpha$.

6. In Problem 5, assume that the samples have different variances σ_1^2 and σ_2^2. Discuss the problem of obtaining confidence intervals for the ratio σ_1^2/σ_2^2.

7. (a) Suppose that $C(R)$ is a *confidence set* for $\gamma(\theta)$ with confidence coefficient $\geq 1 - \alpha$; that is,

$$P_\theta\{\gamma(\theta) \in C(R)\} \geq 1 - \alpha \quad \text{for all } \theta \in N$$

Consider the hypothesis-testing problem

$$H_0: \gamma(\theta) = k$$
$$H_1: \gamma(\theta) \neq k$$

and the following test.

$$\varphi_k(x) = 1 \quad \text{if } k \notin C(x)$$
$$= 0 \quad \text{if } k \in C(x)$$

[Thus $C(x)$ is the *acceptance region* of φ_k.] Show that φ_k is a test at level α.

(b) Suppose that for all k in the range of γ there is a nonrandomized test φ_k [i.e., $\varphi_k(x) = 0$ or 1 for all x] at level α for $H_0: \gamma(\theta) = k$ versus $H_1: \gamma(\theta) \neq k$. Let $C(x)$ be the set $\{k: \varphi_k(x) = 0\}$. Show that $C(R)$ is a confidence set for $\gamma(\theta)$ with confidence coefficient $\geq 1 - \alpha$.

This result allows the confidence interval examples in this section to be translated into the language of hypothesis testing.

*8.7 THE MULTIDIMENSIONAL GAUSSIAN DISTRIBUTION

If $R'_1, \ldots, R'_n$ are independent, normally distributed random variables and we define random variables $R_1, \ldots, R_n$ by $R_i = \sum_{j=1}^{n} a_{ij} R'_j + b_j$, $i = 1, \ldots, n$, the R_i have a distribution of considerable importance in many aspects of probability and statistics. In this section we examine the properties of this distribution and make an application to the problem of prediction.

Let $R = (R_1, \ldots, R_n)$ be a random vector. The *characteristic function* of R (or the *joint characteristic function* of $R_1, \ldots, R_n$) is defined by

$$M(u_1, \ldots, u_n) = E[i(u_1 R_1 + \cdots + u_n R_n)], \qquad u_1, \ldots, u_n \text{ real}$$

$$= \int_{-\infty}^{\infty} \cdots \int_{-\infty}^{\infty} \exp\left(i \sum_{k=1}^{n} u_k x_k\right) dF(x_1, \ldots, x_n)$$

where F is the distribution function of R. It will be convenient to use a vector-matrix notation. If $u = (u_1, \ldots, u_n) \in E^n$, $\mathbf{u}$ will denote the column vector with components $u_1, \ldots, u_n$. Similarly we write $\mathbf{x}$ for col $(x_1, \ldots, x_n)$ and $\mathbf{R}$ for col $(R_1, \ldots, R_n)$. A superscript t will indicate the transpose of a matrix.

Just as in one dimension, it can be shown that the characteristic function determines the distribution function uniquely.

DEFINITION. The random vector $R = (R_1, \ldots, R_n)$ is said to be *Gaussian* (or $R_1, \ldots, R_n$ are said to be *jointly Gaussian*) iff the characteristic function of R is

$$M(u_1, \ldots, u_n) = \exp[i\mathbf{u}^t\mathbf{b}] \exp[-\tfrac{1}{2}\mathbf{u}^t K \mathbf{u}]$$

$$= \exp\left[i \sum_{r=1}^{n} u_r b_r - \frac{1}{2} \sum_{r,s=1}^{n} u_r K_{rs} u_s\right] \qquad (8.7.1)$$

where $b_1, \ldots, b_n$ are arbitrary real numbers and K is an arbitrary real symmetric nonnegative definite n by n matrix. (Nonnegative definite means that $\sum_{r,s=1}^{n} a_r K_{rs} a_s$ is real and ≥ 0 for all real numbers $a_1, \ldots, a_n$.)

We must show that there is a random vector with this characteristic function. We shall do this in the proof of the next theorem.

Theorem 1. *Let R be a random n-vector. R is Gaussian iff $\mathbf{R}$ can be expressed as $W\mathbf{R}' + \mathbf{b}$, where $b = (b_1, \ldots, b_n) \in E^n$, W is an n by n matrix, and $R_1', \ldots, R_n'$ are independent normal random variables with 0 mean.*

The matrix K of (8.7.1) is given by WDW^t, where $D = \operatorname{diag}(\lambda_1, \ldots, \lambda_n)$ is a diagonal matrix with entries $\lambda_j = \operatorname{Var} R_j'$, $j = 1, \ldots, n$. (To avoid having to treat the case $\lambda_j = 0$ separately, we agree that normal with expectation m and variance 0 will mean degenerate at m.)

Furthermore, the matrix W can be taken as orthogonal.

PROOF. If $\mathbf{R} = W\mathbf{R}' + \mathbf{b}$, then

$$E[\exp(i\mathbf{u}^t\mathbf{R})] = \exp[i\mathbf{u}^t\mathbf{b}] \, E[\exp(i\mathbf{u}^t W\mathbf{R}')]$$

But

$$E[\exp(i\mathbf{v}^t\mathbf{R}')] = E\left[\prod_{k=1}^{n} \exp(iv_k R_k')\right]$$

$$= \prod_{k=1}^{n} E[\exp(iv_k R_k')] = \exp\left[-\frac{1}{2}\sum_{k=1}^{n} \lambda_k v_k^2\right]$$

$$= \exp[-\tfrac{1}{2}\mathbf{v}^t D\mathbf{v}]$$

Set $\mathbf{v} = W^t\mathbf{u}$ to obtain

$$E[\exp(i\mathbf{u}^t\mathbf{R})] = \exp[i\mathbf{u}^t\mathbf{b} - \tfrac{1}{2}\mathbf{u}^t K\mathbf{u}]$$

where $K = WDW^t$. K is clearly symmetric, and is also nonnegative definite, since $\mathbf{u}^t K\mathbf{u} = \mathbf{v}^t D\mathbf{v} = \sum_{k=1}^{n} \lambda_k v_k^2 \geq 0$, where $\mathbf{v} = W^t\mathbf{u}$. Thus R is Gaussian. (Notice also that if K is symmetric and nonnegative definite, there is an orthogonal matrix W such that $W^t KW = D$, where D is the diagonal matrix of eigenvalues of K. Thus $K = WDW^t$, so that it is always possible to construct a Gaussian random vector corresponding to a prescribed K and b.)

Conversely, let R have characteristic function $\exp[i\mathbf{u}^t\mathbf{b} - (1/2)(\mathbf{u}^t K\mathbf{u})]$, where K is symmetric and nonnegative definite. Let W be an orthogonal matrix such that $W^t KW = D = \operatorname{diag}(\lambda_1, \ldots, \lambda_n)$, where the λ_j are the eigenvalues of K. Let $\mathbf{R}' = W^t(\mathbf{R} - \mathbf{b})$. Then

$$E[\exp(i\mathbf{u}^t\mathbf{R}')] = \exp(-i\mathbf{u}^t W^t\mathbf{b}) E[\exp(i\mathbf{u}^t W^t\mathbf{R})]$$

$$= \exp[-\tfrac{1}{2}\mathbf{v}^t K\mathbf{v}] \quad \text{where} \quad \mathbf{v} = W\mathbf{u}$$

$$= \exp[-\tfrac{1}{2}\mathbf{u}^t D\mathbf{u}] = \exp\left[-\frac{1}{2}\sum_{k=1}^{n} \lambda_k u_k^2\right]$$

It follows that $R'_1, \ldots, R'_n$ are independent, with R_j normal $(0, \lambda_j)$. Since W is orthogonal, $W^t = W^{-1}$; hence $\mathbf{R} = W\mathbf{R}' + \mathbf{b}$.

The matrix K has probabilistic significance, as follows.

Theorem 2. *In Theorem 1 we have* $E(\mathbf{R}) = \mathbf{b}$, *that is,* $E(R_j) = b_j$, $j = 1, \ldots, n$, *and K is the covariance matrix of the R_j, that is,* $K_{rs} = Cov(R_r, R_s)$, r, $s = 1, \ldots, n$.

PROOF. Since the R'_j have finite second moments, so do the R_j. $E(\mathbf{R}) = \mathbf{b}$ follows immediately by linearity of the expectation. Now the covariance matrix of the R_j is

$$[Cov(R_r, R_s)] = [E((R_r - b_r)(R_s - b_s))] = E[(\mathbf{R} - \mathbf{b})(\mathbf{R} - \mathbf{b})^t]$$

where $E(A)$ for a matrix A means the matrix $[E(A_{rs})]$. Thus the covariance matrix is

$$E[W\mathbf{R}'(W\mathbf{R}')^t] = WE(\mathbf{R}'\mathbf{R}'^t)W^t = WDW^t = K$$

since D is the covariance matrix of the R'_j.

The representation of Theorem 1 yields many useful properties of Gaussian vectors.

Theorem 3. *Let R be Gaussian with representation* $\mathbf{R} = W\mathbf{R}' + \mathbf{b}$, W *orthogonal, as in Theorem 1.*

1. *If K is nonsingular, then the random variables* $R_j^* = R_j - b_j$ *are linearly independent; that is, if* $\sum_{j=1}^n a_j R_j^* = 0$ *with probability 1, then all $a_j = 0$. In this case R has a density given by*

$$f(x) = (2\pi)^{-n/2} (det\ K)^{-1/2} exp\left[-\tfrac{1}{2}(\mathbf{x} - \mathbf{b})^t K^{-1}(\mathbf{x} - \mathbf{b})\right]$$

2. *If K is singular, the R_j^* are linearly dependent. If, say, $\{R_1^*, \ldots, R_r^*\}$ is a maximal linearly independent subset of $\{R_1^*, \ldots, R_n^*\}$, then $(R_1, \ldots, R_r)$ has a density of the above form, with K replaced by $K_r =$ the first r rows and columns of K. $R_{r+1}^*, \ldots, R_n^*$ can be expressed (with probability 1) as linear combinations of $R_1^*, \ldots, R_r^*$.*

PROOF.
1. If K is nonsingular, all λ_j are > 0; hence R' has density

$$f'(y) = (2\pi)^{-n/2}(\lambda_1 \cdots \lambda_n)^{-1/2} exp\left[-\frac{1}{2}\sum_{k=1}^n \frac{y_k^2}{\lambda_k}\right]$$

$$= (2\pi)^{-n/2}(det\ K)^{-1/2} exp\left[-\tfrac{1}{2}\mathbf{y}^t D^{-1}\mathbf{y}\right]$$

The Jacobian of the transformation $\mathbf{x} = W\mathbf{y} + \mathbf{b}$ is det $W = \pm 1$; hence R has density

$$f(\mathbf{x}) = f'(W^t(\mathbf{x} - \mathbf{b}))$$
$$= (2\pi)^{-n/2} (\det K)^{-1/2} \exp\left[-\tfrac{1}{2}(\mathbf{x} - \mathbf{b})^t W D^{-1} W^t(\mathbf{x} - \mathbf{b})\right]$$

Since $K = WDW^t$, we have $K^{-1} = WD^{-1}W^t$, which yields the desired expression for the density.

Now if $\sum_{j=1}^{n} a_j R_j^* = 0$ with probability 1,

$$0 = E\left[\left|\sum_{j=1}^{n} a_j R_j^*\right|^2\right] = \sum_{r,s=1}^{n} a_r E(R_r^* R_s^*) a_s$$
$$= \sum_{r,s=1}^{n} a_r K_{rs} a_s$$

Since K is nonsingular, it is positive rather than merely nonnegative definite, and thus all $a_r = 0$.

2. If K is singular, then $\sum_{r,s=1}^{n} a_r K_{rs} a_s$ will be 0 for some $a_1, \ldots, a_n$, not all 0. (This follows since $\mathbf{u}^t K \mathbf{u} = \sum_{k=1}^{n} \lambda_k v_k^2$, where $\mathbf{v} = W^t \mathbf{u}$; if K is singular, then some λ_j is 0.) But by the analysis of case 1, $E[|\sum_{j=1}^{n} a_j R_j^*|^2] = 0$; hence $\sum_{j=1}^{n} a_j R_j^* = 0$ with probability 1, proving linear dependence. The remaining statements of 2 follow from 1.

REMARK.   The result that K is singular iff the R_j^* are linearly dependent is true for arbitrary random variables with finite second moments, as the above argument shows.

▶ **Example 1.**   Let (R_1, R_2) be Gaussian. Then

$$K = \begin{bmatrix} \sigma_1^2 & \sigma_{12} \\ \sigma_{12} & \sigma_2^2 \end{bmatrix}$$

where $\sigma_1^2 = \mathrm{Var}\, R_1$, $\sigma_2^2 = \mathrm{Var}\, R_2$, $\sigma_{12} = \mathrm{Cov}\,(R_1, R_2)$. Also, $\det K = \sigma_1^2 \sigma_2^2 (1 - \rho_{12}^2)$, where ρ_{12} is the correlation coefficient between R_1 and R_2. Thus K is singular iff $|\rho_{12}| = 1$. In the nonsingular case we have

$$K^{-1} = (\sigma_1^2 \sigma_2^2 (1 - \rho_{12}^2))^{-1} \begin{bmatrix} \sigma_2^2 & -\sigma_{12} \\ -\sigma_{12} & \sigma_1^2 \end{bmatrix}$$

and the density is given by

$$f(x, y) = \frac{1}{2\pi\sigma_1\sigma_2(1 - \rho_{12}^2)^{1/2}}$$
$$\times \exp\left[-\frac{\sigma_2^2(x - a)^2 - 2\sigma_{12}(x - a)(y - b) + \sigma_1^2(y - b)^2}{2\sigma_1^2\sigma_2^2(1 - \rho_{12}^2)}\right]$$

where $a = E(R_1)$, $b = E(R_2)$. The characteristic function of (R_1, R_2) is
$M(u_1, u_2) = \exp\ [i(au_1 + bu_2)] \exp\ [-\frac{1}{2}(\sigma_1^2 u_1^2 + 2\sigma_{12} u_1 u_2 + \sigma_2^2 u_2^2)]$.
Notice that if $n = 1$, the multidimensional Gaussian distribution reduces to
the ordinary Gaussian distribution. For in this case we have

$$K = [\sigma^2], \qquad M(u) = e^{ium} e^{-u^2\sigma^2/2}, \qquad f(x) = \frac{1}{\sqrt{2\pi}\ \sigma}\ e^{-(x-m)^2/2\sigma^2}$$

where $m = E(R)$. ◄

Theorem 4. *If R_1 is a Gaussian n-vector and $R_2 = AR_1$, where A is an m
by n matrix, then R_2 is a Gaussian m-vector.*

PROOF.   Let $\mathbf{R}_1 = W\mathbf{R}' + \mathbf{b}$ as in Theorem 1. Then $\mathbf{R}_2 = AW\mathbf{R}' + A\mathbf{b}$,
and hence R_2 is Gaussian by Theorem 1.

COROLLARY.
  (a) If $R_1, \ldots, R_n$ are jointly Gaussian, so are $R_1, \ldots, R_m$, $m \le n$.
  (b) If $R_1, \ldots, R_n$ are jointly Gaussian, then $a_1 R_1 + \cdots + a_n R_n$ is a
Gaussian random variable.

PROOF.   For (a) take $A = [I \quad 0]$, where I is an m by m identity matrix.
For (b) take $A = [a_1 a_2 \ldots a_n]$.

Thus we see that if $R_1, \ldots, R_n$ are jointly Gaussian, then the R_i are
(individually) Gaussian. The converse is not true, however. It is possible to
find Gaussian random variables R_1, R_2 such that (R_1, R_2) is not Gaussian,
and in addition $R_1 + R_2$ is not Gaussian.
  For example, let R_1 be normal $(0, 1)$ and define R_2 as follows. Let R_3
be independent of R_1, with $P\{R_3 = 0\} = P\{R_3 = 1\} = 1/2$. If $R_3 = 0$, let
$R_2 = R_1$; if $R_3 = 1$, let $R_2 = -R_1$. Then $P\{R_2 \le y\} = (1/2)P\{R_1 \le y\} +
(1/2)\ P\{-R_1 \le y\} = P\{R_1 \le y\}$, so that R_2 is normal $(0, 1)$. But if $R_3 = 0$,
then $R_1 + R_2 = 2R_1$, and if $R_3 = 1$, then $R_1 + R_2 = 0$. Therefore
$P\{R_1 + R_2 = 0\} = 1/2$; hence $R_1 + R_2$ is not Gaussian. By corollary (b)
to Theorem 4, (R_1, R_2) is not Gaussian.
  Notice that if $R_1, \ldots, R_n$ are *independent* and each R_i is Gaussian, then
the R_i are jointly Gaussian (with $K =$ the diagonal matrix of variances of the
R_i).

Theorem 5. *If $R_1, \ldots, R_n$ are jointly Gaussian and uncorrelated, that is,
if $K_{ij} = 0$ for $i \ne j$, they are independent.*

PROOF.   Let $\sigma_j^2 = \text{Var } R_j$. We may assume all $\sigma_j^2 > 0$; if $\sigma_j^2 = 0$, then R_j is constant with probability 1 and may be deleted. Now $K = \text{diag}$ $(\sigma_1^2, \ldots, \sigma_n^2)$; hence $K^{-1} = \text{diag } (1/\sigma_1^2, \ldots, 1/\sigma_n^2)$, and so, by Theorem 3, $R_1, \ldots, R_n$ have a joint density given by

$$f(x_1, \ldots, x_n) = (2\pi)^{-n/2}(\sigma_1 \cdots \sigma_n)^{-1} \exp\left[-\frac{1}{2} \sum_{j=1}^{n} \frac{(x_j - b_j)^2}{\sigma_j^2} \right]$$

Thus $R_1, \ldots, R_n$ are independent, with R_j normal (b_j, σ_j^2).

We now consider the following prediction problem. Let $R_1, \ldots, R_{n+1}$ be jointly Gaussian. We observe $R_1 = x_1, \ldots, R_n = x_n$ and then try to predict the value of R_{n+1}. If the predicted value is $\psi(x_1, \ldots, x_n)$ and the actual value is x_{n+1}, we assume a quadratic loss $(x_{n+1} - \psi(x_1, \ldots, x_n))^2$. In other words, we are trying to minimize the mean square difference between the true value and the predicted value of R_{n+1}. This is simply a problem of Bayes estimation with quadratic loss function, as considered in Section 8.3; in this case R_{n+1} plays the role of the state of nature and $(R_1, \ldots, R_n)$ the observable. It follows that the best estimate is

$$\psi(x_1, \ldots, x_n) = E(R_{n+1} \mid R_1 = x_1, \ldots, R_n = x_n)$$

We now show that in the jointly Gaussian case ψ is a linear function of $x_1, \ldots, x_n$. Thus the optimum predictor assumes a particularly simple form.

Say $\{R_1, \ldots, R_r\}$ is a maximal linearly independent subset of $\{R_1, \ldots, R_n\}$. If $R_1, \ldots, R_r, R_{n+1}$ are linearly dependent, there is nothing to prove; if $R_1, \ldots, R_r, R_{n+1}$ are linearly independent, we may replace $R_1, \ldots, R_n$ by $R_1, \ldots, R_r$ in the problem. Thus we may as well assume $R_1, \ldots, R_{n+1}$ linearly independent. Then $(R_1, \ldots, R_{n+1})$ has a density, and the conditional density of R_{n+1} given $R_1 = x_1, \ldots, R_n = x_n$ is

$$h(x_{n+1} \mid x_1, \ldots, x_n) = \frac{f(x_1, \ldots, x_{n+1})}{\int_{-\infty}^{\infty} f(x_1, \ldots, x_{n+1}) \, dx_{n+1}}$$

$$= \frac{(2\pi)^{-(n+1)/2}(\det K)^{-1/2} \exp\left[-\frac{1}{2} \sum_{r,s=1}^{n+1} x_r q_{rs} x_s \right]}{(2\pi)^{-(n+1)/2}(\det K)^{-1/2} \int_{-\infty}^{\infty} \exp\left[-\frac{1}{2} \sum_{r,s=1}^{n+1} x_r q_{rs} x_s \right] dx_{n+1}}$$

where K is the covariance matrix of $R_1, \ldots, R_{n+1}$ and $Q = [q_{rs}] = K^{-1}$.

Thus

$$h(x_{n+1} \mid x_1, \ldots, x_n) = \frac{1}{B(x_1, \ldots, x_n)}$$

$$\times \exp\left[-\frac{1}{2}\left(\sum_{r,s=1}^{n} x_r q_{rs} x_s + x_{n+1}\sum_{s=1}^{n} q_{n+1,s} x_s\right.\right.$$

$$\left.\left.+ x_{n+1}\sum_{r=1}^{n} x_r q_{r,n+1} + q_{n+1,n+1}x_{n+1}^2\right)\right]$$

$$= \frac{A(x_1, \ldots, x_n)}{B(x_1, \ldots, x_n)} \exp\left[-(Cx_{n+1}^2 + Dx_{n+1})\right]$$

where

$$C = \tfrac{1}{2}q_{n+1,n+1}, \qquad D = D(x_1, \ldots, x_n) = \sum_{r=1}^{n} q_{n+1,r} x_r$$

Therefore the conditional density can be expressed as

$$\frac{A}{B} \exp\left[\frac{D^2}{4C}\right] \exp\left[-C\left(x_{n+1} + \frac{D}{2C}\right)^2\right]$$

Thus, given $R_1 = x_1, \ldots, R_n = x_n$, R_{n+1} is normal with mean $-D/2C$ and variance $1/2C = 1/q_{n+1,n+1}$. Hence

$$E(R_{n+1} \mid R_1 = x_1, \ldots, R_n = x_n) = -\frac{1}{q_{n+1,n+1}}\sum_{r=1}^{n} q_{n+1,r} x_r$$

Tables

Common Density Functions and Their Properties

Type	Density	Parameters
Uniform on $[a, b]$	$\dfrac{1}{b-a}, \quad a \leq x \leq b$	a, b real, $a < b$
Normal	$\dfrac{1}{\sqrt{2\pi}\sigma} e^{-(x-\mu)^2/2\sigma^2}$	μ real, $\sigma > 0$
Gamma	$\dfrac{x^{\alpha-1}e^{-x/\beta}}{\Gamma(\alpha)\beta^\alpha}, \quad x \geq 0$	$\alpha, \beta > 0$
Beta	$\dfrac{x^{r-1}(1-x)^{s-1}}{\beta(r, s)}, \quad 0 \leq x \leq 1$	$r, s > 0$
Exponential ($=$ gamma with $\alpha = 1, \beta = 1/\lambda$)	$\lambda e^{-\lambda x}, \quad x \geq 0$	$\lambda > 0$
Chi-square ($=$ gamma with $\alpha = n/2, \beta = 2$)	$\dfrac{1}{2^{n/2}\Gamma(n/2)} x^{(n/2)-1}e^{-x/2}, \quad x \geq 0$	$n = 1, 2, \ldots$
t	$\dfrac{\Gamma[(n+1)/2]}{\sqrt{n\pi}\,\Gamma(n/2)} \dfrac{1}{(1 + x^2/n)^{(n+1)/2}}$	$n = 1, 2, \ldots$
F	$\dfrac{(m/n)^{m/2}}{\beta(m/2, n/2)} \dfrac{x^{(m/2)-1}}{(1 + (m/n)x)^{(m+n)/2}}, \quad x \geq 0$	$m, n = 1, 2, \ldots$
Cauchy	$\dfrac{\theta}{\pi(x^2 + \theta^2)}$	$\theta > 0$

Common Density Functions (continued)

Type	Mean	Variance	Generalized Characteristic Function (If Easily Computable)
Uniform on $[a, b]$	$\dfrac{a + b}{2}$	$\dfrac{(b - a)^2}{12}$	$\dfrac{1}{b - a}\dfrac{e^{-sa} - e^{-sb}}{s}$, all s
Normal	μ	σ^2	$e^{-s\mu}e^{s^2\sigma^2/2}$, all s
Gamma	$\alpha\beta$	$\alpha\beta^2$	$\left(\dfrac{1/\beta}{s + 1/\beta}\right)^{\alpha}$, $\mathrm{Re}\,s > -1/\beta$
Beta	$\dfrac{r}{r + s}$	$\dfrac{rs}{(r + s)^2(r + s + 1)}$	
Exponential	$\dfrac{1}{\lambda}$	$\dfrac{1}{\lambda^2}$	$\dfrac{\lambda}{s + \lambda}$, $\mathrm{Re}\,s > -\lambda$
Chi-square	n	$2n$	$(2s + 1)^{-n/2}$, $\mathrm{Re}\,s > -1/2$
t	0 if $n > 1$; does not exist if $n = 1$	$\dfrac{n}{n - 2}$ if $n > 2$; ∞ if $n = 2$	
F	$\dfrac{n}{n - 2}$ if $n > 2$; ∞ if $n = 1$ or 2	$\dfrac{2n^2(m + n - 2)}{m(n - 2)^2(n - 4)}$ if $n > 4$; ∞ if $n = 3$ or 4	
Cauchy	Does not exist	Does not exist	$e^{-\theta|u|}$, $s = iu$, u real

Common Probability Functions and Their Properties

Type	Probability $p(k)$	Parameters
Discrete uniform	$\dfrac{1}{N}$, $\quad k = 1, 2, \ldots, N$	$N = 1, 2, \ldots$
Bernoulli	$\begin{aligned}p(1) &= p\\ p(0) &= q\end{aligned}$	$0 \le p \le 1, q = 1 - p$
Binomial	$\binom{n}{k}p^k q^{n-k}$, $\quad k = 0, 1, \ldots, n$	$0 \le p \le 1, q = 1 - p$, $n = 1, 2, \ldots$
Poisson	$e^{-\lambda}\lambda^k/k!$, $\quad k = 0, 1, \ldots$	$\lambda > 0$
Geometric	$q^{k-1}p$, $\quad k = 1, 2, \ldots$	$0 < p \le 1, q = 1 - p$
Negative binomial	$\binom{k-1}{r-1}p^r q^{k-r} = \binom{-r}{k-r}p^r(-q)^{k-r}$, $k = r, r + 1, \ldots$	$0 < p \le 1, q = 1 - p$, $r = 1, 2, \ldots$

Common Probability Functions (continued)

Type	Mean	Variance	Generalized Characteristic Function
Discrete uniform	$\dfrac{N+1}{2}$	$\dfrac{N^2-1}{12}$	$\dfrac{e^{-s}(1-e^{-sN})}{N(1-e^{-s})}$, all s
Bernoulli	p	$p(1-p)$	$q + pe^{-s}$, all s
Binomial	np	$np(1-p)$	$(q + pe^{-s})^n$, all s
Poisson	λ	λ	$\exp[\lambda(e^{-s}-1)]$, all s
Geometric	$\dfrac{1}{p}$	$\dfrac{1-p}{p^2}$	$\dfrac{pe^{-s}}{1-qe^{-s}}$, $\lvert qe^{-s}\rvert < 1$
Negative binomial	$\dfrac{r}{p}$	$\dfrac{r(1-p)}{p^2}$	$\left(\dfrac{pe^{-s}}{1-qe^{-s}}\right)^r$, $\lvert qe^{-s}\rvert < 1$

Selected Values of the Standard Normal Distribution Function

$F(x) = (2\pi)^{-1/2} \int_{-\infty}^{x} e^{-t^2/2}\, dt,\qquad F(-x) = 1 - F(x)$

x	$F(x)$	x	$F(x)$	x	$F(x)$	x	$F(x)$
.0	.500	.9	.816	1.64	.950	2.33	.990
.1	.540	1.0	.841	1.7	.955	2.4	.992
.2	.579	1.1	.864	1.8	.964	2.5	.994
.3	.618	1.2	.885	1.9	.971	2.6	.995
.4	.655	1.28	.900	1.96	.975	2.7	.996
.5	.691	1.3	.903	2.0	.977	2.8	.997
.6	.726	1.4	.919	2.1	.982	2.9	.998
.7	.758	1.5	.933	2.2	.986	3.0	.999
.8	.788	1.6	.945	2.3	.989		

A Brief Bibliography

An excellent source of examples and applications of basic probability is the classic work of W. Feller, *Introduction to Probability Theory and Its Applications* (John Wiley, Vol. 1, 1950; Vol. 2, 1966). Another good source is *Modern Probability Theory* by E. Parzen (John Wiley, 1960).

Many properties of Markov chains and other stochastic processes are given in *A First Course in Stochastic Processes* by S. Karlin (Academic Press, 1966). A. Papoulis' *Probability, Random Variables, and Stochastic Processes* (McGraw-Hill, 1967) is a treatment of stochastic processes that is directed toward engineers.

A comprehensive treatment of basic statistics is given in *Introduction to Mathematical Statistics* by R. Hogg and A. Craig (Macmillan, 1965). A more advanced work emphasizing the decision-theory point of view is *Mathematical Statistics, A Decision Theoretic Approach* by T. Ferguson (Academic Press, 1967).

The student who wishes to take more advanced work in probability will need a course in measure theory. H. Royden's *Real Analysis* (Macmillan, 1963) is a popular text for such a course. For those with a measure theory background, J. Lamperti's *Probability* (W. A. Benjamin, 1966) gives the flavor of modern probability theory in a relatively light and informal way. A more systematic account is given in *Probability* by L. Breiman (Addison-Wesley, 1968).

Solutions to Problems

CHAPTER 1

Section 1.2

1. $A \cap B \cap C = \{4\}$, $A \cup (B \cap C^c) = \{0, 1, 2, 3, 4, 5, 7\}$,
   $(A \cup B) \cap C^c = \{0, 1, 3, 5, 7\}$, $(A \cap B) \cap [(A \cup C)^c] = \varnothing$

3. W = registered Whigs
   F = those who approve of Fillmore
   E = those who favor the electoral college

$$x + y + z + 100 = 550$$
$$x + y + 25 = 325$$

Thus $z + 100 - 25 = 550 - 325 = 225$, so $z = 150$

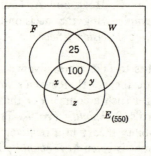

PROBLEM 1.2.3

5. If $a < x < b$ then $x \le b - 1/n$ for some n, hence $x \in \bigcup_{n=1}^{\infty} (a, b - 1/n]$.
Conversely, if $x \in \bigcup_{n=1}^{\infty} (a, b - 1/n]$, then $a < x \le b - 1/n$ for some n, hence $x \in (a, b)$. The other arguments are similar.

9. $x \in A \cap (\bigcup_i B_i)$ iff $x \in A$ and $x \in B_i$ for at least one i
   iff $x \in A \cap B_i$ for at least one i
   iff $x \in \bigcup_i (A \cap B_i)$.

Section 1.3

1. (a) Let $A \subset \Omega$, and let $\mathscr{F}$ consist of $\varnothing$, Ω, A and A^c.

   (b) Let $A_1, A_2, \ldots$ be disjoint sets whose union is Ω. Let $\mathscr{F}$ consist of all finite or countable unions of the A_i (including $\varnothing$ and Ω).

Section 1.4

1. .432

2. (a) The sequence of face values may be chosen in 9 ways (the high card may be anything from an ace to a six); the suit of each card in the straight may be chosen in 4 ways. Thus the probability of a straight is $9(4^5)/\binom{52}{5}$.

   (b) Select the face value to appear three times (13 choices); select the two other face values [$\binom{12}{2}$ choices]. Select three suits out of four for the face value appearing three times, and select one suit for each of the two odd cards. Thus $p = 13\binom{12}{2}\binom{4}{3}(16)/\binom{52}{5}$.

   (c) Select two face values for the pairs [$\binom{13}{2}$ possibilities]; then choose two suits out of four for each of the pairs. Finally, choose the odd card from the 44 cards that remain when the two face values are removed from the deck. Thus $p = \binom{13}{2}\binom{4}{2}\binom{4}{2}(44)/\binom{52}{5}$.

3. 149/432

4. (a) $(52)(48) \cdots (20)(16)/52^{10}$
   (b) $[4\binom{13}{9}39 + 4\binom{13}{10}]/\binom{52}{10}$

5. We must have exactly one number > 8 (2 choices) and exactly 3 numbers < 8 [$\binom{7}{3}$ possibilities]. Thus $p = 2\binom{7}{3}/\binom{10}{5}$.

6. $(m + 1)/\binom{m+w}{w}$

7. $1 - \binom{75}{15}/\binom{100}{15}$

8. $[2\binom{4}{3}\binom{48}{5} - \binom{4}{3}\binom{4}{3}\binom{44}{2}]/\binom{52}{8}$

9. Let A_i be the event that the ticket numbered i appears at the ith drawing. The probability of at least one match is $P(A_1 \cup \cdots \cup A_n)$. Now $P(A_1) = (n - 1)!/n! = 1/n$ (the first ticket must have number 1, the second may be any one of $n - 1$ remaining possibilities, the third one of $n - 2$, etc.) By symmetry, $P(A_i) = 1/n$ for all i. Similarly, $P(A_i \cap A_j) = (n - 2)!/n! = 1/n(n - 1)$, $i < j$, $P(A_i \cap A_j \cap A_k) = (n - 3)!/n! = 1/n(n - 1)(n - 2)$, $i < j < k$, etc. By the expansion formula (1.4.5) for the probability of a union, $P(A_1 \cup \cdots \cup$

$A_n) = n(1/n) - \binom{n}{2}(n-2)!/n! + \binom{n}{3}(n-3)!/n! - \cdots + (-1)^{n-1}\binom{n}{n}0!/n! = 1 - 1/2! + 1/3! - \cdots + (-1)^{n-1}/n!.$

11. $(365)_r/365^r$

12. The number of arrangements with occupancy numbers $r_1 = r_2 = 4$, $r_3 = r_4 = r_5 = 2$, $r_6 = 0$, is $\binom{14}{4}\binom{10}{4}\binom{6}{2}\binom{4}{2}\binom{2}{2} = 14!/4!\,4!\,2!\,2!\,2!\,0!$. We must select 2 boxes out of 6 to receive 4 balls, then 3 boxes of the remaining 4 to receive 2; the remaining box receives 0. This can be done in $\binom{6}{2}\binom{4}{3}\binom{1}{1} = 6!/2!\,3!\,1!$ ways. Thus the total number is $14!\,6!/(4!)^2(2!)^4 3!$.

Section 1.5

4. (a) This is a multinomial problem. The probability is

$$\frac{30!}{10!\,10!\,10!}\left(\frac{50}{100}\right)^{10}\left(\frac{30}{100}\right)^{10}\left(\frac{20}{100}\right)^{10}$$

(b) This is a binomial problem. The probability is

$$\binom{30}{12}\left(\frac{30}{100}\right)^{12}\left(\frac{70}{100}\right)^{18}$$

5. The probability that there will be exactly 3 ones and no two (and 3 from $\{3, 4, 5, 6\}$) is, by the multinomial formula, $(6!/3!\,0!\,3!)(\frac{1}{6})^3(\frac{1}{6})^0(\frac{2}{3})^3$. The probability of exactly 4 ones and 1 two is $(6!/4!\,1!\,1!)(\frac{1}{6})^4(\frac{1}{6})^1(\frac{2}{3})^1$. The sum of these expressions is the desired probability.

Section 1.6

1.

$$(7p^4q^6 + 6p^5q^5 + 5p^6q^4)\bigg/ \sum_{k=4}^{6}\binom{10}{k}p^kq^{10-k}$$

2.

$$\frac{1 - q^n - npq^{n-1} - \binom{n}{2}p^2q^{n-2}}{1 - q^n}$$

3. We have the following tree diagram:

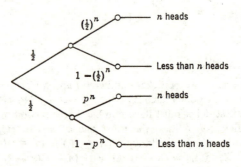

PROBLEM 1.6.3

The desired probability is

$$P\{\text{unbiased coin used and } n \text{ heads obtained}\} \over P\{n \text{ heads obtained}\}$$

$$= \frac{(1/2)^{n+1}}{(1/2)^{n+1} + (1/2)p^n}$$

4. Let $A = \{\text{heads}\}$. By the theorem of total probability,

$$P(A) = \sum_{n=1}^{\infty} P\{I = n\}P(A \mid I = n) = \sum_{n=1}^{\infty} (1/2)^n e^{-n}$$

$$= (1/2)e^{-1}/(1 - \tfrac{1}{2}e^{-1})$$

REMARK.   To formalize this problem, we may take $\Omega = $ all pairs (n, i), $n = 1$, $2, \ldots , i = 0, 1$ [where $(n, 1)$ indicates that $I = n$ and the coin comes up heads, and $(n, 0)$ indicates that $I = n$ and the coin comes up tails].

We assign $p(n, i) = (1/2)^n e^{-n}$ if $i = 1$, and $(1/2)^n(1 - e^{-n})$ if $i = 0$. Alternatively, a tree diagram can be constructed; a typical path is indicated in the diagram.

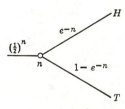

PROBLEM 1.6.4

5. (a) $\binom{6}{3}\binom{20}{10}/\binom{26}{13} = .36$

   (b) $[\binom{6}{2}\binom{20}{11} + \binom{6}{4}\binom{20}{9}]/\binom{26}{13} = 2\binom{6}{2}\binom{20}{11}/\binom{26}{13} = .48$

   (c) $2\binom{6}{1}\binom{20}{12}/\binom{26}{13} = .15$

   (d) $2\binom{20}{13}/\binom{26}{13} = .01$

7. By the theorem of total probability, if X_n is the result of the toss at $t = n$, then (see diagram)

$$y_{n+1} = P\{X_{n+1} = H\} = P\{X_n = H\}P\{X_{n+1} = H \mid X_n = H\}$$
$$+ P\{X_n = T\}P\{X_{n+1} = H \mid X_n = T\} = y_n(1/2) + (1 - y_n)(3/4)$$

Thus $y_{n+1} + (1/4)y_n = 3/4$. The solution to the homogeneous equation $y_{n+1} + (1/4)y_n = 0$ is $y_n = A(-1/4)^n$. Since the "forcing function" 3/4 is constant, we assume as a "particular solution" $y_n = c$. Then $(5c)/4 = 3/4$, or $c = 3/5$. Thus the general solution is $y_n = A(-1/4)^n + 3/5$. Since y_0 is given as 1/2, we have $A = 1/2 - 3/5 = -1/10$. Thus $y_n = 3/5 - 1/10(-1/4)^n$.

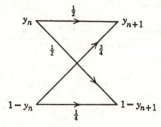

PROBLEM 1.6.7

9. We have (see diagram) $P(D \mid R) = P(D \cap R)/P(R) = .2(.9)(.25)/.2(.9)(.25) +$
.8(.3)(.25) = 3/7

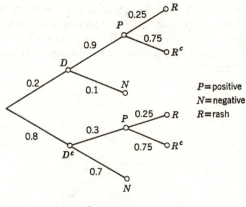

$P =$ positive
$N =$ negative
$R =$ rash

PROBLEM 1.6.9

CHAPTER 2

Section 2.2

1. (a) $\{\omega: R(\omega) \in E^1\} = \Omega \in \mathscr{F}$, hence $E^1 \in \mathscr{C}$. If $B_1, B_2, \ldots \in \mathscr{C}$, then $\{\omega: R(\omega) \in \bigcup_{n=1}^{\infty} B_n\} = \bigcup_{n=1}^{\infty} \{\omega: R(\omega) \in B_n\} \in \mathscr{F}$ since each $B_n \in \mathscr{C}$; thus $\bigcup_{n=1}^{\infty} B_n \in \mathscr{C}$. If $B \in \mathscr{C}$, then $\{\omega: R(\omega) \in B^c\} = \{\omega: R(\omega) \in B\}^c \in \mathscr{F}$, hence $B^c \in \mathscr{C}$.

   (b) $\mathscr{C}$ is a sigma field containing the intervals, hence is at least as large as the smallest sigma-field containing the intervals. Thus all Borel sets belong to $\mathscr{C}$.

Section 2.3

1. (a) $F_R(x) = \frac{1}{2}e^x$, $x \leq 0$; $F_R(x) = 1 - \frac{1}{2}e^{-x}$, $x \geq 0$

   (b) 1. $P\{|R| \leq 2\} = P\{-2 \leq R \leq 2\} = \int_{-2}^{2} f_R(x) \, dx = F_R(2) - F_R(-2) = 1 - e^{-2}$

2. $P\{|R| \leq 2 \text{ or } R \geq 0\} = P\{R \geq -2\} = 1 - F_R(-2) = 1 - \frac{1}{2}e^{-2}$

3. $P\{|R| \leq 2 \text{ and } R \leq -1\} = P\{-2 \leq R \leq -1\} = \frac{1}{2}(e^{-1} - e^{-2})$

4. $P\{|R| + |R - 3| \leq 3\} = P\{0 \leq R \leq 3\} = \frac{1}{2}(1 - e^{-3})$

5. $P\{R^3 - R^2 - R - 2 \leq 0\} = P\{(R - 2)(R^2 + R + 1) \leq 0\} = P\{R \leq 2\}$
   (since $R^2 + R + 1$ is always > 0) $= 1 - \frac{1}{2}e^{-2}$

6. $P\{e^{\sin \pi R} \geq 1\} = P\{\sin \pi R \geq 0\} = P\{0 \leq R \leq 1\} + P\{2 \leq R \leq 3\}$
$$+ P\{4 \leq R \leq 5\} + \cdots$$
$$+ P\{-2 \leq R \leq -1\} + P\{-4 \leq R \leq -3\} + \cdots$$

But $P\{2n - 1 \leq R \leq 2n\} = P\{-2n \leq R \leq -2n + 1\}$ since f_R is an even function. Thus

$$P\{e^{\sin \pi R} \geq 1\} = \sum_{n=0}^{\infty} P\{2n \leq R \leq 2n + 1\} + \sum_{n=1}^{\infty} P\{2n - 1 \leq R \leq 2n\}$$
$$= P\{R \geq 0\} = \frac{1}{2}$$

7. $P\{R \text{ irrational}\} = 1 - P\{R \text{ rational}\}$. The rationals are a countable set, say $\{x_1, x_2, \ldots\}$. Hence, $P\{R \text{ rational}\} = \sum_{i=1}^{\infty} P\{R = x_i\} = 0$, so $P\{R \text{ irrational}\} = 1$.

2. By direct enumeration,
$$P\{R = 0\} = p^5 + p^4q + p^3q^2 + p^2q^3 + pq^4 + q^5$$
$$P\{R = 1\} = 4p^4q + 6p^3q^2 + 6p^2q^3 + 4pq^4$$
$$P\{R = 2\} = 3p^3q^2 + 3p^2q^3$$

Section 2.4

1. If $0 < y < 1$,
$$F_2(y) = P\{R_1 \leq y\} + P\left\{R_1 \geq \frac{1}{y}\right\} = \int_0^y e^{-x}\,dx + \int_{1/y}^{\infty} e^{-x}\,dx$$
$$= 1 - e^{-y} + e^{-1/y}$$
$$F_2(y) = 0, y \leq 0; \quad F_2(y) = 1, y \geq 1$$

F_2 is the integral of
$$f_2(y) = \frac{d}{dy} F_2(y) = e^{-y} + \frac{1}{y^2} e^{-1/y}, 0 < y < 1$$
$$= 0 \text{ elsewhere}$$

Hence R_2 is absolutely continuous.

2.
$$f_2(y) = \frac{1}{2y}, e^{-1} < y < e$$
$$= 0 \text{ elsewhere}$$

3.
$$f_2(y) = 2y^{-2}, \quad 2 < y < 4$$
$$= \tfrac{1}{2}y^{-3/2}, \quad y > 4$$
$$= 0, \quad y < 2$$

4. If $2 \le y \le 4$, $F_2(y) = P\{R_2 \le y\} = \int_1^{y/2} \dfrac{1}{x^2}\,dx = 1 - \dfrac{2}{y}$

If $4 \le y < 5$, $F_2(y) = \tfrac{1}{2}$

If $y \ge 5$, $F_2(y) = 1$

$P\{R_2 = 5\} = P\{R_1 > 2\} = \tfrac{1}{2}$, so R_2 is not absolutely continuous.

7. Consider the graph of $y = g(x)$. The horizontal line at height y will intersect

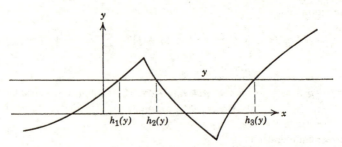

PROBLEM 2.4.7

the graph at points (say) $x_{i_1}, \ldots, x_{i_k}$, with $x_j \in I_j$. If we choose a sufficiently small interval (a, b) about y, we have (except for finitely many y, which lead to intersections at the endpoints of intervals)

$$F_2(b) - F_2(a) = P\{a < R_2 \le b\} = \sum_{j=1}^{n} P\{c_j < R_1 \le d_j\}$$

where $c_j = h_j(a)$ and $d_j = h_j(b)$ if $j \in \{i_1, \ldots, i_k\}$ and h_j is increasing at x_j
  $c_j = h_j(b)$ and $d_j = h_j(a)$ if $j \in \{i_1, \ldots, i_k\}$ and h_j is decreasing at x_j
  $c_j = $ (say) 1 and $d_j = 0$ if $j \notin \{i_1, \ldots, i_k\}$. Thus

$$F_2(b) - F_2(a) = \sum_{j=1}^{n} \int_{c_j}^{d_j} f_1(x)\,dx = \sum_{j=1}^{n} [F_1(d_j) - F_1(c_j)]$$

where $F_1(d_j) - F_1(c_j)$ is interpreted as 0 if $c_j > d_j$. Differentiate with respect to b to obtain $f_2(b) = \sum_{j=1}^{n} f_1[h_j(b)]\,|h_j'(b)|$, as desired.

Section 2.5

1. (a) $\tfrac{1}{3}$
   (b) $\tfrac{1}{3}$
   (c) $\tfrac{1}{3} + F_R(1.5) - F_R(.5) = \tfrac{1}{3} + \tfrac{1}{3} - \tfrac{1}{3}(\tfrac{1}{4}) = 7/12$
   (d) $F_R(3) - F_R(.5) = \tfrac{5}{6} - \tfrac{1}{12} = \tfrac{3}{4}$

Section 2.6

4. (a) $\frac{23}{32}$

   (b) $\frac{23}{32}$

   (c) $\dfrac{5 + \ln 4}{8}$

   (d) $\frac{5}{8}$

   (e) $\frac{1}{4}$

   (f) $\frac{1}{2}$

   (g) $1 - \frac{1}{2}e^{-1}$

In each case, the probability is $\frac{1}{4}$ (shaded area).

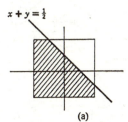

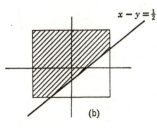

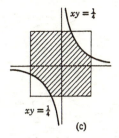

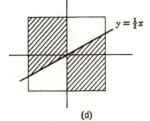

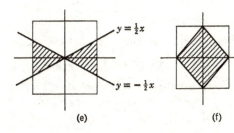

PROBLEM 2.6.4

Section 2.7

1. (a) $f_1(x) = 6x^2 - 4x^3 (0 \leq x \leq 1)$,

     $f_2(y) = 2y(0 \leq y \leq 1)$

   (b) $3/8$

2. $f_1(x) = 2e^{-x} - 2e^{-2x}, x \geq 0$
   $f_2(y) = 2e^{-2y}, y \geq 0$

3.

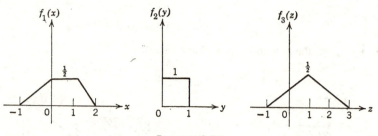

PROBLEM 2.7.3

8. (a) $P\{R_1^2 + R_2 \leq 1\} = \int_{x=0}^{1} \int_{y=0}^{1-x^2} \frac{1}{8}(x + y) \, dy \, dx = \dfrac{31}{480}$

(b) Let $A = \{R_1 \leq 1, R_2 \leq 1\}$, $B = \{R_1 \leq 1, R_2 > 1\}$, $C = \{R_1 > 1, R_2 \leq 1\}$

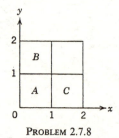

PROBLEM 2.7.8

The probability that at least one of the random variables is ≤ 1 is

$$P(A \cup B \cup C) = P(A) + P(B) + P(C) = \tfrac{1}{8} + \tfrac{1}{4} + \tfrac{1}{4} = \tfrac{5}{8}$$

The probability that exactly one of the random variables is ≤ 1 is $P(B \cup C) = P(B) + P(C) = \tfrac{1}{2}$. If $D = \{$exactly one random variable $\leq 1\}$ and $E = \{$at least one random variable $\leq 1\}$, then

$$P(D \mid E) = \frac{P(D \cap E)}{P(E)} = \frac{P(D)}{P(E)} = \frac{1/2}{5/8} = \frac{4}{5}$$

$\left(\text{Notice that, for example, } P(B) = \int_{x=0}^{1} \int_{y=1}^{2} \tfrac{1}{8}(x + y) \, dy \, dx = \tfrac{1}{4}, \text{ etc.} \right)$

(c) $P\{R_1 \leq 1, R_2 \leq 1\} = \tfrac{1}{8} \neq P\{R_1 \leq 1\}P\{R_2 \leq 1\} = (3/8)^2$, hence the random variables are not independent.

Section 2.8

1. $f_3(z) = \frac{1}{3}$, $0 \leq z \leq 1$; $f_3(z) = \frac{1}{3}z^{-3/2}$, $z > 1$; $f_3(z) = 0$, $z < 0$

2. (a) $f_3(z) = ze^{-z}$, $z \geq 0$; $f_3(z) = 0$, $z < 0$

   (b) $f_3(z) = \dfrac{1}{(z+1)^2}$, $z \geq 0$, $f_3(z) = 0$, $z < 0$

3. $f_3(z) = \dfrac{1}{\pi(1+z^2)}$, all z

4. $f_3(z) = \dfrac{1}{z^2}$, $z \geq 1$, $f_3(z) = 0$, $z < 1$

5. $3\sqrt{3}/8$

6. $2/3$

7. $8/27$

9. $2/33$

10. Let R_1 = arrival time of the man, R_2 = arrival time of the woman. Then the probability that they will meet is

$$P\{|R_1 - R_2| \leq z\} = \frac{\text{shaded area}}{\text{total area}} = 1 - (1-z)^2 = z(2-z)$$

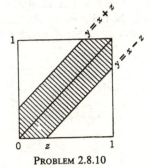

PROBLEM 2.8.10

11. $\left[1 - \dfrac{(n-1)d}{L}\right]^n$ if $(n-1)d \leq L$, and 0 if $(n-1)d > L$

13. $f_{R_0}(r) = \dfrac{1}{b^2}re^{-r^2/2b^2}$, $r > 0$; $f_{\theta_0}(\theta) = \dfrac{1}{2\pi}$, $0 < \theta < 2\pi$.

Section 2.9

2. (a) $n \geq 4600$
   (b) $1 - 5e^{-2}$

3. (a) No.

(b). 9

4. $P\{R_1 + R_2 = k\} = \sum_{i=0}^{k} P\{R_1 = i, R_2 = k - i\}$

$$= \sum_{i=0}^{k} \binom{n}{i} p^i q^{n-i} \binom{m}{k-i} p^{k-i} q^{m-k+i}$$

$$= p^k q^{n+m-k} \sum_{i=0}^{k} \binom{n}{i} \binom{m}{k-i}. \text{ But}$$

(a) $\sum_{i=0}^{k} \binom{n}{i} \binom{m}{k-i} = \binom{n+m}{k}.$

[We may select k positions out of $n + m$ in $\binom{n+m}{k}$ ways. The number of selections in which exactly i positions are chosen from the first n is $\binom{n}{i}\binom{m}{k-i}$. Sum over i to obtain (a).] Thus $P\{R_1 + R_2 = k\} = \binom{n+m}{k} p^k q^{n+m-k}$, $k = 0, 1, \ldots, n + m$. (Intuitively, $R_1 + R_2$ is the number of successes in $n + m$ Bernoulli trials, with probability of success p on a given trial.) Now $P\{R_1 = j, R_1 + R_2 = k\} = P\{R_1 = j, R_2 = k - j\} = \binom{n}{j} p^j q^{n-j} \binom{m}{k-j} p^{k-j} q^{m-k+j} = \binom{n}{j}\binom{m}{k-j} p^k q^{n+m-k}$, $j = 0, 1, \ldots, n, k = j, j+1, \ldots, n + m$

Thus $P\{R_1 = j \mid R_1 + R_2 = k\} = \dfrac{\binom{n}{j}\binom{m}{k-j}}{\binom{n+m}{k}}$, the hypergeometric probability function (see Problem 7 of Section 1.5). Intuitively, given that k successes have occurred in $n + m$ trials, the positions for the successes may be chosen in $\binom{n+m}{k}$ ways. The number of such selections in which j successes occur in the first n trials is $\binom{n}{j}\binom{m}{k-j}$.

CHAPTER 3

Section 3.2

2. (a) $E([R_1]) = \displaystyle\int_{-\infty}^{\infty} [x] f_1(x)\, dx = \int_{0}^{\infty} [x] e^{-x}\, dx$

$$= \int_{1}^{2} e^{-x}\, dx + \int_{2}^{3} 2e^{-x}\, dx + \int_{3}^{4} 3e^{-x}\, dx + \cdots + \int_{n}^{n+1} n e^{-x}\, dx + \cdots$$

$$= e^{-1} - e^{-2} + 2(e^{-2} - e^{-3}) + 3(e^{-3} - e^{-4}) + \cdots$$

$$= e^{-1} + e^{-2} + e^{-3} + \cdots = e^{-1}(1 + e^{-1} + e^{-2} + \cdots) = \frac{e^{-1}}{1 - e^{-1}}$$

(b) $P\{R_2 = n\} = P\{n \leq R_1 < n + 1\} = \int_n^{n+1} e^{-x}\, dx = e^{-n} - e^{-(n+1)}$

$$E(R_2) = \sum_{n=0}^{\infty} nP\{R_2 = n\} = \sum_{n=0}^{\infty} n[e^{-n} - e^{-(n+1)}]$$

$$= e^{-1} - e^{-2} + 2(e^{-2} - e^{-3}) + 3(e^{-3} - e^{-4}) + \cdots$$

$$= \frac{e^{-1}}{1 - e^{-1}} \text{ as above.}$$

3. (a) 1   (b) 0   (c) 1

4. 1/3

5. $2 + 30e^{-3}$

8. $E[R(R-1)\cdots(R-r+1)] = \sum_{k=0}^{\infty} \dfrac{k(k-1)\cdots(k-r+1)}{k!} e^{-\lambda}\lambda^k$

$$= \lambda^r \sum_{k=r}^{\infty} \frac{1}{(k-r)!} e^{-\lambda}\lambda^{k-r} = \lambda^r e^{-\lambda} e^{\lambda} = \lambda^r$$

Set $r = 1$ to obtain $E(R) = \lambda$; set $r = 2$ to obtain $E(R^2 - R) = \lambda^2$, hence $E(R^2) = \lambda + \lambda^2$. It follows that Var $R = \lambda$.

Section 3.3

1. This is immediate from Theorem 2 of Section 2.7.

2. $E[(R - m)^n] = 0$, n odd

$$= \sigma^n(n-1)(n-3)\cdots(5)(3)(1), \; n \text{ even}$$

Section 3.4

1. Let $a(R_1 - ER_1) + b(R_2 - ER_2) = 0$ (with probability 1). If, say, $b \neq 0$ then we may write $R_2 - ER_2 = c(R_1 - ER_1)$. Thus $\sigma_2{}^2 = c^2\sigma_1{}^2$ and Cov $(R_1, R_2) = c\sigma_1{}^2$. Therefore $\rho(R_1, R_2) = \dfrac{c\sigma_1{}^2}{|c|\,\sigma_1{}^2}$, hence $|\rho| = 1$.

4. In (a), let R take the values $1, 2, \ldots, n$, each with probability $1/n$, and set $R_1 = g(R)$, $R_2 = h(R)$, where $g(i) = a_i$, $h(i) = b_i$. Then

$$E(R_1R_2) = \sum_{i=1}^{n} \frac{1}{n} a_ib_i, \quad E(R_1{}^2) = \sum_{i=1}^{n} \frac{1}{n} a_i{}^2, \quad E(R_2{}^2) = \sum_{i=1}^{n} \frac{1}{n} b_i{}^2$$

In (b) let R be uniformly distributed between a and b, and set $R_1 = g(R)$, $R_2 = h(R)$. Then

$$E(R_1R_2) = \int_a^b \frac{g(x)h(x)}{b-a}\, dx, \quad E(R_1{}^2) = \int_a^b \frac{g^2(x)}{b-a}\, dx, \quad E(R_2{}^2) = \int_a^b \frac{h^2(x)}{b-a}\, dx$$

In each case the result follows from Theorem 2 of Section 3.4.

5. This follows from the argument of property 7, Section 3.3.

Section 3.5

3. $P\{R_1 = j, R_2 = k\} = \dfrac{n!}{j!\,k!\,(n - j - k)!}\,(\tfrac{1}{6})^{j+k}(\tfrac{2}{3})^{n-j-k}$

   $j, k = 0, 1, \ldots, n, j + k \le n$ (see Example 1, Section 2.9).

4. $(n - 1)p(1 - p)$.

5. Let $A_i = \{$trial i results in success and trial $i + 1$ in failure$\}$. Then $R_0 = \sum_{i=1}^{n-1} I_{A_i}$
   and $R_0{}^2 = \sum_{i=1}^{n-1} I_{A_i}{}^2 + 2 \sum_{i<j} I_{A_i} I_{A_j}$. $I_{A_i} I_{A_{i+1}} \equiv 0$, and if $j \ge i + 2$, I_{A_i} and I_{A_j} are
   independent, with $E(I_{A_i} I_{A_j}) = P(A_i)P(A_j) = (pq)^2$. Thus

   $$E(R_0{}^2) = (n - 1)pq + 2\sum_{i=1}^{n-3} \sum_{j=i+2}^{n-1} (pq)^2$$
   $$= (n - 1)pq + 2(pq)^2 \sum_{i=1}^{n-3} (n - 2 - i)$$
   $$= (n - 1)pq + 2(pq)^2[1 + 2 + \cdots + (n - 3)]$$
   $$= (n - 1)pq + (pq)^2(n - 3)(n - 2).$$

   Therefore, Var $R_0 = E(R_0{}^2) - [E(R_0)]^2 = (n - 1)pq + (n - 2)(n - 3)(pq)^2 - (n - 1)^2(pq)^2$, $q = 1 - p$ (assuming $n \ge 2$).

6. $50(49/50)^{100}$

Section 3.6

1. (a) .532

   (b) -2.84

Section 3.7

1. $m = \int_0^\infty xe^{-x}\,dx = 1$, $E(R^2) = \int_0^\infty x^2 e^{-x}\,dx = 2$, hence $\sigma^2 = 1$. $P\{|R - m| \ge k\sigma\} = P\{|R - 1| \ge k\}$. If $0 < k \le 1$, this is $\int_0^{1-k} e^{-x}\,dx + \int_{1+k}^\infty e^{-x}\,dx = 1 - e^{-(1-k)} + e^{-(1+k)}$. When $k > 1$, it becomes $\int_{1+k}^\infty e^{-x}\,dx = e^{-(1+k)}$. Notice that the Chebyshev bound is vacuous when $k \le 1$, and for $k > 1$, $e^{-(1+k)}$ approaches zero much more rapidly than $1/k^2$.

CHAPTER 4

Section 4.2

1. $\Gamma(r) = \int_0^\infty t^{r-1}e^{-t}\,dt = $ (with $t = x^2$) $2\int_0^\infty x^{2r-1}e^{-x^2}\,dx$.

   Thus $\Gamma(r)\Gamma(s) = 4\int_0^\infty \int_0^\infty x^{2r-1}y^{2s-1}e^{-(x^2+y^2)}\,dx\,dy$

   $= $ (in polar coordinates) $4\int_0^{\pi/2} d\theta \int_0^\infty (\cos\theta)^{2r-1}(\sin\theta)^{2s-1}e^{-\rho^2}\rho^{2r+2s-1}\,d\rho$.

   Now $\int_0^\infty \rho^{2r+2s-1}e^{-\rho^2}\,d\rho = $ (set $u = \rho^2$) $\tfrac{1}{2}\int_0^\infty u^{r+s-1}e^{-u}\,du = \tfrac{1}{2}\Gamma(r + s)$

   Therefore $\dfrac{\Gamma(r)\Gamma(s)}{2\Gamma(r + s)} = \int_0^{\pi/2} (\cos\theta)^{2r-1}(\sin\theta)^{2s-1}\,d\theta$

Let $z = \cos^2 \theta$ so that $1 - z = \sin^2 \theta$, $dz = -2 \cos \theta \sin \theta \, d\theta$,

$$d\theta = -\frac{dz}{2z^{1/2}(1-z)^{1/2}}$$

Thus

$$\frac{\Gamma(r)\Gamma(s)}{2\Gamma(r+s)} = -\tfrac{1}{2}\int_1^0 z^{r-1}(1-z)^{s-1}\,dz = \tfrac{1}{2}\beta(r,s)$$

3. $p_1(e^{-3} - e^{-5}) + p_2(e^{-4} - e^{-8}) + p_3(e^{-3} - e^{-9}) + p_4(1 - e^{-8}) + p_5(1 - e^{-5})$

5. $f_2(y) = \tfrac{1}{2}, \quad 0 \le y \le 1$

$$= \frac{1}{2y^2}, \quad y > 1$$

Section 4.3

1. $h(y \mid x) = e^{x-y}$, $0 \le x \le y$, and 0 elsewhere; $P\{R_2 \le y \mid R_1 = x\} = 1 - e^{x-y}$, $y \ge x$, and 0 elsewhere.

2. $f_1(x) = \displaystyle\int_{-1}^{x} kx\,dy = kx(x+1), 0 \le x \le 1$

$$= \int_{-1}^{x} -kx\,dy = -kx(x+1), -1 \le x \le 0$$

$f_2(y) = \displaystyle\int_{y}^{1} kx\,dx = \tfrac{1}{2}k(1-y^2), 0 \le y \le 1$

$$= \int_{y}^{0} -kx\,dx + \int_0^1 kx\,dx = \tfrac{1}{2}k(1+y^2), -1 \le y \le 0$$

Since $\int_{-1}^1 f_1(x)\,dx = \int_{-1}^1 f_2(y)\,dy = k$, we must have $k = 1$.
The conditional density of R_2 given R_1 is

$$h_2(y \mid x) = \frac{f(x,y)}{f_1(x)} = \frac{1}{x+1}, -1 \le x \le 1, -1 \le y \le x$$

The conditional density of R_1 given R_2 is

$$h_1(x \mid y) = \frac{f(x,y)}{f_2(y)} = \frac{2x}{1-y^2}, 0 \le y \le 1, y \le x \le 1$$

$$= \frac{2x}{1+y^2}, -1 \le y \le 0, 0 \le x \le 1$$

$$= -\frac{2x}{1+y^2}, -1 \le y \le 0, y \le x \le 0$$

4. The conditional density of R_3 given $R_1 = x$ is $h(z \mid x) = e^{-z}$, $z \ge 0$, $x \ge 0$,

$$P\{1 \le R_3 \le 2 \mid R_1 = x\} = e^{-1} - e^{-2}, x \ge 0$$

5. $P\{g(R_1, R_2) \leq z \mid R_1 = x\} = P\{g(x, R_2) \leq z \mid R_1 = x\} = \displaystyle\int_{\{y:\, g(x,y)\leq z\}} h(y \mid x)\, dy$

Section 4.4

1. (a) $h_2(y \mid x) = \dfrac{f(x, y)}{f_1(x)} = \dfrac{8xy}{4x^3} = \dfrac{2y}{x^2}, \qquad 0 \leq y \leq x \leq 1$

$h_1(x \mid y) = \dfrac{f(x, y)}{f_2(y)} = \dfrac{8xy}{4y(1 - y^2)} = \dfrac{2x}{1 - y^2}, \qquad 0 \leq y \leq x \leq 1$

Thus $E(R_2 \mid R_1 = x) = \displaystyle\int_{-\infty}^{\infty} y h_2(y \mid x)\, dy = \int_0^x y \left(\dfrac{2y}{x^2}\right) dy = \tfrac{2}{3}x,\ 0 \leq x \leq 1$

$E(R_1 \mid R_2 = y) = \displaystyle\int_{-\infty}^{\infty} x h_1(x \mid y)\, dx = \int_y^1 x \left(\dfrac{2x}{1 - y^2}\right) dx = \dfrac{2}{3}\dfrac{(1 - y^3)}{(1 - y^2)},$

$$0 \leq y \leq 1$$

(b) $E(R_2^4 \mid R_1 = x) = \displaystyle\int_{-\infty}^{\infty} y^4 h_2(y \mid x)\, dy = \int_0^x y^4 \left(\dfrac{2y}{x^2}\right) dy = \tfrac{1}{3}x^4$

(c) The conditional density of (R_1, R_2) given A is

$$f(x, y \mid A) = \dfrac{f(x, y)}{P(A)} = \dfrac{8xy}{\displaystyle\int_0^{1/2} dx \int_0^x 8xy\, dy} = 128xy, \qquad 0 \leq y \leq x \leq \tfrac{1}{2}$$

The conditional density of R_2 given A is

$$f_2(y \mid A) = \displaystyle\int_{-\infty}^{\infty} f(x, y \mid A)\, dx = \int_y^{1/2} 128xy\, dx = 16y - 64y^3, \quad 0 \leq y \leq \tfrac{1}{2}$$

$$= 0 \text{ elsewhere}$$

$$E(R_2 \mid A) = \displaystyle\int_{-\infty}^{\infty} y f_2(y \mid A)\, dy = \int_0^{1/2} y(16y - 64y^3)\, dy$$

$$= \tfrac{2}{3} - \tfrac{2}{5} = \tfrac{4}{15}$$

Alternatively, $E(R_2 \mid A) = \dfrac{E(R_2 I_A)}{P(A)} = \dfrac{\displaystyle\int_0^{1/2} \int_0^x y(8xy)\, dy\, dx}{\displaystyle\int_0^{1/2} \int_0^x 8xy\, dy\, dx} = \dfrac{1/60}{1/16} = 4/15$

Alternatively, $E(R_2 \mid A) = \int_{-\infty}^{\infty} f_1(x \mid A) E(R_2 \mid R_1 = x)\, dx$, where $f_1(x \mid A) = f_1(x)/P(A)$ if $0 \leq x \leq \tfrac{1}{2}$, and 0 elsewhere. Thus

$$E(R_2 \mid A) = \displaystyle\int_0^{1/2} \dfrac{4x^3}{1/16} \tfrac{2}{3}x\, dx = 4/15$$

2. $\dfrac{(1 + x)^n}{n!}, \qquad x = x_1 + \cdots + x_n$

3. $E(R_2 \mid R_1 = x) = x/2, \quad 0 \le x \le 1$
$$= 1/2, \quad 1 \le x \le 2$$
$$= (x - 1)/2, \quad 2 \le x \le 3$$

4. $\frac{1}{5}(n - k)$

5. (a) 3/7
   (b) 19/21

7. $E(R) = 1P\{R = 1\} + 2P\{R = 2\} + 3P\{R = 3\}$
   where

$$P\{R = 1\} = \frac{2}{3} \frac{30!}{10!\,10!\,10!} (.5)^{10}(.3)^{10}(.2)^{10} + \frac{1}{3} \frac{\binom{50}{10}\binom{30}{10}\binom{20}{10}}{\binom{100}{30}}$$

$$P\{R = 2\} = \tfrac{2}{3}\binom{30}{12}(.3)^{12}(.7)^{18}$$

$$P\{R = 3\} = 1 - P\{R = 1\} - P\{R = 2\}$$

8. (a) $P\{R_3 \le z\} = P\{R_3 \le z, R_1^2 + R_2^2 \le 1\} + P\{R_3 \le z, R_1^2 + R_2^2 > 1\}$. If $0 \le z \le 1$, then $R_1^2 + R_2^2 > 1$ implies $R_3 = 2 > z$, so $R\{R_3 \le z, R_1^2 + R_2^2 > 1\} = 0$. Thus

$$F_3(z) = P\{R_3 \le z, R_1^2 + R_2^2 \le 1\}$$

$$= \int\limits_{\substack{x^2+y^2 \le 1, \\ x \le z}} f(x, y)\, dx\, dy = \int_0^z dx \int_0^{(1-x^2)^{1/2}} dy = \int_0^z (1 - x^2)^{1/2}\, dx$$

$$= \tfrac{1}{2}[z(1 - z^2)^{1/2} + \arcsin z],\ 0 \le z \le 1$$

If $1 \le z < 2$, $P\{R_3 \le z, R_1^2 + R_2^2 > 1\}$ is still 0, and $F_3(z) = P\{R_3 \le z, R_1^2 + R_2^2 \le 1\} = P\{R_1^2 + R_2^2 \le 1\} = \pi/4$. If $z \ge 2, P\{R_3 \le z\} = 1$.

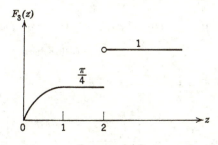

PROBLEM 4.4.8

By (4.4.6), $E(R_3) = 2P\{R_3 = 2\} + \int_0^1 z(1 - z^2)^{1/2}\, dz$

$$= 2\left(1 - \frac{\pi}{4}\right) + \frac{1}{3} = \frac{7}{3} - \frac{\pi}{2}.$$

(b)
$$\int_{-\infty}^{\infty} \int_{-\infty}^{\infty} g(x, y) f(x, y) \, dx \, dy = \iint_{x^2+y^2 \leq 1} x f_1(x) f_2(y) \, dx \, dy$$

$$+ \iint_{x^2+y^2 > 1} 2 f_1(x) f_2(y) \, dx \, dy$$

$$= \int_0^1 x \, dx \int_0^{(1-x^2)^{1/2}} dy + 2 \iint_{\substack{0 \leq x \leq 1, 0 \leq y \leq 1, \\ x^2+y^2 > 1}} dx \, dy$$

$$= \int_0^1 x(1 - x^2)^{1/2} \, dx + 2\left(1 - \frac{\pi}{4}\right)$$

$$= \tfrac{1}{3} + 2\left(1 - \frac{\pi}{4}\right) \text{ as before.}$$

(c) Since $R_3 = 2$ when $R_1^2 + R_2^2 > 1$, $E(R_3 \mid R_1^2 + R_2^2 > 1) = 2$. The conditional density of (R_1, R_2) given $A = \{R_1^2 + R_2^2 \leq 1\}$ is

$$f(x, y \mid A) = \frac{f(x, y)}{P(A)} = \frac{4}{\pi}, \quad x^2 + y^2 \leq 1, x, y \geq 0$$

$$= 0 \text{ elsewhere}$$

$$E(R_3 \mid A) = \frac{E(R_3 I_A)}{P(A)} = \frac{E(R_1 I_A)}{P(A)} \text{ since } R_1^2 + R_2^2 \leq 1 \text{ implies } R_3 = R_1$$

$$= E(R_1 \mid A) = \int_{-\infty}^{\infty} \int_{-\infty}^{\infty} x f(x, y \mid A) \, dx \, dy = \int_{-\infty}^{\infty} x f_1(x \mid A) \, dx$$

where

$$f_1(x \mid A) = \int_{-\infty}^{\infty} f(x, y \mid A) \, dy = \frac{4}{\pi} (1 - x^2)^{1/2}, \quad 0 \leq x \leq 1$$

$$= 0 \text{ elsewhere}$$

Thus $E(R_3 \mid A) = \int_0^1 \frac{4}{\pi} x(1 - x^2)^{1/2} \, dx = \frac{4}{3\pi}$

Alternatively, $\dfrac{E(R_1 I_A)}{P(A)} = \dfrac{\displaystyle\int_0^1 \int_0^1 x I_{\{x^2+y^2 \leq 1\}} \, dx \, dy}{\pi/4} = \dfrac{\displaystyle\int_0^1 x \, dx \int_0^{(1-x^2)^{1/2}} dy}{\pi/4}$

$$= \frac{4}{3\pi}$$

Now $E(R_3) = P(A)E(R_3 \mid A) + P(A^c)E(R_3 \mid A^c)$

$$= \frac{\pi}{4}\left(\frac{4}{3\pi}\right) + \left(1 - \frac{\pi}{4}\right)2 = \frac{7}{3} - \frac{\pi}{2} \text{ as before.}$$

10. Let $R_3 = R_1 + R_2$, $R_4 = R_1 - R_2$. The joint density of R_3 and R_4 is

$$f_{34}(u, v) = f_{12}(x, y) \left| \frac{\partial(x, y)}{\partial(u, v)} \right|$$

where $u = x + y$, $v = x - y$, $x = \frac{1}{2}(u + v)$, $y = \frac{1}{2}(u - v)$ (see Section 2.8, problem 12). Thus $f_{34}(u, v) = \frac{1}{2}$, $0 \le x \le 1$, $0 \le y \le 1$. But $0 \le x \le 1$, $0 \le y \le 1$ corresponds to $(u, v) \in D$ (see diagram).

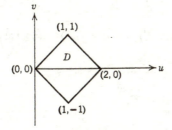

PROBLEM 4.4.10

The density of R_4 is $f_4(v) = \displaystyle\int_{-\infty}^{\infty} f_{34}(u, v)\, du = 1 - |v|$, $\qquad |v| \le 1$

$$= 0 \text{ elsewhere}$$

Therefore $h_3(u \mid v) = \dfrac{f_{34}(u, v)}{f_4(v)} = \dfrac{1}{2(1 - v)}$, $0 \le v \le 1$, $v \le u \le 2 - v$

$$= \frac{1}{2(1 + v)}, \quad -1 \le v \le 0, \; -v \le u \le 2 + v$$

If $0 \le v \le 1$, $E(R_3^2 \mid R_4 = v) = \dfrac{1}{2(1 - v)} \displaystyle\int_v^{2-v} u^2\, du = \dfrac{4 - 2v + v^2}{3}$

If $-1 \le v \le 0$, $E(R_3^2 \mid R_4 = v) = \dfrac{1}{2(1 + v)} \displaystyle\int_{-v}^{2+v} u^2\, du = \dfrac{4 + 2v + v^2}{3}$

Thus $E[(R_1 + R_2)^2 \mid R_1 - R_2 = v] = \frac{1}{3}[4 - 2|v| + v^2]$, $-1 \le v \le 1$.

11. $x^2 + 7/6$

14. $f_R(\lambda \mid R_1 = x_1, \ldots, R_n = x_n) = \lambda^x (1 - \lambda)^{n-x} / \beta(1 + x, n - x + 1)$, $0 \le \lambda \le 1$,

$$\text{where } x = x_1 + \cdots + x_n$$

$E(R \mid R_1 = x_1, \ldots, R_n = x_n) = (x + 1)/(n + 2)$

16. $(np - npq^{n-1})/(1 - q^n - npq^{n-1})$

17. $(\frac{12}{10} + \frac{1}{30}(8 - 2^{3/2}))/(\frac{6}{10} - \frac{1}{10}\sqrt{2})$

18. $f_{12}(x, y \mid z) = \dfrac{1}{\pi z^2}$, $x^2 + y^2 < z^2$, and 0 elsewhere

(a) $E(D \mid R = z) = \displaystyle\int_{-\infty}^{\infty} \int_{-\infty}^{\infty} (x^2 + y^2)^{1/2} f_{12}(x, y \mid z) \, dx \, dy$

$$= \frac{1}{\pi z^2} \iint\limits_{x^2+y^2<z^2} (x^2 + y^2)^{1/2} \, dx \, dy$$

$$= \frac{1}{\pi z^2} \int_0^{2\pi} d\theta \int_0^z r^2 \, dr = \tfrac{2}{3} z. \text{ Thus } E(D) = \int_0^\infty \tfrac{2}{3} z e^{-z} \, dz = \tfrac{2}{3}$$

(b) $h(z \mid x, y) = f(x, y, z)/f_{12}(x, y) = f_R(z) f_{12}(x, y \mid z) \Big/ \displaystyle\int_{-\infty}^{\infty} f(x, y, z) \, dz$

$$= \frac{e^{-z}/\pi z^2}{\displaystyle\int_{(x^2+y^2)^{1/2}}^{\infty} \frac{e^{-z}}{\pi z^2} \, dz}, \; z > (x^2 + y^2)^{1/2}$$

19. (b)

$$d(x) = -1, \qquad -3 \le x < -1$$
$$= 0, \qquad -1 \le x \le 1$$
$$= 1, \qquad 1 < x \le 3$$
$$E_{\min}[(\theta^* - \theta)^2] = 1/2$$

20. $\tfrac{1}{2}(x + 1)$

CHAPTER 5

Section 5.2

1. $N_0(s) = N_1(s) N_2(s) N_3(s) = \left[\dfrac{1}{2s} (e^s - e^{-s}) \right]^3$

$$= \frac{1}{8s^3} (e^{3s} - 3e^s + 3e^{-s} - e^{-3s}), \text{ all } s$$

$f_0(x) = \tfrac{1}{16} [(x + 3)^2 u(x + 3) - 3(x + 1)^2 u(x + 1) + 3(x - 1)^2 u(x - 1)$
$\qquad - (x - 3)^2 u(x - 3)]$

Thus

$$f_0(x) = \tfrac{1}{16}(x + 3)^2, \; -3 \le x \le -1$$
$$= \frac{3 - x^2}{8}, \; -1 \le x \le 1$$
$$= \tfrac{1}{16}(x - 3)^2, \; 1 \le x \le 3$$
$$= 0 \text{ elsewhere.}$$

2. $f_0(x) = \frac{1}{9}[(x + 2)u(x + 2) + 2(x + 1)u(x + 1) - 3xu(x) - 4(x - 1)u(x - 1)$

$+4(x - 2)u(x - 2)]$

3. $N_{R_i^2}(s) = E(e^{-sR_i^2}) = \int_{-\infty}^{\infty} e^{-sx^2} \dfrac{1}{\sqrt{2\pi}} e^{-x^2/2}\, dx = [\text{with } y = (s + \frac{1}{2})^{1/2}x]$

$\int_{-\infty}^{\infty} [2\pi(s + \frac{1}{2})]^{-1/2} e^{-y^2}\, dy = [2(s + \frac{1}{2})]^{-1/2}, \text{ Re } s > -\frac{1}{2}$

$N_R(s) = \displaystyle\prod_{i=1}^{n} N_{R_i^2}(s) = 2^{-n/2}(s + \frac{1}{2})^{-n/2}, \text{ Re } s > -\frac{1}{2}$

and the result follows from Table 5.1.1.

4. $N_R(s) = \dfrac{1}{\Gamma(\alpha)\beta^\alpha} \displaystyle\int_0^{\infty} x^{\alpha-1} e^{-(s+\beta-1)x}\, dx$

$= \dfrac{1}{\Gamma(\alpha)\beta^\alpha} \displaystyle\int_0^{\infty} \dfrac{y^{\alpha-1}e^{-y}}{(s + \beta^{-1})^\alpha}\, dy = \dfrac{1}{(1 + \beta s)^\alpha}, \text{ Re } s > -\dfrac{1}{\beta}$

If $R_0 = R_1 + R_2$ where R_1 and R_2 are independent and R_i has the gamma distribution with parameters α_i and β, $i = 1, 2$, then

$$N_0(s) = N_1(s)N_2(s) = \left(\dfrac{1}{1 + \beta s}\right)^{\alpha_1+\alpha_2}, \text{ Re } s > -1/\beta$$

so R_0 has the gamma distribution with parameters $\alpha_1 + \alpha_2$ and β.

5. $f_0(x) = \lambda^n x^{n-1} e^{-\lambda x} u(x)/(n - 1)!$

9. Integrate $\dfrac{e^{-iuz}}{\pi(1 + z^2)}$ around contour (a) if $u \geq 0$, and around contour (b) if $u \leq 0$. (Notice that $|e^{-iu(x+iy)}|$ is bounded if $u \geq 0$, as long as $y \leq 0$.) If $u \geq 0$,

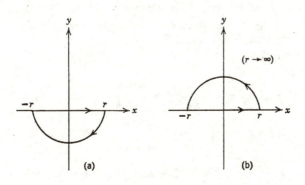

(a)   (b)

PROBLEM 5.2.9

the integral is $-2\pi i \left[\text{residue of } \dfrac{e^{-iuz}}{\pi(1+z^2)} \text{ at } z = -i \right] = -2\pi i \dfrac{[e^{-iu(-i)}]}{\pi(-2i)} = e^{-u}$.

If $u \leq 0$, the integral is $2\pi i \left[\text{residue of } \dfrac{e^{-iuz}}{\pi(1+z^2)} \text{ at } z = i \right] = \dfrac{2\pi i [e^{-iu(i)}]}{\pi(2i)} = e^{u}$.

The result follows since the integral around the semicircle of radius r approaches 0 as $r \to \infty$.

Section 5.3

4. $M_R(u) = \displaystyle\sum_{n=-\infty}^{\infty} p_n e^{-iu(a+nd)}$ where $p_n = P\{R = a + nd\}$. Thus $M_R(u) = e^{-iua}M(u)$ where M is periodic with period $2\pi/d$; the numbers p_n are the coefficients of the Fourier series of M.

5. $N_R(s) = \exp[\lambda(e^{-s} - 1)]$, hence $\dfrac{dN_R(s)}{ds} = -\lambda e^{-s} \exp[\lambda(e^{-s} - 1)] = -\lambda$ when $s = 0$. $\dfrac{d^2 N_R(s)}{ds^2} = (\lambda e^{-s})^2 \exp[\lambda(e^{-s} - 1)] + \lambda e^{-s} \exp[\lambda(e^{-s} - 1)] = \lambda^2 + \lambda$ when $s = 0$. By (5.3.3), $E(R) = \lambda, E(R^2) = \lambda^2 + \lambda$, hence Var $R = \lambda$.

Section 5.4

1. (a) $P\{R_n \leq x\} = P\{R_n \leq x, R > x + \varepsilon\} + P\{R_n \leq x, R \leq x + \varepsilon\}$
$$\leq P\{|R_n - R| \geq \varepsilon\} + P\{R \leq x + \varepsilon\}$$
since $R_n \leq x, R > x + \varepsilon$ implies $|R_n - R| \geq \varepsilon$ and
$R_n \leq x, R \leq x + \varepsilon$ implies $R \leq x + \varepsilon$.
Similarly $P\{R \leq x - \varepsilon\} = P\{R \leq x - \varepsilon, R_n > x\} + P\{R \leq x - \varepsilon, R_n \leq x\}$
$$\leq P\{R_n - R \mid \geq \varepsilon\} + P\{R_n \leq x\}$$
Thus $F(x - \varepsilon) - P\{|R_n - R| \geq \varepsilon\} \leq F_n(x) \leq P\{|R_n - R| \geq \varepsilon\} + F(x + \varepsilon)$
(b) Given $\delta > 0$, choose $\varepsilon > 0$ so small that

$$F(x + \varepsilon) < F(x) + \frac{\delta}{2}, \qquad F(x - \varepsilon) > F(x) - \frac{\delta}{2}.$$

(This is possible since F is continuous at x.) For large enough n, $P\{|R_n - R| \geq \varepsilon\} < \delta/2$ since $R_n \xrightarrow{P} R$. By (a), $F(x) - \delta < F_n(x) < F(x) + \delta$ for large enough n. Thus $F_n(x) \to F(x)$.

5. (a) $n \geq 1{,}690{,}000$
   (b) $n \geq 9604$

7. $P\{|R - \tfrac{1}{2}n| > .005n\} = P\left\{\dfrac{|R - \tfrac{1}{2}n|}{\tfrac{1}{2}\sqrt{n}} > \dfrac{.005n}{\tfrac{1}{2}\sqrt{n}}\right\} \sim P\{|R^*| > .01\sqrt{n}\}$

$$= 2P\{R^* > .01\sqrt{n}\} = \frac{2}{\sqrt{2\pi}} \int_{.01\sqrt{n}}^{\infty} e^{-t^2/2}\, dt \leq \frac{2}{\sqrt{2\pi}} \frac{1}{.01\sqrt{n}} e^{-.0001n/2}$$

$$= \frac{200}{\sqrt{2\pi n}} e^{-(1/2)10^{-4}n}. \text{ For example, if } n = 10^6, \text{ this is } \frac{.2}{\sqrt{2\pi}} e^{-50}$$

8. .91

CHAPTER 6

Section 6.2

1. $f_1(x) = \lambda_1^x$ and $f_2(x) = \lambda_2^x$ [or $f_1(x) = \lambda^x, f_2(x) = x\lambda^x$ in the repeated root case] are linearly independent solutions. For if $c_1 f_1 + c_2 f_2 \equiv 0$ then

$$c_1 \lambda_1^x + c_2 \lambda_2^x = 0$$

$$c_1 \lambda_1^{x+1} + c_2 \lambda_2^{x+1} = 0$$

If c_1 and c_2 are not both 0, then

$$\begin{vmatrix} \lambda_1^x & \lambda_2^x \\ \lambda_1^{x+1} & \lambda_2^{x+1} \end{vmatrix} = 0$$

hence

$$\begin{vmatrix} 1 & 1 \\ \lambda_1 & \lambda_2 \end{vmatrix} = 0, \text{ a contradiction}$$

If f is any solution then for some constants A and C,

$$\begin{bmatrix} f(0) \\ f(1) \end{bmatrix} = A \begin{bmatrix} f_1(0) \\ f_1(1) \end{bmatrix} + C \begin{bmatrix} f_2(0) \\ f_2(1) \end{bmatrix}$$

since three vectors in a two-dimensional space are linearly dependent. But then

$$f(2) = d_1 f(0) + d_2 f(1), \ d_1 = \frac{1}{p}, \ d_2 = \frac{-q}{p}$$

$$= d_1[A f_1(0) + C f_2(0)] + d_2[A f_1(1) + C f_2(1)]$$

$$= A f_1(2) + C f_2(2)$$

Recursively, $f(x) = A f_1(x) + C f_2(x)$. Thus all solutions are of the form $A f_1(x) + C f_2(x)$. But we have shown in the text that A and C are uniquely determined by the boundary conditions at 0 and b; the result follows.

2. By the theorem of total expectation, if R is the duration of the game and $A = \{\text{win on trial 1}\}$ then

$$E(R) = P(A)E(R \mid A) + P(A^c)E(R \mid A^c)$$

Thus

$$D(x) = p[1 + D(x + 1)] + q[1 + D(x - 1)], \ x = 1, 2, \ldots, b - 1$$

since if we win on trial 1, the game has already lasted for one trial, and the average number of trials remaining after the first is $D(x + 1)$. [Notice that this argument, just as the one leading to (6.2.1), is intuitive rather than formal.]

3. In standard form, $pD(x + 2) - D(x + 1) + qD(x) = -p - q = -1, \ D(0) = D(b) = 0$.

CASE 1. $p \neq q$.

The homogeneous equation is the same as (6.2.1), with solution $A + C(q/p)^x$. To find a particular solution, notice that the "forcing function" -1 already satisfies the homogeneous equation, so try $D(x) = kx$. Then

$$k[p(x + 2) - (x + 1) + qx] = (2p - 1)k = (p - q)k = -1.$$

Thus

$$D(x) = A + C(q/p)^x + \frac{x}{q - p}.$$

Set $D(0) = D(b) = 0$ to solve for A and C.

CASE 2. $p = q = 1/2$.

The homogeneous solution is $A + Cx$. Since polynomials of degree 0 and 1 already satisfy the homogeneous equation, try as a particular solution $D(x) = kx^2$. Then

$$k[\tfrac{1}{2}(x + 2)^2 - (x + 1)^2 + \tfrac{1}{2}x^2] = k = -1$$

Thus

$$D(x) = A + Cx - x^2$$

$$D(0) = A = 0, \ D(b) = b(C - b) = 0 \text{ so that } C = b.$$

Therefore $D(x) = x(b - x)$.

If we let $b \to \infty$ we obtain

$$D(x) = \infty \text{ if } p \geq q; \ D(x) = \frac{x}{q - p} \text{ if } p < q$$

Section 6.3

1. $P\{S_1 \geq 0, \ldots, S_{2n-1} \geq 0, S_{2n} = 0\}$ is the number of paths from $(0, 0)$ to $(2n, 0)$ lying on or above the axis, times $(pq)^n$. These paths are in one-to-one correspondence with the paths from $(-1, -1)$ to $(2n, 0)$ lying above -1 [connect $(-1, -1)$ to $(0, 0)$ to establish the correspondence]. Thus the number of paths is the same as the number from $(0, 0)$ to $(2n + 1, 1)$ lying above 0, namely

$$\binom{2n + 1}{a} \frac{a - b}{2n + 1} \text{ where } a + b = 2n + 1, a - b = 1, \text{ that is, } a = n + 1, b = n$$

Thus the desired probability is

$$\binom{2n + 1}{n + 1} \frac{1}{2n + 1} (pq)^n = \frac{(2n)!}{n! \, (n + 1)!} (pq)^n = \frac{u_{2n}}{n + 1}$$

5. $u_{2n} = \binom{2n}{n} (pq)^n = \frac{(2n)!}{n! \, n!} (pq)^n \sim \frac{(2n)^{2n} \sqrt{2\pi 2n}}{(n^n \sqrt{2\pi n})^2} (pq)^n = \frac{(4pq)^n}{\sqrt{n\pi}} = \frac{1}{\sqrt{n\pi}}$ if

$$p = q = \tfrac{1}{2}$$

By Problem 2,

$$h_{2n} = \frac{u_{2n-2}}{2n} \sim \frac{1}{2n} \frac{1}{\sqrt{(n-1)\pi}} \sim \frac{1}{2\sqrt{\pi}\, n^{3/2}}$$

Let T be the time required to return to 0. Then

$$P\{T = 2n\} = h_{2n}, \quad n = 1, 2, \ldots \text{ where } \sum_{n=1}^{\infty} h_{2n} = 1$$

$$E(T) = \sum_{n=1}^{\infty} 2nP\{T = 2n\} = \sum_{n=1}^{\infty} 2nh_{2n}$$

But $2nh_{2n} \sim K/\sqrt{n}$ and $\Sigma\, 1/\sqrt{n} = \infty$, hence $E(T) = \infty$.

6. The probability that both players will have k heads is $[\binom{n}{k}(\frac{1}{2})^n]^2$; sum from $k = 0$ to n to obtain the desired result.

7. The probability that both players will receive the same number of heads = the probability that the number of heads obtained by player 1 = the number of tails obtained by player 2 (since $p = q = 1/2$), and this is the probability of being at 0 after $2n$ steps of a simple random walk, namely $\binom{2n}{n}(\frac{1}{2})^{2n}$. Comparing this expression with the result of Problem 6, we obtain the desired conclusion. (Alternatively, we may use the formula of Section 2.9, Problem 4, with $m = k = n$.)

Section 6.4

1. $(1 - 4pqz^2)^{1/2} = \sum_{n=0}^{\infty} \binom{1/2}{n}(-4pqz^2)^n$

$$= \sum_{n=0}^{\infty} \binom{1/2}{n}(-4pq)^n z^{2n}$$

Thus

$$H(z) = 1 - (1 - 4pqz^2)^{1/2} = -\sum_{n=1}^{\infty} \binom{1/2}{n}(-4pq)^n z^{2n}$$

Thus

$$h_{2n} = (-1)^{n+1}\binom{1/2}{n}(4pq)^n, \quad n = 1, 2, \ldots$$

But

$$(-1)^{n+1}\binom{1/2}{n} = (-1)^{n+1} \frac{(1/2)(-1/2)(-3/2) \cdots [(2n-3)/2]}{n!}$$

$$= \frac{1 \cdot 3 \cdot 5 \cdots (2n-3)}{2^n n!} = \frac{(2n-2)!}{2^n n!\, 2 \cdot 4 \cdots (2n-2)}$$

$$= \frac{(2n-2)!}{2^n n(n-1)!\, 2^{n-1}(n-1)!} = \frac{2}{n}\binom{2n-2}{n-1}\left(\frac{1}{2}\right)^{2n}$$

Therefore $h_{2n} = (2/n)\binom{2n-2}{n-1}(pq)^n$, in agreement with (6.3.5).

2.

$$\sum_{n=0}^{\infty} a_{n+1}z^n - 3\sum_{n=0}^{\infty} a_n z^n = 4\sum_{n=0}^{\infty} z^n = \frac{4}{1-z}$$

If

$$A(z) = \sum_{n=0}^{\infty} a_n z^n, \text{ then } \frac{1}{z}[A(z) - a_0] - 3A(z) = \frac{4}{1-z}$$

or

$$A(z)(z^{-1} - 3) = \frac{4}{1-z} + \frac{a_0}{z}$$

Thus

$$A(z) = \frac{4z}{(1-z)(1-3z)} + \frac{a_0}{1-3z}$$

$$= -\frac{2}{1-z} + \frac{2+a_0}{1-3z} = -2\sum_{n=0}^{\infty} z^n + (2+a_0)\sum_{n=0}^{\infty} 3^n z^n$$

Thus $a_n = (2+a_0)3^n - 2$. Notice that $(2+a_0)3^n$ is the homogeneous solution, -2 the particular solution.

6. (a) $P\{N_r = k\} = P\{$the first $k-1$ trials result in exactly $r-1$ successes, and trial k results in a success$\} = \binom{k-1}{r-1}p^{r-1}q^{k-r}p$, $k = r, r+1, \ldots$. Now

$$\binom{k-1}{r-1} = \binom{k-1}{k-r} = (k-1)(k-2)\cdots(r+1)r/(k-r)!$$

$$= (-1)^{k-r}(-r)(-r-1)(-r-2)\cdots[-r-(k-2-r)]$$

$$\times [-r-(k-1-r)]/(k-r)!$$

$$= (-1)^{k-r}\binom{-r}{k-r}, \text{ and the result follows.}$$

Note that if $j = k - r$, this computation shows that $(-1)^j\binom{-r}{j} = \binom{j+r-1}{r-1}$, $r = 1, 2, \ldots, j = 0, 1, \ldots$.

(b) We show that T_1 and T_2 are independent. The argument for $T_1, \ldots, T_r$ is similar, but the notation becomes cumbersome.

$$P\{T_1 = j, T_2 = k\} = P\{R_1 = \cdots = R_{j-1} = 0, R_j = 1,$$
$$R_{j+1} = \cdots = R_{j+k-1} = 0, R_{j+k} = 1\}$$
$$= p^2 q^{j+k-2}, j, k = 1, 2, \ldots$$

Now $P\{T_1 = j\} = q^{j-1}p$ by Problem 5, and

$$P\{T_2 = k\} = \sum_{j=1}^{\infty} P\{T_1 = j, T_2 = k\} = \left(\sum_{j=1}^{\infty} q^{j-1}\right)p^2 q^{k-1}$$

$$= \frac{p^2 q^{k-1}}{1-q} = pq^{k-1}$$

Hence $P\{T_1 = j, T_2 = k\} = P\{T_1 = j\}P\{T_2 = k\}$ and the result follows.

(c) $E(N_r) = r/p$, Var $N_r = r[(1 - p)/p^2]$ since $N_r = T_1 + \cdots + T_r$ and the T_i are independent. The generalized characteristic function of N_r is

$$\left(\frac{pe^{-s}}{1 - qe^{-s}}\right)^r, \qquad |qe^{-s}| < 1$$

Set $s = iu$ to obtain the characteristic function, $z = e^{-s}$ to obtain the generating function.

7. $P\{R = k\} = p^k q + q^k p$, $k = 1, 2, \ldots$; $E(R) = pq^{-1} + qp^{-1}$

8. $1/\sqrt{2}$

Section 6.5

2. $P\{\text{an even number of customers arrives in } (t, t + \tau]\} = \sum_{k=0,2,4,\ldots}^{\infty} e^{-\lambda\tau}(\lambda\tau)^k/k!$

$$= \tfrac{1}{2}\sum_{k=0}^{\infty} e^{-\lambda\tau}(\lambda\tau)^k/k! + \tfrac{1}{2}\sum_{k=0}^{\infty} e^{-\lambda\tau}(-\lambda\tau)^k/k!$$

$$= \tfrac{1}{2}e^{-\lambda\tau}(e^{\lambda\tau} + e^{-\lambda\tau}) = \tfrac{1}{2}(1 + e^{-2\lambda\tau})$$

$P\{\text{an odd number of customers arrives in } (t, t + \tau]\} = \sum_{k=1,3,5,\ldots}^{\infty} e^{-\lambda\tau}(\lambda\tau)^k/k!$

$$= \tfrac{1}{2}\sum_{k=0}^{\infty} e^{-\lambda\tau}(\lambda\tau)^k/k! - \tfrac{1}{2}\sum_{k=0}^{\infty} e^{-\lambda\tau}(-\lambda\tau)^k/k!$$

$$= \tfrac{1}{2}e^{-\lambda\tau}(e^{\lambda\tau} - e^{-\lambda\tau}) = \tfrac{1}{2}(1 - e^{-2\lambda\tau})$$

[Alternatively, we may note that $\sum_{k=0,2,4,\ldots}^{\infty} \dfrac{(\lambda\tau)^k}{k!} = \cosh \lambda\tau$ and $\sum_{k=1,3,5,\ldots}^{\infty} \dfrac{(\lambda\tau)^k}{k!} = \sinh \lambda\tau$.]

3. (a) $P\{R_t = 1, R_{t+\tau} = 1\} = P\{R_t = -1, R_{t+\tau} = -1\} = \tfrac{1}{4}(1 + e^{-2\lambda\tau})$

   $P\{R_t = 1, R_{t+\tau} = -1\} = P\{R_t = -1, R_{t+\tau} = 1\} = \tfrac{1}{4}(1 - e^{-2\lambda\tau})$

   (b) $K(t, \tau) = e^{-2\lambda\tau}$

Section 6.6

4. (a) is immediate from Theorem 1.

   (b) If $\omega \in A_n$ for infinitely many n, then $\omega \in A$, which in turn implies that $\omega \in A_n$ eventually. Thus lim sup $A_n \subset A \subset$ lim inf $A_n \subset$ lim sup A_n, so all these sets are equal.

   (c) Let $A_n = [1 - 1/n, 2 - 1/n]$; lim sup $A_n =$ lim inf $A_n = [1, 2)$ (*Another example.* If the A_n are disjoint, lim sup $A_n =$ lim inf $A_n = \varnothing$.)

   (d) $\bigcap_{k=n}^{\infty} A_k \subset A_n \subset \bigcup_{k=n}^{\infty} A_k$, and $B_n = \bigcap_{k=n}^{\infty} A_k$ expands to $\lim_n$ inf $A_n = A$, $C_n = \bigcup_{k=n}^{\infty} A_k$ contracts to $\lim_n$ sup $A_n = A$. Thus $P(A_n)$ is boxed between $P(B_n)$ and $P(C_n)$, each of which approaches $P(A)$.

   (e) $(\lim_n$ inf $A_n)^c = (\bigcup_{n=1}^{\infty} \bigcap_{k=n}^{\infty} A_k)^c = \bigcap_{n=1}^{\infty} \bigcup_{k=n}^{\infty} A_k^c = \lim_n$ sup A_n^c by the DeMorgan laws; $(\lim_n$ sup $A_n)^c = \lim_n$ inf A_n^c similarly.

6. lim inf $A_n = \{x, y\colon x^2 + y^2 < 1\}$,

   lim sup $A_n = \{(x, y)\colon x^2 + y^2 \le 1\} - \{(0, 1), (0, -1)\}$.

8. $P(\lim_n \sup A_n) = \lim_{n\to\infty} P(\bigcup_{k=n}^\infty A_k)$ by definition of lim sup, hence

$$P(\lim_n \sup A_n) = \lim_{n\to\infty} \lim_{m\to\infty} P(\bigcup_{k=n}^m A_k)$$

Now

$$P\left(\bigcup_{k=n}^m A_k\right)^c = P\left(\bigcap_{k=n}^m A_k{}^c\right) = \prod_{k=n}^m P(A_k{}^c) \text{ by independence}$$

$$\le \prod_{k=n}^m e^{-P(A_k)} \text{ since } P(A_k{}^c) = 1 - P(A_k) \le e^{-P(A_k)}$$

$$= e^{-\sum_{k=n}^m P(A_k)} \to 0 \text{ as } m \to \infty \text{ since } \sum_n P(A_n) = \infty$$

The result follows.

9. If $\sum_{n=1}^\infty P\{|R_n - c| \ge \varepsilon\} < \infty$ for every $\varepsilon > 0$, $R_n \xrightarrow{a.s.} c$ by Theorem 5. Thus assume that $\sum_{n=1}^\infty P\{|R_n - c| \ge \varepsilon\} = \infty$ for some $\varepsilon > 0$. Then by the second Borel-Cantelli lemma, $P\{|R_n - c| \ge \varepsilon$ for infinitely many $n\} = 1$. But $|R_n - c| \ge \varepsilon$ for infinitely many n implies that $R_n \not\to c$, hence $P\{R_n \not\to c\} = 1$, that is, $P\{R_n \to c\} = 0$.

12. Let $S_n = (R_1 + \cdots + R_n)/n$; then $E(S_n/n)^2 = (1/n^2) \operatorname{Var} S_n$

$$= \frac{1}{n^2} \sum_{k=1}^n \operatorname{Var} R_k \le \frac{M}{n^2} \sum_{k=1}^n \frac{1}{k}$$

But

$$\frac{1}{2} + \frac{1}{3} + \cdots + \frac{1}{n} \le \int_1^n \frac{1}{x}\, dx = \ln n \left(\text{notice } \frac{1}{k} \le \frac{1}{x} \text{ if } k - 1 \le x \le k\right)$$

Hence

$$E\left(\frac{S_n}{n}\right)^2 \le \frac{M}{n^2}(1 + \ln n), \text{ so that}$$

$$\sum_{n=1}^\infty E\left(\frac{S_n}{n}\right)^2 < \infty. \text{ By Theorem 6, } \frac{S_n}{n} \xrightarrow{a.s.} 0$$

CHAPTER 7

Section 7.1

1. $P\{R_n = j \mid R_0 = r\} = \dfrac{P\{R_n = j, R_0 = r\}}{P\{R_0 = r\}}$

$$= \frac{\sum_{i_1,\ldots,i_{n-1}} P_r P_{ri_1} \cdots P_{i_{n-2}i_{n-1}} P_{i_{n-1}j}}{P_r} = p_{rj}^{(n)}$$

$$P\{T_n = j\} = \sum_{i_0,\ldots,i_{n-1}} P\{T_0 = i_0, \ldots, T_{n-1} = i_{n-1}, T_n = j\}$$

$$= \sum_{i_1,\ldots,i_{n-1}} q_r P_{ri_1} \cdots P_{i_{n-2}i_{n-1}} P_{i_{n-1}j} = p_{rj}^{(n)}$$

4. $P\{R_{n+1} = i_1, \ldots, R_{n+k} = i_k \mid R_n = i\}$

$$= P\{R_n = i, R_{n+1} = i_1, \ldots, R_{n+k} = i_k\}/P\{R_n = i\}$$

$$= \frac{\sum\limits_{j_0, \ldots, j_{n-1}} P_{j_0} P_{j_0 j_1} \cdots P_{j_{n-2} j_{n-1}} P_{j_{n-1} i} P_{i i_1} \cdots P_{i_{k-1} i_k}}{\sum\limits_{j_0, \ldots, j_{n-1}} P_{j_0} P_{j_0 j_1} \cdots P_{j_{n-2} j_{n-1}} P_{j_{n-1} i}} = P_{i i_1} \cdots P_{i_{k-1} i_k}$$

Section 7.2

1. The desired probability is

$$\frac{P[D \cap \{R_n = i, R_{n+1} = i_1, \ldots, R_{n+k} = i_k\}]}{P[D \cap \{R_n = i\}]}$$

D must be of the form $(R_0, \ldots, R_n) \in B$ for some $B \subset S^{n+1}$, hence D is a countable union of sets of the form $\{R_0 = j_0, \ldots, R_n = j_n\}$. Now

$$P\{R_0 = j_0, \ldots, R_n = j_n, R_n = i, \ldots, R_{n+k} = i_k\} = 0 \qquad \text{if} \qquad j_n \neq i$$

$$= P\{R_0 = j_0, \ldots, R_{n-1} = j_{n-1}, R_n = i\} P\{R_{n+1} = i_1, \ldots, R_{n+k} = i_k \mid R_n = i\}$$

if $j_n = i$

Thus the numerator is simply

$$P[D \cap \{R_n = i\}] P\{R_{n+1} = i_1, \ldots, R_{n+k} = i_k \mid R_n = i\}$$

and the result follows.

Section 7.3

1. (a) If $n_1 \in A$, then n_1 is a multiple of d, say $n_1 = r_1 d$. If $r_1 = 1$ then $\{n_1\}$ is the required set. If not choose $n_2 \in A$ such that n_1 does not divide n_2 (if n_2 does not exist then $d = n_1$, hence $r_1 = 1$). If $n_2 = r_1' d$ then $gcd(n_1, n_2) = gcd(r_1 d, r_1' d) = r_2 d$ for some positive integer $r_2 < r_1$; (if $r_2 = r_1$, then n_1 divides n_2). If $r_2 = 1$, $\{n_1, n_2\}$ is the required set, if not, find $n_3 \in A$ such that $r_2 d$ does not divide n_3. [If n_3 does not exist, then $r_2 d$ divides everything in A including n_1 and n_2. But $r_2 d = gcd(n_1, n_2)$ so that $r_2 d = d$, hence $r_2 = 1$, a contradiction.] If $n_3 = r_2' d$ then $gcd(n_1, n_2, n_3) = gcd(r_2 d, r_2' d) = r_3 d$, $r_3 < r_2$. If $r_3 = 1$, $\{n_1, n_2, n_3\}$ is the desired set. If not, find $n_4 \in A$ such that $r_3 d$ does not divide n_4, and continue in this fashion. We obtain a decreasing sequence of positive integers $r_1 > r_2 > \cdots$; the sequence must terminate in a finite number of steps, thus yielding a finite set whose gcd is d.

   (b) By (a) there are integers $n_1, \ldots, n_k \in A$, such that $gcd(n_1, \ldots, n_k) = d$. Thus for some integers $c_1, \ldots, c_k$ we have $c_1 n_1 + \cdots + c_k n_k = d$. Collect positive and negative terms to obtain, since A is closed under addition, $md, nd \in A$ such that $md - nd = d$, that is, $m - n = 1$. Now let $q = cd$, $c \geq n(n-1)$.

Write $c = an + b, a \geq n - 1, 0 \leq b \leq n - 1$. Then $q = cd = [(a - b)n + (bn + b)]d = [(a - b)n + bm]d \in A$, since $a - b \geq 0$ and A is closed under addition. The result follows.

(c) Let $A = \{n \geq 1 : p_{ii}^{(n)} > 0\}$, $B = \{n \geq 1 : f_{ii}^{(n)} > 0\}$. Then $d_i = gcd(A)$. If $e_i = gcd(B)$, then since $B \subset A$ we have $e_i \geq d_i$. To show that $d_i \geq e_i$ we show that e_i divides all elements of A. If this is not the case, let n be the smallest positive integer such that $p_{ii}^{(n)} > 0$ and e_i does not divide n. Write $n = ae_i + b, 0 < b < e_i$. Then

$$p_{ii}^{(n)} = \sum_{k=1}^{n} f_{ii}^{(k)} p_{ii}^{(n-k)} = \sum_{r=1}^{a} f_{ii}^{(re_i)} p_{ii}^{(ae_i+b-re_i)}$$

Now e_i does not divide $(a - r)e_i + b$, so by minimality of n,

$$p_{ii}^{(ae_i+b-re_i)} = 0 \text{ for all } r = 1, 2, \ldots, a.$$

But then $p_{ii}^{(n)} = 0$, a contradiction. The result follows.

4. (a) S forms a single aperiodic equivalence class. Starting from 1, the probability of returning to 1 is 1 if $p \leq q$, and $q + p(q/p) = 2q$ if $p > q$ (see 6.2.6). Thus S is recurrent if $p \leq q$, transient if $p > q$.

(b) and (c) S forms a single aperiodic equivalence class. By Theorem 6, S is recurrent.

(d) The equivalence classes are $C = \{1\}$ and $D = \{2, 3\}$. C is not closed, hence is transient; D is closed, hence recurrent. C and D are aperiodic.

Section 7.4

1. For each n so that $f_n > 0$, form the following system of states (always originating from a fixed state i). It is clear from the construction that $f_{ii}^{(n)} = f_n$ for all n.

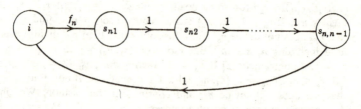

PROBLEM 7.4.1

Also, $p_{ii}^{(n)} = u_n$ by induction, using the First Entrance Theorem. *Note.* we need not have $gcd\{j : f_j > 0\} = 1$ for this construction.

2. Construct a Markov chain such that for some i,

$$f_{ii}^{(n)} = P\{T_1 = n\}, \qquad n = 1, 2, \ldots$$

We claim that $G(n) = p_{ii}^{(n)}$ for all $n = 1, 2, \ldots$. For $G(1) = P\{T_1 = 1\} = f_{ii}^{(1)} = p_{ii}^{(1)}$, and if $G(r) = p_{ii}^{(r)}$ for $r = 1, 2, \ldots, n$, then

$$G(n + 1) = \sum_{k=1}^{\infty} P\{T_1 + \cdots + T_k = n + 1\}$$

$$= \sum_{k=1}^{\infty} \sum_{l=1}^{n+1} P\{T_1 = l\} P\{T_2 + \cdots + T_k = n + 1 - l\}$$

$$= \sum_{l=1}^{n+1} f_{ii}^{(l)} G(n + 1 - l), \text{ if we define } G(0) = 1$$

$$= \sum_{l=1}^{n+1} f_{ii}^{(l)} p_{ii}^{(n+1-l)} \text{ by induction hypothesis}$$

$$= p_{ii}^{(n+1)} \text{ by the First Entrance Theorem}$$

Now state i is recurrent since $\sum_{n=1}^{\infty} P\{T_1 = n\} = 1$, and has period d by hypothesis. Thus by Theorem 2(c), $\lim_{n \to \infty} G(nd) = d/\mu$.

Since a renewal can only take place at times nd, $n = 1, 2, \ldots$, $G(nd)$ is the probability that a renewal takes place in the interval $[nd, (n + 1) d)$. If the average length of time between renewals is μ, for large n it is reasonable that one renewal should take place every μ seconds, hence there should be, on the average, d/μ renewals in a time interval of length d. Thus we expect intuitively that $G(nd) \to d/\mu$.

4. (a) Let the initial state be i. Then $V_{ij} = \sum_{n=0}^{\infty} I_{\{R_n = j\}}$, and the result follows.

(b) By (a), $N = \sum_{n=0}^{\infty} Q^n$ so that $QN = \sum_{n=1}^{\infty} Q^n = N - I$. (In particular, QN is finite.)

(c) By (b), $(I - Q)N = I$. But $N = \sum_{n=0}^{\infty} Q^n$ so that $QN = NQ$, hence $N(I - Q) = I$.

Section 7.5

2. (a) j_0 is recurrent since $p_{ij_0}^{(n)} \geq \delta$ for all $n \geq N$ (see Problem 2, Section 7.1), hence

$$\sum_{n=1}^{\infty} p_{j_0 j_0}^{(n)} = \infty$$

If i is any recurrent state then since $p_{ij_0}^{(N)} \geq \delta > 0$, i leads to j_0. By Theorem 5 of Section 7.3, j_0 leads to i, so that there can only be one recurrent class. Since $p_{ij_0}^{(n)} \geq \delta > 0$ for all $n \geq N$, the class is aperiodic, so that $\lim_n p_{j_0 j_0}^{(n)} = 1/\mu_{j_0}$. But then $1/\mu_{j_0} \geq \delta > 0$, hence $\mu_{j_0} < \infty$ and the class is positive.

(Note also that if i is any state and C is the equivalence class of j_0, then, for $n \geq N$, $P\{R_n \notin C \mid R_0 = i\} \leq 1 - \delta$, hence $P\{R_{kn} \notin C \mid R_0 = i\} \leq (1 - \delta)^k \to 0$ as $k \to \infty$. Thus $f_{ij_0} = 1$, and it follows that a steady state distribution exists.)

(b) If $\prod^N$ has a positive column then by (a) there is exactly one recurrent class, which is (positive and) aperiodic, and therefore a steady state distribution exists. Conversely, let $\{v_j\}$ be a steady state distribution. Pick j_0 so that $v_{j_0} [= \lim_{n \to \infty} p_{ij_0}^{(n)}] > 0$. Since the chain is finite, $p_{ij_0}^{(n)} > 0$ for all i if n is sufficiently large, say $n \geq N$. But then $\prod^N$ has a positive column.

3.1. If $p \neq q$, the chain is transient, so $p_{ij}^{(n)} \to 0$ for all i, j. If $p = q$ the chain is recurrent. We have observed (Problem 5, Section 6.3) that the mean recurrence time is infinite, hence the chain is recurrent null, and thus $p_{ij}^{(n)} \to 0$. In either case there is no stationary distribution, hence no steady state distribution. The period is 2.

2. There is one positive recurrent class, namely $\{0\}$; the remaining states form a transient class. Thus there is a unique stationary distribution, given by $v_0 = 1$, $v_j = 0, j \geq 1$. Now starting from $i \geq 1$, the probability of eventually reaching 0 is $\lim_{n \to \infty} p_{i0}^{(n)}$, since the events $\{R_n = 0\}$ expand to $\{0 \text{ is reached eventually}\}$. By (6.2.6),

$$\lim_{n \to \infty} p_{i0}^{(n)} = (q/p)^i \quad \text{if} \quad p > q$$

$$= 1 \quad \text{if} \quad p \leq q$$

(Also $p_{00}^{(n)} \equiv 1$, $p_{ij}^{(n)} \to 0, j \geq 1$). If $p > q$ the limit is not independent of i so there is no steady state distribution.

3. There are two positive recurrent classes $\{0\}$ and $\{b\}$. $\{1, 2, \ldots, b - 1\}$ is a transient class. Thus, there are uncountably many stationary distributions, given by $v_0 = p_1$, $v_0 = p_2$, $v_i = 0, 1 \leq i \leq b - 1$, where $p_1, p_2 \geq 0, p_1 + p_2 = 1$. There is no steady state distribution. By (6.2.3) and (6.2.4),

$$\lim_{n \to \infty} p_{i0}^{(n)} = \frac{(q/p)^i - (q/p)^b}{1 - (q/p)^b} \quad \text{if} \quad p \neq q$$

$$= 1 - \frac{i}{b} \quad \text{if} \quad p = q$$

$$\lim_{n \to \infty} p_{ib}^{(n)} = 1 - \lim_{n \to \infty} p_{i0}^{(n)}$$

$$\lim_{n \to \infty} p_{ij}^{(n)} = 0, \quad 1 \leq j \leq b - 1$$

4. The chain is aperiodic. If $p > q$ then $f_{i1} = (q/p)^{i-1} < 1, i > 1$, hence the chain is transient. Therefore $p_{ij}^{(n)} \to 0$ as $n \to \infty$ for all i, j, and there is no stationary

or steady state distribution. Now if $p \leq q$ then $f_{i1} = 1$ for $i > 1$, hence $f_{11} = q + pf_{i1} = 1$, and the chain is recurrent. The equations $V\Pi = V$ become

$$v_1 q + v_2 q = v_1$$
$$v_1 p + v_3 q = v_2$$
$$v_2 p + v_4 q = v_3$$
$$\vdots$$

This may be reduced to $v_j = (p/q)v_{j-1}$, $j = 2, 3, \ldots$. If $p = q$ then all v_j are equal, hence $v_j \equiv 0$ and there is no stationary or steady state distribution. Thus the chain is recurrent null. If $p < q$, the condition $\sum_{j=1}^{\infty} v_j = 1$ yields the unique solution

$$v_j = \frac{(q-p)}{q}\left(\frac{p}{q}\right)^{j-1}, \qquad j = 1, 2, \ldots$$

Thus there is a unique stationary distribution, so that the chain is recurrent positive; $\{v_j\}$ is also the steady state distribution.

5. The chain forms a recurrent positive aperiodic class, hence $p_{ij}^{(n)} \to v_j$ where the v_j form the unique stationary distribution and the steady state distribution. The equations $V\Pi = V$, $\sum_j v_j = 1$ yield

$$v_j = \frac{(p/q)^{j-1}}{\sum_{j=1}^{l}(p/q)^{j-1}}$$

6. The chain forms a recurrent positive aperiodic class. Since Π^2 has identical rows $(p^2, pq, qp, q^2) = V$, there is a steady state distribution ($=$ the unique stationary distribution), namely V.

7. The chain forms a recurrent positive aperiodic class, hence $p_{ij}^{(n)} \to v_j$ where the v_j form the unique stationary distribution and the steady state distribution. The equations $V\Pi = V$, $\sum_j v_j = 1$ yield

$$v_1 = \tfrac{7}{24}, \qquad v_2 = \tfrac{1}{3}, \qquad v_3 = \tfrac{1}{24}, \qquad v_4 = \tfrac{1}{3}$$

8. There is a single positive recurrent class $\{2, 3\}$, which is aperiodic, hence $p_{ij}^{(n)} \to v_j$ where the v_j form the unique stationary distribution and the steady state distribution. We find that $v_1 = 0$, $v_2 = 3/7$, $v_3 = 4/7$.

9. We may take $p_{ij} = P\{R_n = j\}$ for all i, j (with initial distribution $p_j = P\{R_n = j\}$ also). The chain forms a recurrent class since from any initial state, $P\{R_n$ never $= j\} = \prod_{n=1}^{\infty} P\{R_n = j\} = \prod_{n=1}^{\infty} p_j = 0$. The class is aperiodic. Clearly $v_j = p_j$ is a stationary distribution, so that the chain is recurrent positive and the stationary distribution is unique and coincides with the steady state distribution.

10. The chain forms a positive recurrent class of period 3 (see Section 7.3, Example 2). Thus there is a unique stationary distribution given by

$$v_1 = \tfrac{1}{9}, \quad v_2 = \tfrac{2}{9}, \quad v_3 = \tfrac{1}{9}, \quad v_4 = \tfrac{2}{9}, \quad v_5 = \tfrac{1}{12}, \quad v_6 = \tfrac{1}{36}, \quad v_7 = \tfrac{2}{9}$$

Now the cyclically moving subclasses are $C_0 = \{1, 2\}$, $C_1 = \{3, 4\}$, $C_2 = \{5, 6, 7\}$. By Theorem 2(c) of Section 7.4, if $i \in C_r, j \in C_{r+a}$ then $p_{ij}^{(3n+a)} \to 3v_j$. Thus

$$\Pi^{3n} \to \begin{array}{c} \\ 1 \\ 2 \\ 3 \\ 4 \\ 5 \\ 6 \\ 7 \end{array} \begin{array}{ccccccc} 1 & 2 & 3 & 4 & 5 & 6 & 7 \\ \left[\begin{array}{ccccccc} \tfrac{1}{3} & \tfrac{2}{3} & 0 & 0 & 0 & 0 & 0 \\ \tfrac{1}{3} & \tfrac{2}{3} & 0 & 0 & 0 & 0 & 0 \\ 0 & 0 & \tfrac{1}{3} & \tfrac{2}{3} & 0 & 0 & 0 \\ 0 & 0 & \tfrac{1}{3} & \tfrac{2}{3} & 0 & 0 & 0 \\ 0 & 0 & 0 & 0 & \tfrac{1}{4} & \tfrac{1}{12} & \tfrac{2}{3} \\ 0 & 0 & 0 & 0 & \tfrac{1}{4} & \tfrac{1}{12} & \tfrac{2}{3} \\ 0 & 0 & 0 & 0 & \tfrac{1}{4} & \tfrac{1}{12} & \tfrac{2}{3} \end{array}\right] \end{array}$$

$$\Pi^{3n+1} \to \begin{array}{c} \\ 1 \\ 2 \\ 3 \\ 4 \\ 5 \\ 6 \\ 7 \end{array} \begin{array}{ccccccc} 1 & 2 & 3 & 4 & 5 & 6 & 7 \\ \left[\begin{array}{ccccccc} 0 & 0 & \tfrac{1}{3} & \tfrac{2}{3} & 0 & 0 & 0 \\ 0 & 0 & \tfrac{1}{3} & \tfrac{2}{3} & 0 & 0 & 0 \\ 0 & 0 & 0 & 0 & \tfrac{1}{4} & \tfrac{1}{12} & \tfrac{2}{3} \\ 0 & 0 & 0 & 0 & \tfrac{1}{4} & \tfrac{1}{12} & \tfrac{2}{3} \\ \tfrac{1}{3} & \tfrac{2}{3} & 0 & 0 & 0 & 0 & 0 \\ \tfrac{1}{3} & \tfrac{2}{3} & 0 & 0 & 0 & 0 & 0 \\ \tfrac{1}{3} & \tfrac{2}{3} & 0 & 0 & 0 & 0 & 0 \end{array}\right] \end{array}$$

$$\Pi^{3n+2} \to \begin{array}{c} \\ 1 \\ 2 \\ 3 \\ 4 \\ 5 \\ 6 \\ 7 \end{array} \begin{array}{ccccccc} 1 & 2 & 3 & 4 & 5 & 6 & 7 \\ \left[\begin{array}{ccccccc} 0 & 0 & 0 & 0 & \tfrac{1}{4} & \tfrac{1}{12} & \tfrac{2}{3} \\ 0 & 0 & 0 & 0 & \tfrac{1}{4} & \tfrac{1}{12} & \tfrac{2}{3} \\ \tfrac{1}{3} & \tfrac{2}{3} & 0 & 0 & 0 & 0 & 0 \\ \tfrac{1}{3} & \tfrac{2}{3} & 0 & 0 & 0 & 0 & 0 \\ 0 & 0 & \tfrac{1}{3} & \tfrac{2}{3} & 0 & 0 & 0 \\ 0 & 0 & \tfrac{1}{3} & \tfrac{2}{3} & 0 & 0 & 0 \\ 0 & 0 & \tfrac{1}{3} & \tfrac{2}{3} & 0 & 0 & 0 \end{array}\right] \end{array}$$

CHAPTER 8

Section 8.2

1. (a) $L(x) = 2e^{-2x}/e^{-x} = 2e^{-x}$, so $L(x) > \lambda$ iff $x < c = -\ln \lambda/2$. Thus

$$\alpha = \int_0^c e^{-x}\, dx = 1 - e^{-c}$$

$$\beta = \int_c^\infty 2e^{-2x}\, dx = e^{-2c} = (1 - \alpha)^2$$

Hence as in Example 1 of the text, $S_A = \{(\alpha, (1 - \alpha)^2), 0 \leq \alpha \leq 1\}$ and $S = \{(\alpha, \beta): 0 \leq \alpha \leq 1, (1 - \alpha)^2 \leq \beta \leq 1 - \alpha^2\}$.

(b) $e^{-c} = 1 - \alpha = .95$, so that $c = .051$. Thus we reject H_0 if $x < .051$, accept H_0 if $x > .051$. We have $\beta = (1 - \alpha)^2 = .9025$, which indicates that tests based on a single observation are not very promising here.

(c) Set $\alpha = \beta = (1 - \alpha)^2$; thus $\alpha = (3 - \sqrt{5})/2 = .38 = 1 - e^{-c}$, so that $c = .477$.

3. (a)

x	1	2	3	4	5	6
$p_0(x)$	$\frac{1}{6}$	$\frac{1}{6}$	$\frac{1}{6}$	$\frac{1}{6}$	$\frac{1}{6}$	$\frac{1}{6}$
$p_1(x)$	$\frac{1}{4}$	$\frac{1}{4}$	$\frac{1}{8}$	$\frac{1}{8}$	$\frac{1}{8}$	$\frac{1}{8}$
$L(x)$	$\frac{3}{2}$	$\frac{3}{2}$	$\frac{3}{4}$	$\frac{3}{4}$	$\frac{3}{4}$	$\frac{3}{4}$

LRT	Rejection Region	Acceptance Region	α	β
$0 \leq \lambda < \frac{3}{4}$	all x	empty	1	0
$\frac{3}{4} < \lambda < \frac{3}{2}$	$x = 1, 2$	$x = 3, 4, 5, 6$	$\frac{1}{3}$	$\frac{1}{2}$
$\frac{3}{2} < \lambda \leq \infty$	empty	all x	0	1

The admissible risk points are given in the diagram.

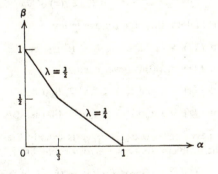

PROBLEM 8.2.3(a)

(b) Reject with probability a if $x = 1$ or 2, accept if $x = 3, 4, 5$ or 6, where $a/3 = .1$, that is, $a = .3$. We have $1 - \beta = .3(1/2) = .15$, or $\beta = .85$.

(c) $\lambda = pc_1/(1 - p)c_2 = 3/2$, so reject with probability a if $x = 1$ or 2, accept if $x = 3, 4, 5, 6$, where a is any number in $[0, 1]$. Thus there are uncountably many Bayes solutions.

4. $n \geq 13$.

5. By Problem 2d, the test is of the form

$$\varphi(x) = 1 \text{ if } \sum_{k=1}^{n} x_k{}^2 > c$$

$$= 0 \text{ if } \sum_{k=1}^{n} x_k{}^2 < c$$

$$= \text{anything if } \sum_{k=1}^{n} x_k{}^2 = c$$

Now if the true variance is θ, $\sum_{k=1}^{n} (R_k{}^2)/\theta$ has the chi-square density h_n with n degrees of freedom (see Problem 3, Section 5.2), hence

$$P_\theta\left\{x: \sum_{k=1}^{n} x_k{}^2 > c\right\} = \int_{c/\theta}^{\infty} h_n(x)\, dx = A_n(c/\theta)$$

(The numbers A_n are tabulated in most statistics books.) Thus the error probabilities are given by

$$\alpha = A_n(c/\theta_0)$$

$$1 - \beta = A_n(c/\theta_1)$$

For a given value of n, the specification of α determines c, which in turn determines β. In practice, one must keep trying larger values of n until β is reduced to the desired figure.

6. First consider $\theta = \theta_0$ versus $\theta = \theta_1$, $\theta_1 > \theta_0$.

$$L(x) = f_{\theta_1}(x)/f_{\theta_0}(x) = (\theta_0/\theta_1)^n \text{ if } 0 \leq t(x) = \max x_i \leq \theta_0$$

$$= \infty \text{ if } \theta_0 < t(x) \leq \theta_1$$

Let $\lambda = (\theta_0/\theta_1)^n$ and consider the following test

$$\varphi(x) = 1 \text{ if } L(x) > \lambda, \text{ that is, if } \theta_0 < t(x) \leq \theta_1$$

$$= 0 \text{ if } L(x) < \lambda \text{ (this never occurs)}$$

$$= 1 \text{ if } L(x) = \lambda \text{ and } t(x) \leq \theta_0\alpha^{1/n}, \text{ that is, if } 0 \leq t(x) \leq \theta_0\alpha^{1/n}$$

$$= 0 \text{ if } L(x) = \lambda \text{ and } t(x) > \theta_0\alpha^{1/n}, \text{ that is, if } \theta_0\alpha^{1/n} < t(x) \leq \theta_0$$

Since $t(x)$ can never be < 0 or $> \theta_1$, φ is exactly the test proposed in the statement of the problem. Its type 1 error probability is

$$P_{\theta_0}\{x: \max x_i \leq \theta_0\alpha^{1/n}\} = (\theta_0\alpha^{1/n}/\theta_0)^n = \alpha$$

Since φ is a LRT, it is most powerful at level α. But φ does not depend on θ_1, hence is UMP for $\theta = \theta_0$ versus $\theta > \theta_0$.

Now let $\theta_1 < \theta_0$. Then

$$L(x) = (\theta_0/\theta_1)^n \text{ if } 0 \le t(x) \le \theta_1;$$
$$= 0 \text{ if } \theta_1 < t(x) \le \theta_0$$

Let $\lambda = (\theta_0/\theta_1)^n$ and consider the following test.

$$\varphi'(x) = 1 \text{ if } L(x) > \lambda \text{ (this never occurs)}$$
$$= 0 \text{ if } L(x) < \lambda$$
$$= 1 \text{ if } L(x) = \lambda \text{ and } t(x) \le \theta_0 \alpha^{1/n}$$
$$= 0 \text{ if } L(x) = \lambda \text{ and } t(x) > \theta_0 \alpha^{1/n}$$

Since $t(x)$ cannot be $> \theta_0$ in this case, $\varphi' \equiv \varphi$. Again, φ is UMP for $\theta = \theta_0$ versus $\theta < \theta_0$, and the result follows. The power function is (see diagram)

$$Q(\theta) = E_\theta \varphi = 1 \text{ if } 0 < \theta \le \theta_0 \alpha^{1/n}$$
$$= (\theta_0 \alpha^{1/n}/\theta)^n = \alpha(\theta_0/\theta)^n,\ \theta_0 \alpha^{1/n} \le \theta \le \theta_0$$
$$= 1 - P_\theta\{x: \theta_0 \alpha^{1/n} < t(x) \le \theta_0\}$$
$$= 1 - [(\theta_0/\theta)^n - (\theta_0 \alpha^{1/n}/\theta)^n]$$
$$= 1 - (1 - \alpha)(\theta_0/\theta)^n,\ \theta > \theta_0$$

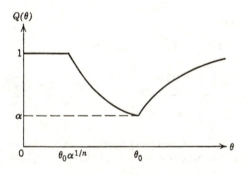

PROBLEM 8.2.6

7. The risk set is $\{(\alpha, \beta): 0 \le \alpha \le 1,\ (1 - \alpha)2^{-n} \le \beta \le 1 - \alpha 2^{-n}\}$, and the set of admissible risk points is $\{(\alpha, (1 - \alpha)2^{-n}): 0 \le \alpha \le 1\}$.

10. If $\alpha(\varphi) = \alpha < \alpha_0$, let $\varphi' \equiv 1$ and $\varphi_t = (1 - t)\varphi + t\varphi'$, $0 \le t \le 1$. Then

$$\alpha(\varphi_t) = (1 - t)\alpha(\varphi) + t\alpha(\varphi')$$
$$\beta(\varphi_t) = (1 - t)\beta(\varphi) + t\beta(\varphi')$$

Since $\alpha(\varphi) < \alpha_0$, $\alpha(\varphi_t)$ will be $< \alpha_0$ for some $t \in (0, 1)$. But $\beta(\varphi') = 0$ and $\beta(\varphi) > 0$ hence $\beta(\varphi_t) < \beta(\varphi)$, contradicting the assumption that φ is most powerful at level α_0.

For the counterexample, let R be uniformly distributed between a and b, and let H_0: $a = 0$, $b = 1$, H_1: $a = 2$, $b = 3$. Let $\varphi_t(x) = 1$ if $x \geq t$, $\varphi_t(x) = 0$ otherwise, where $0 \leq t \leq 1$. Then $\beta(\varphi_t) = 0$, $\alpha(\varphi_t) = 1 - t$. For $t < 1$, φ_t is most powerful at level $\alpha_0 = 1$, but is of size <1.

14. (a) φ_n is Bayes with $c_1 = c_2 = 1$, $p = 1/2$ (hence $\lambda = 1$), and $L(x) = f_{\theta_1}(x)/f_{\theta_0}(x)$ where $f_\theta(x) = \prod_{i=1}^n h_\theta(x_i)$; the result follows.

(b) $P_{\theta_0}\{x: g_n(x) > 1\} = P_{\theta_0}\{g_n(R) > 1\} \leq E_{\theta_0}[g_n(R)^{1/2}]$ by Chebyshev's inequality

$$= \prod_{i=1}^n E_{\theta_0} t(R_i) = [E_{\theta_0} t(R_1)]^n$$

since all R_i have the same density.

(c) $\text{Var}_{\theta_0} t(R_1) = E_{\theta_0} t^2(R_1) - [E_{\theta_0} t(R_1)]^2 > 0$ (assuming $h_{\theta_1} \neq h_{\theta_0}$, so $E_{\theta_0} t(R_1) < [E_{\theta_0} t^2(R_1)]^{1/2}$. But

$$E_{\theta_0} t^2(R_1) = E_{\theta_0}[h_{\theta_1}(R_1)/h_{\theta_0}(R_1)] = \int_{\{x_i: h_{\theta_0}(x_i) > 0\}} \frac{h_{\theta_1}(x_i)}{h_{\theta_0}(x_i)} h_{\theta_0}(x_i)\, dx_i \leq 1$$

Section 8.3

1. (a) $\hat\theta = -n / \sum_{i=1}^n \ln x_i$ (b) $\hat\theta = \bar x$ (c) $\hat\theta = \max(x_1, \ldots, x_n)$

2. $\hat\theta = |x|$

3. $\hat\theta = r/x$

4. $\rho(\theta) = \theta(1 - \theta)/n$

5. By (8.3.2) with $g(\theta) = 1$, $0 \leq \theta \leq 1$, we have

$$\psi(x) = \frac{\int_0^1 \binom{n}{x} \theta^{x+1}(1 - \theta)^{n-x}\, d\theta}{\int_0^1 \binom{n}{x} \theta^x (1 - \theta)^{n-x}\, d\theta} = \frac{\beta(x + 2, n - x + 1)}{\beta(x + 1, n - x + 1)} = \frac{x + 1}{n + 2}$$

6. For each x, we wish to minimize $\int_0^1 \binom{n}{x} \theta^x (1 - \theta)^{n-x}[(\theta - \psi(x))^2/\theta(1 - \theta)]\, d\theta$ [see (8.3.1)]. In the same way that we derived (8.3.2), we find that

$$\psi(x) = \frac{\int_0^1 \theta^x (1 - \theta)^{n-x-1}\, d\theta}{\int_0^1 \theta^{x-1}(1 - \theta)^{n-x-1}\, d\theta} = \frac{\beta(x + 1, n - x)}{\beta(x, n - x)} = \frac{x}{n}$$

The risk function is

$$\rho_\psi(\theta) = E_\theta\left[\frac{[(R/n) - \theta]^2}{\theta(1 - \theta)}\right] = \frac{1}{n^2\theta(1 - \theta)} \text{Var}_\theta R = \frac{1}{n} = \text{constant}$$

Section 8.4

1. $f_\theta(x_1, \ldots, x_n) = (\theta_2 - \theta_1)^{-n} \prod_{i=1}^n I_{[\theta_1, \theta_2]}(x_i)$, where I is an indicator function

$$= (\theta_2 - \theta_1)^{-n} I_{[\theta_1, \infty)}(\min x_i) I_{(-\infty, \theta_2]}(\max x_i)$$

Thus in (a), $(\min R_i, \max R_i)$ is sufficient; in (b), $\max R_i$ is sufficient; in (c) $\min R_i$ is sufficient

2.

$$f_\theta(x_1, \ldots, x_n) = \frac{\left(\prod_{i=1}^n x_i \right)^{\theta_1 - 1} e^{-\sum_{i=1}^n x_i/\theta_2}}{[\Gamma(\theta_1)\theta_2^{\theta_1}]^n}$$

hence if θ_1, θ_2 are both unknown, $(\prod_{i=1}^n R_i, \sum_{i=1}^n R_i)$ is sufficient; if θ_1 is known, $\sum_{i=1}^n R_i$ is sufficient; if θ_2 is known, $\prod_{i=1}^n R_i$ is sufficient.

3. $[\prod_{i=1}^n R_i, \prod_{i=1}^n (1 - R_i)]$ is sufficient if θ_1 and θ_2 are unknown; if θ_1 is known, $\prod_{i=1}^n (1 - R_i)$ is sufficient, and if θ_2 is known, $\prod_{i=1}^n R_i$ is sufficient.

Section 8.5

1. $(1 - 1/n)^T$.

2. (a) T has density $f_T(y) = n y^{n-1}/\theta^n$, $0 \le y \le \theta$ (Example 3, Section 2.8), so $E_\theta g(T) = \int_0^\theta g(y) f_T(y) \, dy = (n/\theta^n) \int_0^\theta y^{n-1} g(y) \, dy$. If $E_\theta g(T) = 0$ for all $\theta > 0$ then $y^{n-1} g(y) = 0$, hence $g(y) = 0$, for all y (except on a set of Lebesgue measure 0). Thus

$$P_\theta\{g(T) = 0\} = \int_{\{y : g(y) = 0\}} f_T(y) \, dy = 1$$

(b) If $g(T)$ is an unbiased estimate of $\gamma(\theta)$ then

$$E_\theta g(T) = \frac{n}{\theta^n} \int_0^\theta y^{n-1} g(y) \, dy = \gamma(\theta)$$

Assuming g continuous we have

$$\theta^{n-1} g(\theta) = \frac{d}{d\theta} \left[\frac{\theta^n \gamma(\theta)}{n} \right] \quad \text{or} \quad g(\theta) = \gamma(\theta) + \frac{\theta}{n} \gamma'(\theta)$$

Conversely, if g satisfies this equation then $n\theta^{n-1}g(\theta) = d/d\theta[\theta^n \gamma(\theta)]$ hence $n \int_0^\theta y^{n-1} g(y) \, dy = \theta^n \gamma(\theta)$, assuming $\theta^n \gamma(\theta) \to 0$ as $\theta \to 0$. Thus a UMVUE of $\gamma(\theta)$ is given by $g(T) = \gamma(T) + (T/n)\gamma'(T)$. For example, if $\gamma(\theta) = \theta$ then $g(T) = T + T/n = [(n + 1)/n]T$; if $\gamma(\theta) = 1/\theta$ then $g(T) = 1/T + (T/n)(-1/T^2) = (1/T)[1 - (1/n)]$, assuming that $n > 1$.

4. We have

$$P_N\{R_1 = x_1, \ldots, R_n = x_n\} = \frac{1}{N^n} \prod_{i=1}^{n} I_{\{1,2 \ldots\}}(x_i) I_{\{1,2, \ldots, N\}}(\max x_i)$$

hence $T = \max R_i$ is sufficient. Now

$$P_N\{T \le k\} = \left(\frac{k}{N}\right)^n, \qquad k = 1, 2, \ldots, N;$$

therefore

$$P_N\{T = k\} = \frac{k^n - (k-1)^n}{N^n}, \qquad k = 1, 2, \ldots, N$$

Thus

$$E_N g(T) = \sum_{k=1}^{N} g(k) \left[\frac{k^n - (k-1)^n}{N^n}\right]$$

If $E_N g(T) = 0$ for all $N = 1, 2, \ldots$, take $N = 1$ to conclude that $g(1) = 0$.
If $g(k) = 0$ for $k = 1, \ldots, N-1$, then $E_N g(T) = 0$ implies that

$$g(N) \left[\frac{N^n - (N-1)^n}{N^n}\right] = 0$$

hence $g(N) = 0$. By induction, $g \equiv 0$ and T is complete. To find a UMVUE of $\gamma(N)$, we must solve the equation

$$\sum_{k=1}^{N} g(k)[k^n - (k-1)^n] = N^n \gamma(N), \qquad N = 1, 2, \ldots$$

or

$$g(N)[N^n - (N-1)^n] = N^n \gamma(N) - (N-1)^n \gamma(N-1)$$

Thus

$$g(N) = \frac{N^n \gamma(N) - (N-1)^n \gamma(N-1)}{N^n - (N-1)^n}, \qquad N = 1, 2, \ldots$$

5. R is clearly sufficient for itself, and

$$E_\theta g(R) = \sum_{k=r}^{\infty} g(k) \binom{k-1}{r-1} (1 - \theta)^r \theta^{k-r}$$

If $E_\theta g(R) = 0$ for all $\theta \in [0, 1)$ then $\sum_{k=r}^{\infty} g(k)\binom{k-1}{r-1}\theta^{k-r} \equiv 0$, so that $g \equiv 0$.
Thus R is complete. The above expression for $E_\theta g(R)$ shows that for a UMVUE to exist, $\gamma(\theta)$ must be expandable in a power series. Conversely, let $\gamma(\theta) = \sum_{i=0}^{\infty} a_i \theta^i$, $0 \le \theta < 1$. We must find g such that

$$\sum_{k=r}^{\infty} g(k) \binom{k-1}{r-1} \theta^{k-r} = (1 - \theta)^{-r} \gamma(\theta) = \sum_{i=0}^{\infty} b_i \theta^i = \sum_{i=r}^{\infty} b_{i-r} \theta^{i-r}$$

Therefore

$$g(i) = \frac{b_{i-r}}{\binom{i-1}{r-1}}, \qquad i = r, r+1, \ldots$$

For example, if $\gamma(\theta) = \theta^k$ then

$$(1 - \theta)^{-r}\gamma(\theta) = \theta^k \sum_{j=0}^{\infty} (-1)^j \binom{-r}{j} \theta^j = \sum_{j=0}^{\infty} \binom{j+r-1}{r-1} \theta^{k+j} \quad \text{(Problem 6a,}$$

Section 6.4)

$$= \sum_{i=k}^{\infty} \binom{i+r-1-k}{r-1} \theta^i$$

Thus $b_i = 0$ for $i < k$, and $b_i = \binom{i+r-1-k}{r-1}$ for $i \geq k$

$$g(i) = \frac{\binom{i-1-k}{r-1}}{\binom{i-1}{r-1}}, \qquad i = r+k, r+k+1, \ldots$$

$$= 0 \text{ otherwise}$$

In particular, if $k = 1$ then

$$g(i) = \frac{\binom{i-2}{r-1}}{\binom{i-1}{r-1}} = \frac{i-r}{i-1}, \quad i \geq r+1$$

Thus a UMVUE of $\theta = 1 - p$ is $(R - r)/(R - 1)$; a UMVUE of $p = 1 - \theta$ is $1 - (R - r)/(R - 1) = (r - 1)/(R - 1)$ (The maximum likelihood estimate of p is r/R, which is biased; see Problem 3, Section 8.3.)

6. (a) $E_\theta \psi(R) = \dfrac{e^{-\theta}}{1 - e^{-\theta}} \displaystyle\sum_{k=1}^{\infty} \psi(k) \dfrac{\theta^k}{k!} = e^{-\theta}$

Thus

$$\sum_{k=1}^{\infty} \psi(k) \frac{\theta^k}{k!} = 1 - e^{-\theta} = \sum_{k=1}^{\infty} (-1)^{k-1} \frac{\theta^k}{k!}$$

The UMVUE is given by

$$\psi(k) = (-1)^{k-1} = -1 \text{ if } k \text{ is even}$$
$$= +1 \text{ if } k \text{ is odd}$$

(b) Since $e^{-\theta}$ is always >0, the estimate found in part (a) looks rather silly. If $\psi'(k) \equiv 1$ then $E_\theta\{[\psi'(R) - e^{-\theta}]^2\} < E_\theta\{[\psi(R) - e^{-\theta}]^2\}$ for all θ, hence ψ is inadmissible.

9.

$$E_\theta\left[\left(\frac{2}{n}\sum_{i=1}^{n} R_i - \theta\right)^2\right] = \mathrm{Var}_\theta\left(\frac{2}{n}\sum_{i=1}^{n} R_i\right) = \frac{4}{n}\mathrm{Var}_\theta\, R_i = \frac{\theta^2}{3n}$$

$$E_\theta\left[\left(\frac{(n+1)}{n} T - \theta\right)^2\right] = \mathrm{Var}_\theta\left[\frac{(n+1)}{n} T\right] = \frac{(n+1)^2}{n^2}\mathrm{Var}_\theta\, T$$

$$E_\theta T = \frac{n}{\theta^n}\int_0^\theta y^n\, dy = \frac{n}{n+1}\,\theta \text{ (see Problem 2b),}$$

and

$$E_\theta T^2 = \frac{n}{\theta^n}\int_0^\theta y^{n+1}\, dy = \frac{n\theta^2}{n+2}$$

Thus

$$\mathrm{Var}_\theta\, T = \theta^2\left[\frac{n}{n+2} - \frac{n^2}{(n+1)^2}\right] = \frac{n\theta^2}{(n+1)^2(n+2)}$$

Therefore

$$E_\theta\left[\frac{(n+1)}{n} T - \theta\right]^2 = \frac{\theta^2}{n(n+2)} < \frac{\theta^2}{3n}$$

if $n > 1$

11. In the inequality $[(a+b)/2]^2 \le (a^2 + b^2)/2$, set $a = \psi_1(R) - \gamma(\theta)$, $b = \psi_2(R) - \gamma(\theta)$, to obtain

$$E_\theta\left[\frac{\psi_1(R) - \gamma(\theta)}{2} + \frac{\psi_2(R) - \gamma(\theta)}{2}\right]^2 \le \tfrac{1}{2}[\rho_{\psi_1}(\theta) + \rho_{\psi_2}(\theta)] = \rho_{\psi_1}(\theta)$$

By minimality, we actually have equality. But the left side is

$$\tfrac{1}{4}[\rho_{\psi_1}(\theta) + \rho_{\psi_2}(\theta) + 2E\{[\psi_1(R) - \gamma(\theta)][\psi_2(R) - \gamma(\theta)]\}$$

$$= \tfrac{1}{2}\rho_{\psi_1}(\theta) + \tfrac{1}{2}\mathrm{Cov}_\theta\,[\psi_1(R), \psi_2(R)]$$

Thus

$$\mathrm{Cov}_\theta\,[\psi_1(R), \psi_2(R)] = \rho_{\psi_1}(\theta) = [\rho_{\psi_1}(\theta)\rho_{\psi_2}(\theta)]^{1/2}$$

$$= [\mathrm{Var}_\theta\,\psi_1(R)\,\mathrm{Var}_\theta\,\psi_2(R)]^{1/2}$$

We therefore have equality in the Schwarz inequality, and it follows that (with probability 1) one of the two random variables $\psi_1(R) - \gamma(\theta)$, $\psi_2(R) - \gamma(\theta)$ is a multiple of the other (Problem 3, Section 3.4). The multiple is $+1$ or -1 since $\psi_1(R)$ and $\psi_2(R)$ have the same variance. If the multiple is $+1$, we are finished, and if it is -1, then $(\psi_1(R) + \psi_2(R))/2 = \gamma(\theta)$. The minimum variance is therefore 0, hence $\psi_1(R) = \psi_2(R) = \gamma(\theta)$, as desired.

Section 8.6

1. $P\left\{\dfrac{R_1}{R_2} \le z\right\} = \displaystyle\int_0^\infty \int_0^{zy} f_{12}(x, y)\, dx\, dy$

$$= \frac{1}{2^{(m+n)/2}\,\Gamma(m/2)\Gamma(n/2)} \int_0^\infty y^{(n/2)-1}e^{-y/2} \int_0^{zy} x^{(m/2)-1}e^{-x/2}\, dx\, dy$$

But $\int_0^{zy} x^{(m/2)-1}e^{-x/2}\, dx =$ (with $x = uy$) $\int_0^z (uy)^{(m/2)-1}e^{-uy/2}y\, du$. Thus

$$P\left\{\frac{R_1}{R_2} \le z\right\} = \int_0^z h(x)\, dx$$

where

$$h(x) = \frac{1}{2^{(m+n)/2}\Gamma(m/2)\Gamma(n/2)}\, x^{(m/2)-1}\int_0^\infty y^{(m+n)/2-1}e^{-(y/2)(1+x)}\, dy$$

$$= \frac{\Gamma[(m+n)/2]x^{(m/2)-1}}{2^{(m+n)/2}\Gamma(m/2)\Gamma(n/2)}\,\frac{2^{(m+n)/2}}{(1+x)^{(m+n)/2}}$$

$$= \frac{1}{\beta(m/2,\, n/2)}\,\frac{x^{(m/2)-1}}{(1+x)^{(m+n)/2}}, \qquad x \ge 0$$

If

$$W = \frac{R_1/m}{R_2/n} = \frac{n}{m}\frac{R_1}{R_2} \qquad \text{then} \qquad f_W(x) = h\left(\frac{m}{n}x\right)\frac{m}{n}$$

$$= f_{mn}(x), \text{ as desired}$$

2. If R is chi-square with n degrees of freedom then $R = R_1^2 + \cdots + R_n^2$ where the R_i are independent and normal $(0, 1)$. Thus $E(R) = n$, and $\text{Var } R = n \text{ Var } R_i^2 = n[E(R_i^4) - [E(R_i^2)]^2] = n(3 - 1) = 2n$. If R has the t distribution with n degrees of freedom, then $E(R) = 0$ by symmetry, unless $n = 1$, in which case R has a Cauchy density and $E(R)$ does not exist. Now in the integral

$$\int_0^\infty \frac{x^2}{[1 + (x^2/n)]^{(n+1)/2}}\, dx, \qquad \text{let } y = \frac{1}{1 + (x^2/n)}$$

so that

$$dy = \frac{-2x/n}{[1 + (x^2/n)]^2}\, dx = \frac{-2xy^2}{n}\, dx$$

But

$$\frac{x^2}{n} = \frac{1}{y} - 1, \qquad \text{hence} \qquad dy = -\frac{2}{\sqrt{n}}\sqrt{\frac{1-y}{y}}\, y^2\, dx$$

The integral becomes

$$\frac{1}{2}\int_0^1 n\left(\frac{1-y}{y}\right)y^{(n+1)/2}\sqrt{n}\sqrt{\frac{y}{1-y}}\,\frac{1}{y^2}\, dy$$

$$= \tfrac{1}{2}n^{3/2}\int_0^1 y^{(n/2)-2}(1-y)^{1/2}\, dy = \tfrac{1}{2}n^{3/2}\beta\left(\frac{n}{2} - 1,\, \frac{3}{2}\right)$$

Thus

$$\text{Var } R = E(R^2) = \frac{\Gamma[(n + 1)/2]}{\sqrt{n\pi}\ \Gamma(n/2)}\, n^{3/2}\, \frac{\Gamma[(n/2) - 1]\Gamma(3/2)}{\Gamma[(n + 1)/2]}$$

$$= \frac{n/2}{(n/2) - 1} = \frac{n}{n - 2}, \qquad n > 2$$

If $n = 2$, the same calculation gives Var $R = \infty$.

A similar calculation shows that if R has the $F(m, n)$ distribution then $E(R) = n/(n - 2)$ if $n > 2$, $E(R) = \infty$ if $n = 1$ or 2, Var $R = [2n^2(m + n - 2)]/[m(n - 2)^2(n - 4)]$ if $n > 4$, Var $R = \infty$ if $n = 3$ or 4.

Index

Variance, 108, 155–118
Venn diagrams, 4

Weak law of large numbers, 128, 169, 171, 207